Geography of the Physical Environment

The *Geography of the Physical Environment* book series provides a platform for scientific contributions in the field of Physical Geography and its subdisciplines. It publishes a broad portfolio of scientific books covering case studies, theoretical and applied approaches as well as novel developments and techniques in the field. The scope is not limited to a certain spatial scale and can cover local and regional to continental and global facets. Books with strong regional focus should be well illustrated including significant maps and meaningful figures to be potentially used as field guides and standard references for the respective area.

The series appeals to scientists and students in the field of geography as well as regional scientists, landscape planners, policy makers, and everyone interested in wide-ranging aspects of modern Physical Geography. Peer-reviewed research monographs, edited volumes, advance and undergraduate level textbooks, and conference proceedings covering the major topics in Physical Geography are included in the series. Submissions to the Book Series are also invited on the theme 'The Physical Geography of…', with a relevant subtitle of the author's/editor's choice.

More information about this series at http://www.springer.com/series/15117

Levan Tielidze
Editor

Geomorphology of Georgia

Springer

Editor
Levan Tielidze
Department of Geomorphology and
Geoecology, Vakhushti Bagrationi
Institute of Geography
Ivane Javakhishvili Tbilisi State
University
Tbilisi
Georgia

and

Department of Earth Sciences
Georgian National Academy of Sciences
Tbilisi
Georgia

ISSN 2366-8865 ISSN 2366-8873 (electronic)
Geography of the Physical Environment
ISBN 978-3-319-77763-4 ISBN 978-3-319-77764-1 (eBook)
https://doi.org/10.1007/978-3-319-77764-1

Library of Congress Control Number: 2018938374

Cover image by Sonja Weber, München

Printed on acid-free paper

This Springer imprint is published by the registered company Springer International Publishing AG part of Springer Nature
The registered company address is: Gewerbestrasse 11, 6330 Cham, Switzerland

We dedicate the presented work to the 100th anniversary of the establishment of Ivane Javakhishvili Tbilisi State University (TSU). The first-ever national university in the Caucasus was opened on January 26 (new style February 8), 1918. The university laid the foundation for a European-type higher school in Georgia, based on Georgian educational traditions.

We also dedicate the presented work to the 85th anniversary of the establishment of Vakhushti Bagrationi Institute of Geography, founded in 1933 at the Tbilisi State University by the leadership of academician Alexander Javakhishvili—the founder of the Georgian geographical school.

Finally we dedicate this work to the Georgian geomorphologist-researchers of previous generation, who mainly worked in Georgia in the last century and created considerable geomorphological works; they were: Academician Alexander Javakhishvili, Professors: Levan Maruashvili, David Tsereteli, Grigol Devdariani, Shalva Kipiani, Nikoloz Astakhov, Nikoloz

Skhirtladze, Simon Nemanishvili, Demur Tabidze, Givi Maisuradze, George Changashvili, Zurab Tintiliozov (Tatashidze), Archil Tsagareli, Karlo Liponava, Ramin Gobejishvili and others. In the twentieth century, the major scientific literature was published only in Georgian and Russian languages (due to the former Soviet Union's "locked system"), and in many cases the names of the authors and their works could not reach the international scientific community, and unfortunately most of them still remain unknown.

Levan Tielidze

Preface

Georgia, located in the Caucasus region at the crossroads of Europe and Asia, is characterized by geological, geomorphological, and geographical diversities. The high snow-icy mountains, medium and high mountainous deeply fragmented landscape, narrow and deep gorges, seaside lowlands, volcanic cones, uplands and mountain ranges, the intermountain plains, hollows, and highlands give the special beauty to the landscape of Georgia.

The peculiarity of the landscape greatly affects the natural conditions. The diversity of climate, hydrographic network, vegetation cover, soils, fauna, natural and anthropogenic landscapes entirely depend on relief. For example, the Greater Caucasus Range hinders the invasion of northern arctic cold air masses in Georgia, which is why the average annual temperature here is higher than in the North Caucasus (Gobejishvili 2011). The presence of the humid and dry subtropical zones, the sharp differentiation of atmospheric precipitation between the west and east is stipulated by the morpho-hypsometry of the Likhi and Achara-Imereti ranges. The diverse spectrum of natural zones and the altitudinal zonality of geodynamic processes are associated with the hypsometry of the relief. The humid air masses blown from the Atlantic Ocean and the Black Sea leave the main amount of precipitation on the barrier-located ranges (Gagra, Bzipi, Kodori, Egrisi, Svaneti, Racha, Likhi, Achara-Imereti, and Arsiani).

The landscape of Georgia has played a major role in the formation of the Georgian nation for centuries. The ethnic zones of our country, located on the southern and northern slopes of the Greater Caucasus, intermountain lowland, and southern Georgia, look similar, but at the same time, they are very different in character, nature, dialect, behavior, and traditions that eventually form a rich and diverse common Georgian ethno-culture.

The first fundamental work of geomorphological content "The Geography of Georgia, Volume I, Geomorphology" was published by Alexander Javakhishvili in 1926. During the twentieth century, geomorphological research was carried out mainly in the Department of Geomorphology of the Institute of Geography. In 1971, as a result of their research, the most important work was created in the Russian language—"Geomorfologia Gruzii" ("Geomorphology of Georgia". Ed. Maruashvili 1971). In the 1980s, more works were published, in which Georgian geomorphologists characterized individual regions or genetic types of the relief (Astakhov 1973; Nemanishvili 1973, 1982; Tskhovrebashvili 1978; Chkheidze 2000, 2004). Since then, numerous geomorphological facts, processes, and events have been documented by a range of authors. Research methods have also been improved, with new geomorphological research in Georgia and therefore highlighting the need for a geomorphological restudy of Georgia.

The present work is the first English language book of geomorphological content about Georgia, driven by the surveys conducted for years in the most part of Georgia by different generations of geomorphologists. We could not conduct expeditions in those regions that are currently temporarily occupied within Russia boundaries (Apkhazeti and Tskhinvali region). The relevant geomorphological materials on each region were collected from our predecessor researchers' works and processed, and they are attached to each chapter of the book in the form of used References.

We would like to express our gratitude to Ms. Nino Chikhradze and Dr. Roger Wheate for supporting us during working on the book.

Tbilisi, Georgia
2017

Levan Tielidze

References

Astakhov NE (1973) Structurnaya Geomorfologia Gruzii (Structural Geomorphology of Georgia). Publ. "Metsniereba". Tbilisi (in Russian)
Chkheidze O (2000) Geomorphology of Imereti, Part I Tbilisi (in Georgian)
Chkheidze O (2004) Geomorphology of Imereti, Part II Tbilisi (in Georgian)
Gobejishvili R (2011) Relief of Georgia. Publ. "Universali", Tbilisi (in Georgian)
Javakhishvili A (1926) Geography of Georgia. Vol 1—geomorphology (with Atlas), Tbilisi (in Georgian)
Maruashvili L (1971) Edit. Geomorphology of Georgia. Pub. House "Metsniereba", Tbilisi (in Russian)
Nemanishvili S (1973) River Terraces of Mountainous areas. Vol 1, Tbilisi (in Georgian)
Nemanishvili S (1982) River Terraces of Mountainous areas. Vol 2, Tbilisi (in Georgian)
Tskhovrebashvili Sh (1978) Geomorphology of Achara-Trialeti Mountain Range, monograph, Tbilisi (in Georgian)

Contents

Editor and Contributors

About the Editor

Levan Tielidze is a Ph.D. of Ivane Javakhishvili Tbilisi State University in the field of geomorphology–glaciology–cartography and a senior research scientist at Vakhushti Bagrationi Institute of Geography. He is also a researcher-scholarship holder of the Department of Earth Sciences of the Georgian National Academy of Sciences. The field of his research is mountain cartography and glacial-geomorphological study of the mountainous areas in the Quaternary (Late Pleistocene and Holocene). He is a member of several international organizations. In 2014/2015, he conducted a research stay at the Climate Change Institute of the University of Maine, USA, and in 2015/2016–2017 at the University of Northern British Columbia, Canada. He is the author of about 40 scientific papers, maps, and eight monographs.

Contributors

Lasha Asanidze Department of Geomorphology and Geoecology, Vakhushti Bagrationi Institute of Geography, Ivane Javakhishvili Tbilisi State University, Tbilisi, Georgia

Nikoloz Astakhov Department of Geomorphology and Geoecology, Vakhushti Bagrationi Institute of Geography, Ivane Javakhishvili Tbilisi State University, Tbilisi, Georgia

Nino Chikhradze Department of Geomorphology and Geoecology, Vakhushti Bagrationi Institute of Geography, Ivane Javakhishvili Tbilisi State University, Tbilisi, Georgia

Lela Gadrani Department of Geomorphology and Geoecology, Vakhushti Bagrationi Institute of Geography, Ivane Javakhishvili Tbilisi State University, Tbilisi, Georgia

Ramin Gobejishvili Department of Geomorphology and Geoecology, Vakhushti Bagrationi Institute of Geography, Ivane Javakhishvili Tbilisi State University, Tbilisi, Georgia; Department of Geography, Faculty of Exact and Natural Sciences, Ivane Javakhishvili Tbilisi State University, Tbilisi, Georgia

Alexander Javakhishvili Department of Geomorphology and Geoecology, Vakhushti Bagrationi Institute of Geography, Ivane Javakhishvili Tbilisi State University, Tbilisi, Georgia; Department of Geography, Faculty of Exact and Natural Sciences, Ivane Javakhishvili Tbilisi State University, Tbilisi, Georgia; Department of Earth Sciences, Georgian National Academy of Sciences, Tbilisi, Georgia

George Kavlashvili Department of Geomorphology and Geoecology, Vakhushti Bagrationi Institute of Geography, Ivane Javakhishvili State University, Tbilisi, Georgia

Roman Kumladze Laboratory of Cartography and GIS, Vakhushti Bagrationi Institute of Geography, Ivane Javakhishvili Tbilisi State University, Tbilisi, Georgia; Geological Mapping Division, Department of Geology, National Environmental Agency, Ministry of Environment Protection and Agriculture of Georgia, Tbilisi, Georgia

Zaza Lezhava Department of Geomorphology and Geoecology, Vakhushti Bagrationi Institute of Geography, Ivane Javakhishvili Tbilisi State University, Tbilisi, Georgia

George Lominadze Department of Geomorphology and Geoecology, Vakhushti Bagrationi Institute of Geography, Ivane Javakhishvili State University, Tbilisi, Georgia

Levan Maruashvili Department of Geomorphology and Geoecology, Vakhushti Bagrationi Institute of Geography, Ivane Javakhishvili Tbilisi State University, Tbilisi, Georgia

Vazha Trapaidze Department of Geography, Faculty of Exact and Natural Sciences, Ivane Javakhishvili Tbilisi State University, Tbilisi, Georgia

Lia Matchavariani Department of Geography, Faculty of Exact and Natural Sciences, Ivane Javakhishvili Tbilisi State University, Tbilisi, Georgia

Irakli Papashvili Gamma Consulting Ltd., Tbilisi, Georgia

Levan Tielidze Department of Geomorphology and Geoecology, Vakhushti Bagrationi Institute of Geography, Ivane Javakhishvili Tbilisi State University, Tbilisi, Georgia; Department of Earth Sciences, Georgian National Academy of Sciences, Tbilisi, Georgia

David Tsereteli Department of Geomorphology and Geoecology, Vakhushti Bagrationi Institute of Geography, Ivane Javakhishvili Tbilisi State University, Tbilisi, Georgia

Kukuri Tsikarishvili Department of Geomorphology and Geoecology, Vakhushti Bagrationi Institute of Geography, Ivane Javakhishvili Tbilisi State University, Tbilisi, Georgia

Mariam Tsitsagi Department of Hydrology and Climatology, Vakhushti Bagrationi Institute of Geography, Ivane Javakhishvili Tbilisi State University, Tbilisi, Georgia

Bejan Tutberidze Department of Geology, Faculty of Exact and Natural Sciences, Ivane Javakhishvili Tbilisi State University, Tbilisi, Georgia

Roger Wheate Natural Resources and Environmental Studies, University of Northern British Columbia (UNBC), Prince George, Canada

Part I

Introduction to the Geomorphology of Georgia

Long-Term Geomorphic History of Georgia

1

Levan Tielidze, Ramin Gobejishvili, Bejan Tutberidze, Levan Maruashvili, Nikoloz Astakhov and Roger Wheate

Abstract

This chapter discusses the location of Georgia, its orographic units, geological and geomorphological structure, zones and zoning. The Proterozoic, Paleozoic, Mesozoic, and Cenozoic sediments; Jurassic and Cretaceous formations; Palaeogenic, Oligocene and Miocene sediments; Pliocene and Quaternary formations (Pleistocene and Holocene) are described. The three main geomorphological zones such as the Greater Caucasus, Georgia's intermountain plain, and Southern highland of Georgia are described. Within Georgia, the Neogene and Quaternary volcanisms (Late Miocene–Early Pliocene, Late Pliocene–Early Pleistocene, Middle and Late Pleistocene, and Holocene) are characterized. We do not provide the description of the Late Pleistocene glaciation of the Georgian Caucasus in this chapter, because a separate monograph directly about the glaciers of Georgia has been published.

Keywords

Geology of Georgia · Geomorphology of Georgia · Volcanism of Georgia

L. Tielidze (✉) · R. Gobejishvili · L. Maruashvili · N. Astakhov
Department of Geomorphology and Geoecology, Vakhushti Bagrationi Institute of Geography, Ivane Javakhishvili Tbilisi State University, Tbilisi, Georgia
e-mail: levan.tielidze@tsu.ge

B. Tutberidze
Department of Geology, Faculty of Exact and Natural Sciences, Ivane Javakhishvili Tbilisi State University, Tbilisi, Georgia

L. Tielidze
Department of Earth Sciences, Georgian National Academy of Sciences, Tbilisi, Georgia

R. Gobejishvili
Department of Geography, Faculty of Exact and Natural Sciences, Ivane Javakhishvili Tbilisi State University, Tbilisi, Georgia

R. Wheate
Natural Resources and Environmental Studies, University of Northern British Columbia (UNBC), Prince George, Canada

1.1 Geographical Information

The territory of Georgia is mainly located in the southern slope of the Greater Caucasus: incorporating the mountainous regions of the Caucasus and southern Georgia, and including Georgia's intermountain plain. Only a small area is located on the northern slope of the Greater Caucasus. It has a distinct natural boundary in the west (the Black Sea) and north (the Greater Caucasus crest) (Fig. 1.1).

L. Tielidze (ed.), *Geomorphology of Georgia*, Geography of the Physical Environment,
https://doi.org/10.1007/978-3-319-77764-1_1

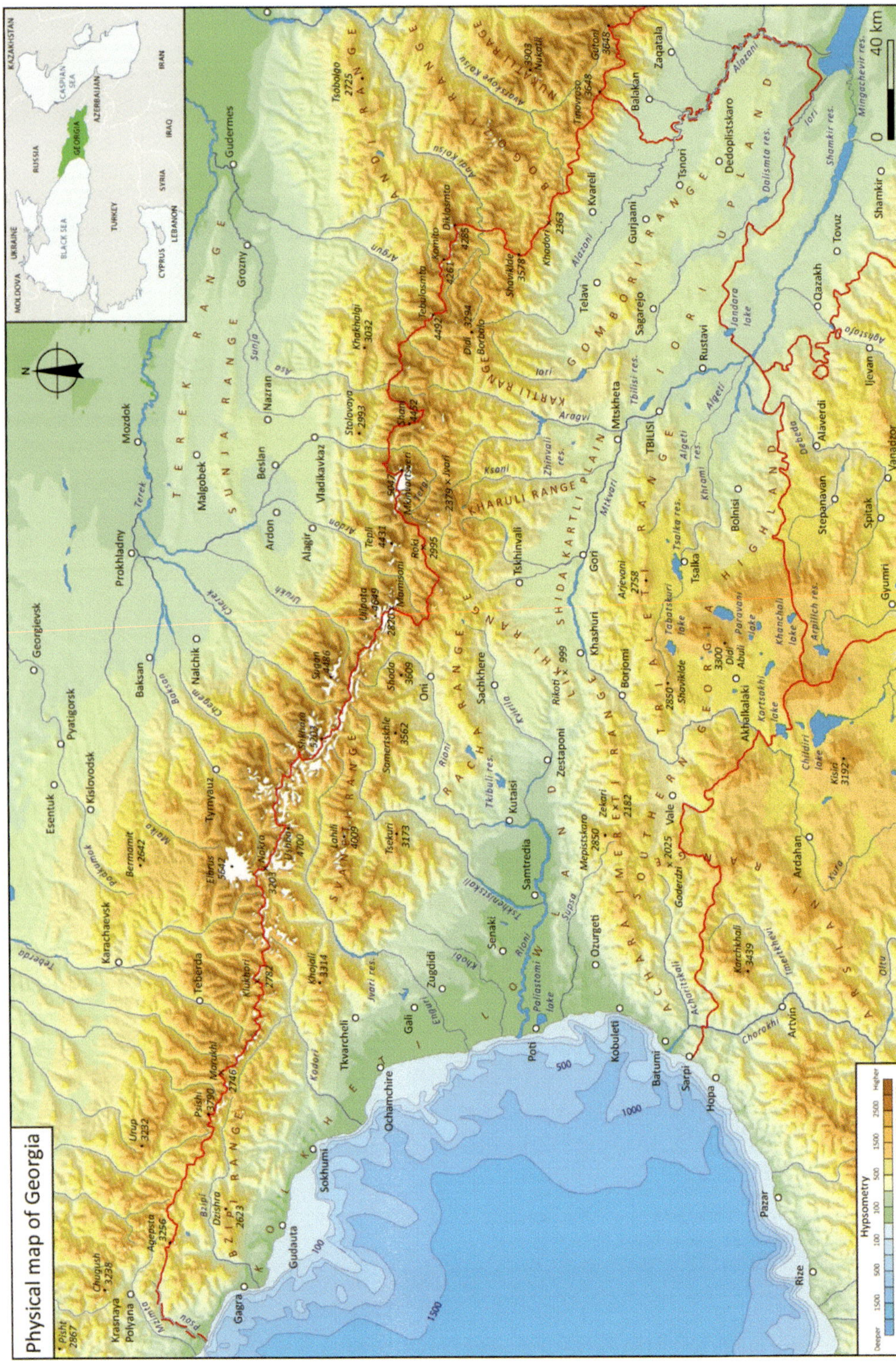

Fig. 1.1 Physical map and location of Georgia

The Caucasus Mountains occupy the entire northern belt of Georgia; the highest peak is Shkhara—5203 m above sea level (asl). The southern slope of the Greater Caucasus Range in Georgia stretches from Mt. Avadhara (2960 m) to Mt. Tinovroso (3374 m) for about 750 km (Tielidze 2016). Parallel and transverse ranges include: Gagra, Bzipi, Chkhalta (Apkhazeti), Kodori, Svaneti, Egrisi (Samegrelo), Lechkhumi, Racha, Shoda-Kedela, Germukhi, Kharuli, Lomisi, Kartli, Kakheti, and others. A short section of the northern slope of the Greater Caucasus intrudes from Mt. Vatsisparsi (3573 m) to Mt. Shaviklde (3578 m) including the Khokhi, Kuro, Shani, Kidegani, Khevsureti, Pirikita, Atsunta, and Mutso subranges. Adjacent to the medium and high mountainous tectogenic-erosive relief, there are widespread glacial, old glacial (Pleistocene and Holocene), volcanic, karst, and gravitational landscape types.

The Psou, Bzipi, Kodori, Enguri, Rioni, Liakhvi, Ksani, Aragvi, Iori, and Alazani rivers flow to the south from the Greater Caucasus, while the Tergi (Terek), Asa, Arghuni, and Tushetis Alazani (Andi Koysu) rivers flow to the north. Many notable rivers originate from the lakes of Didi Ritsa, Patara Ritsa, Amtkeli, Kvedi, Ertso, Keli, and Archvebi.

The Caucasian climate is humid within the western part of Georgia, promoting the development of rich vegetation—deciduous and dark coniferous forests, evergreen and relic species. High mountainous relief is decorated by alpine and subalpine vegetation; the glaciers are distributed over the crest of the Greater Caucasus.

Georgia's intermountain plain is represented by the Kolkheti Lowland, the Imereti Highland, the Shida Kartli, Kvemo Kartli and Alazani plains, the Iori Upland, and some lower ranges.

The important rivers in the Kolkheti Lowland are: Kodori, Mokvi, Rioni, Enguri, Khobi, and Supsa. The Paliastomi Lake is the largest among the lakes. The climate of Kolkheti is warm and humid, frosts are rare, and annual precipitation is 1500–2500 mm. The vegetation cover is mostly modified by human influence, although in some areas it is represented by Colchis type forests—oak (*Quercus*), hornbeam (*Carpinus*), chestnut (*Castanea*), alder (*Alnus*) with abundant evergreen subforest consisting of box tree (*Buxus colchica*), rhododendron (*Rhododendron ponticum*), laurel (*Laurocerasus officinalis*), and holly (*Ilex colchica*). The main part of the Imereti Highland is built from old crystalline rocks (Dzirula massif). The climate here is cool and drier than in the Kolkheti Lowland. Natural vegetation is represented by deciduous forest and poor Colchis subforest (Gobejishvili 2011).

Shida Kartli Plain is located at a height of 500–800 m asl; the climate is moderately warm semi-continental and the annual precipitation is 500–700 mm. Vegetation is represented by forest-valley and secondary valley. The Mtkvari (Kura) River and its tributaries—Prone, Liakhvi, Lekhura, Ksani, and Aragvi flow across the plain. Kvemo Kartli Plain is located at a height of 300–500 m above sea level, but otherwise is similar to the Shida Kartli Plain.

In the Iori Upland, flat plains are mixed with structural hillocks at a height of 1000–1200 m, built of Tertiary sediments. The climate is arid, with annual precipitation 300–400 mm. Chernozems and Kastanozems underlie valley vegetation; this is reduced to a semidesert landscape in the Eldari Lowland.

The Alazani Plain is a sediment-filled continental geosyncline, at 400–700 m elevation. The climate is milder than in the other regions of the eastern Georgian plain. The vegetation is represented by forest and forest-valley; there can be found the elements of mesophilic flora. The Likhi, Kvernakebi, Saguramo-Ialno, and Gombori ranges are located in Georgia's intermountain plain.

The highland of southern Georgia consists of volcanic upland and ranges (Achara-Imereti, Trialeti, Loki, and Shavsheti) built of Tertiary sediments bordering from the north. In the western part of the Trialeti Range, at Borjomi, the young volcanic relief is developed in the form of lava flows, plateaus and cones. In the western part of the Achara-Imereti Range, the climate is humid with abundant vegetation (deciduous and coniferous forests, subalpine with rhododendron shrubs (*Rhododendron caucasicum*) and alpine meadows). The climate in the ranges located in the east becomes dry and losses the Colchis elements. The Mtkvari River (Borjomi gorge) antecedently crosses the Achara-Trialeti Range system.

The volcanic upland of Georgia is constructed from Neogene and Quaternary effusives, covered with boulders; the highest peak is Didi Abuli—3301 m asl. The relief is represented by lava plateaus (Dmanisi, Gomareti, Chochiani, Bedeni, and Kvemo Kartli), folded hillocks and flatlands, and rows of volcanic cones (Samsari and Javakheti ranges). The Javakheti region is rich in lakes: Paravani, Tabatskuri, Kartsakhi, and Saghamo. The climate is cold and mildly humid with 600–1000 mm annual precipitation. Chernozems dominate in the lava plateaus, characterized by the absence of forest cover. There are alpine meadows with rhododendron shrubs (*Rhododendron caucasicum*) and birch (*Betula*) groves in the slopes of Javakheti and Samsari Ranges (above 2100 m).

1.2 Stratigraphy

The territory of Georgia contains geological formations, ranging from Neoproterozoic (~1000–1200 mln years) to modern era (Gamkrelidze 2012) (Fig. 1.2).

The oldest—the Proterozoic and Early-Middle Paleozoic formations, are metamorphic and are represented within the Greater Caucasus main watershed range and Dzirula and Khrami crystalline massifs. These rocks are dated based on both the geologic (interrelation of rocks) and new geo-chronological data. All of these have experienced a polycyclic regional metamorphism. The Laba metamorphic complex of the Greater Caucasus main range as well as the metamorphic slate suites of Dzirula and Loki massifs belong to the Early and Middle Paleozoic.

Upper Paleozoic formations are represented by continental rhyolites and rhyodacites and their pyrocrystals in the north of the Dzirula massif and in the north and east of the Khrami massif. The molasse sediments of sand suite in the Greater Caucasus main range belong to the upper part of the Carboniferous system.

The mainly terrigenous and weakly metamorphic rocks of the Dizi series belong to the Middle and Late Paleozoic. In some areas, the upper parts of the Dizi series also include the Triassic stage. The Dizi series is outcropped on the southern slope of the Greater Caucasus within the Svaneti region.

Mesozoic and Cenozoic sediments are fully represented in the territory of Georgia and are characterized by rich fauna fossils.

Jurassic sediments, which are transgressively located almost everywhere, contain all the stages of the Jurassic age including Tithonian, and are dated mainly by ammonite fauna (Gobejishvili et al. 2011). These sediments are found in all tectonic units of Georgia and they are quite different by character. They are mainly represented by thick clay slates and sandstones of the deep sea, flysch formations, shallow sea limestones, and lime-alkaline and tollite volcanogenic rocks. The volcanogenic suite of the Bajocian age (porphyritic series) is particularly thick, represented by the lime-alkaline basalt-andesite-rhyolite series.

Cretaceous formations, which are often transgressively distributed over the older formations, are also widespread and are represented mainly by carbonate (limestones, marls) and volcanogenic rocks. Thick flysch Cretaceous sediments are widespread in the southern slopes of the Greater Caucasus.

Palaeogenic sediments occur across all tectonic zones of Georgia and are represented by terrigenous, carbonate and sediment-volcanogenic rocks. In many areas, the Paleogene is transgressively located over the older sediments.

The Oligocene and Early Miocene (Maicopian series) are represented by the marine molasse: carbonate clays (Khadum horizon), gypsum clays, conglomerates, and sandstones.

Middle and Late Miocene and Pliocene sediments are developed in orogenic hollows (molasse depressions) and are represented by marine and continental molasses. The Chokrakian age sediments are often located transgressively and with angular discrepancy.

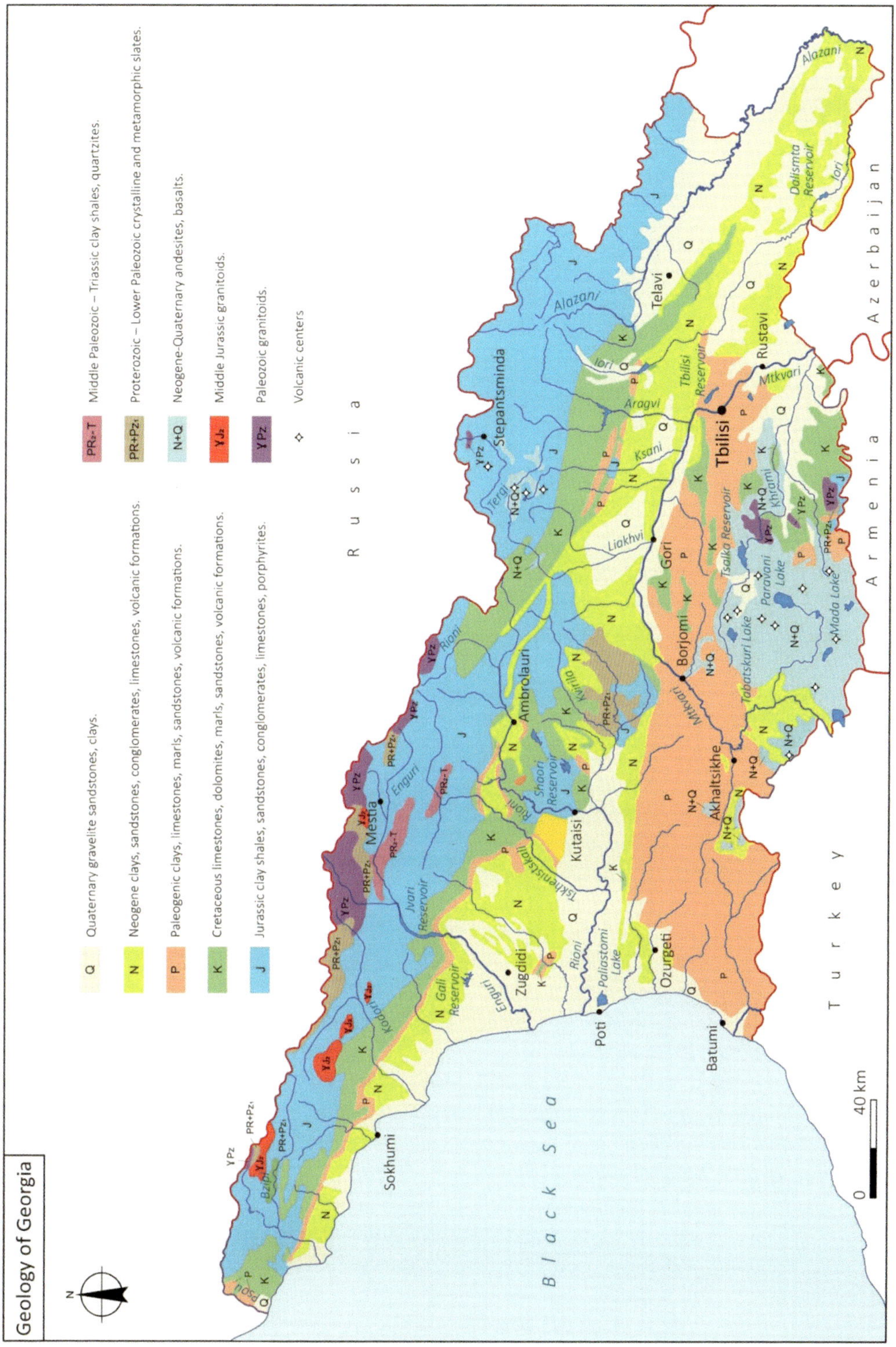

Fig. 1.2 Geological map of Georgia (Gamkrelidze 1992; Khazaradze et al. 2000)

Following the Maeotian transgression, the territory of Georgia has been divided into two sediment provinces—the Black Sea and Caspian Sea provinces (Gamkrelidze 2012).

In the Late Miocene and Early Pliocene, in the southern part of the Achara-Trialeti zone and in the western part of the Artvini-Bolnisi zone, subaerial volcanism (the Goderdzi suite) is present.

Pliocene sediments are widely developed in western Georgia and represented by marine terrigenous rocks, while the Pliocene is represented by continental and marine facies in eastern Georgia and approaching the Caspian Sea basin.

In the Late Pliocene and Early Pleistocene, in the territory of Javakheti Upland, subaerial lime-alkaline, andesite-basaltic, andesite and also doleritic, basaltic, and dacite volcanisms are evident.

Quaternary system is created by marine, alluvial, lacustrine, wetland, glacial, deluvial, proluvial, and cave sediments (Khazaradze et al. 2000; Gobejishvili and Tsereteli 2012a) (Fig. 1.3).

The marine sediments are developed in the sinking zones of western and eastern parts of Georgia (Kolkheti Lowland and Gare Kakheti), in Apkhazeti and Achara. Quaternary marine sediments are constructed with terrigenous facies: with the alternation of cobblestones, sands and clays. In Apkhazeti and Achara, the marine sediments are mostly found on the Black Sea terraces. Under the influence of tectonic movements and Black Sea eustatics, the terraced surfaces of different ages are located at different hypesometric levels. Here, the marine sediments are composed with the alternation of cobblestones, sands, and clays as well.

Lacustrine sediments are widely spread in Javakheti Upland. Their genesis is related to negative forms created by lava flows. The lacustrine sediments are also found in the surroundings of Tbilisi that are composed with the alternation of sandstones and clays.

Wetland sediments are widespread in Kolkheti Lowland; their absolute age is about 6000 years and is associated with the "New Black Sea transgression phase" of the Black Sea. The maximum height of these sediments is about 9 m (Khazaradze et al. 2000). The wetland sediments are formed by peat and sapropel.

Glacial formations of Quaternary age are mainly represented in the high mountainous zone of the Greater Caucasus Range. Here, in the terrain, the traces of glaciation are mainly preserved in the form of cirques and trough valleys. Alongside the southern slope of the Greater Caucasus Range, the lateral and stadial moraines create glacial deposits in the heads of the gorges of all large rivers (Bzipi, Kodori, Enguri, Rioni, Liakhvi, Ksani, and Aragvi). In some gorges, the number of stadial moraines is about 8, which indicates a short-term suspension of movement of the old glaciers. In some areas, the large size erratic boulders brought by glaciers are observed.

Deluvial and proluvial sediments are widespread and are the rock weathering product. Their thick accumulations are observed in the mountainous part of western Georgia. Relatively high temperature, humidity, and high indicators of atmospheric precipitation contribute to the intensity of weathering processes and creation of significant thickness of deluvial and proluvial sediments.

In the second half of Quaternary period, the primitive human often used karst caves for temporal or permanent habitation. Cultural deposits are preserved in many of them that contain the ancient tools of hunting and economic activities of Stone Age humans as well as the remnants of their cuisine wastes (osteological material).

In Georgia, cave dwellings of primitive human are found in many areas—on the southern slope of the Greater Caucasus Range (Kudaro and Tsona), in Dzirula massif (Ortvala Klde, Dzudzuana, Samgle Klde, and Chruchula), in the Okriba-Argveti hill (Tsutskhvati cavern and Sagvarjile cave), in Apkhazeti (Apiancha) and etc. Their age is Middle and Late Pleistocene. The cave sediments are composed primarily with the alternation of limestones, detritus, and loams.

Among the Quaternary sediments of Georgia, the open dwellings of Homo sapiens are found rich in paleontological and anthropological materials (Dmanisi), the absolute age of which is 1.8–2.04 million years (Vekua et al. 2002).

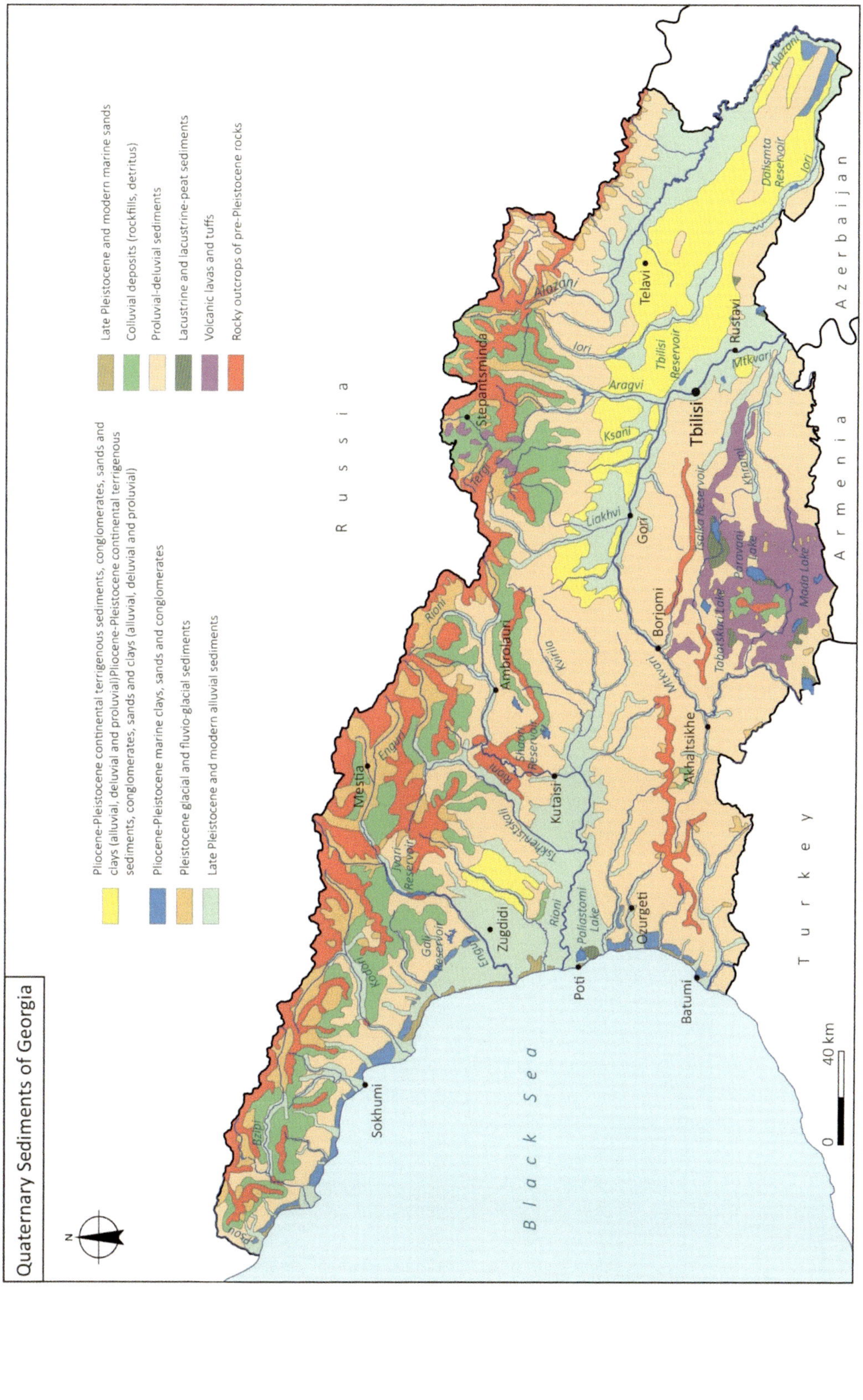

Fig. 1.3 Quaternary sediments of Georgia (Tsereteli 1964; Gobejishvili and Tsereteli 2012a)

Cultural sediments are of polygenetic genesis and are mainly composed of loams and sands, as well as volcanic ash and carbonate sediments.

In the Pleistocene and Holocene, the Volcanites are mainly represented by rocks of moderately acidic (andesites) and acidic (dacites) composition. Within the Achara-Trialeti zone, in the territory of Borjomi-Bakuriani, Early (Middle) Pleistocene andesite and andesite-dacite subaerial volcanism are presented.

1.3 Geomorphological Evolution

The modern appearance of Georgia's landscape is the result of simultaneous interaction of endogenous and exogenous processes and is divided into the three main zones and 12 subzones with multispectral morphological and morphometric conditions (Astakhov et al. 1964; Gobejishvili and Tsereteli 2012b) (Fig. 1.4).

1. The southern slope and crest of the Greater Caucasus is the Horst-anticlinal uprise of the ancient crystalline heart, which helped create the watersheds of the highest part of the system with nival-glacial and erosive-glacial ranges. On their peripheral parts, these ranges are built by the complex of Middle Jurassic volcanogenic rocks and Early Jurassic slates. In the western part of the southern slopes of the Greater Caucasus, there is a continuous belt of limestone ranges and massifs, containing karst landscape. The ranges composed of Jurassic and Cretaceous sediments (carbonate and non-carbonated flysch zone), oriented longitudinally and sublatitudinally in the medium and high mountainous relief, are evident within the central and eastern sections of the main watershed.
   The formation of large volcanic massifs and wide lava plateaus in the central part of the Greater Caucasus crest belt began from the Pliocene, and are mostly present in the Keli Upland and Mt. Mkinvartsveri (Kazbegi).
2. Formation of the relief of Georgia's intermountain plain is closely linked to the surrounding mountain system's formation. The Imereti Highland began forming in the Pliocene, with a denudation of uprising relief and the outcropped old granite ground under the Tertiary sea sediments. An extensive area was freed from the sea forming the foothills of the eastern Georgia, as well as the early sinking areas, where the marine-continental molasse sediments were accumulated. The Kolkheti Lowland experienced intense sinking and powerful Pliocene sea sediments were accumulated. In the background of the general uprising, there was relative sinking in the Quaternary and the erosion-denudation relief of the foothill hillocks began. At the end of the Pleistocene, the relief took a modern appearance. The denudation surface of Imereti Highland experienced erosion fragmentation. The sinking of the Kolkheti Lowland and the Alazani Plain and their filling with alluvial sediments continue today.
3. The Southern Highland of Georgia morphologically is divided into two parts: the high uplands and volcanic mountains of the Javakheti volcanic highland in the south, and the peripheral latitudinal ranges in the north.

The Javakheti Highland is built of Pliocene and Pleistocene volcanic formations that cover relatively older folded strata. On the volcanogenic pedestal along the meridian depth faults, there is a line of volcanic cones of the Samsari Range and the volcanic domes of the Javakheti Range, where there are deep river canyons. Old erosion-glacial forms are preserved in the crests of volcanic ranges with heights exceeding 2500 m.

The peripheral ranges of the southern highland of Georgia are mainly built on volcanogenic formations of Paleocene and Eocene that are resistant to erosion; the peculiarity of the relief is the flattened surfaces. As opposed to the Greater Caucasus, the nival-glacial relief is represented only on the uprisen crest. In the east, among the Khrami-Loki massifs, there is erosion-denudation relief of medium height built of carbonate flysch.

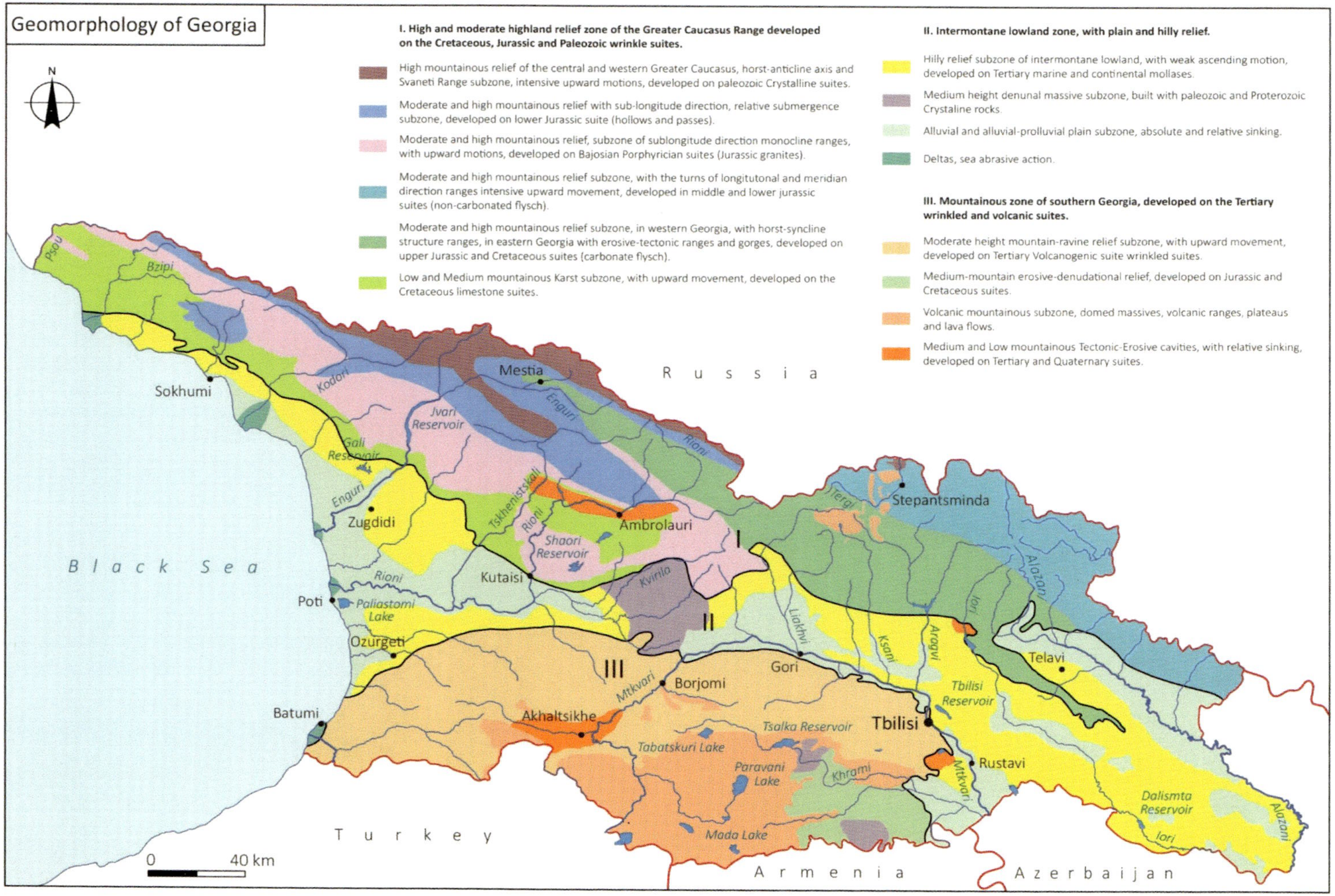

Fig. 1.4 Geomorphological map of Georgia

1.4 Volcanism

In the formation of the modern relief of the territory of Georgia, which started in the Cenozoic Era, the leading role of endogenous relief-originating factors belongs to tectonics and volcanism. In this regard, the role of Neogenic-Anthropogenic volcanism is significant, which is expressed in all major structural-formation units of the first rate of the Georgian territory on a different scale. There are five main stages of Neogene-Anthropogenic volcanic activity in the territory of Georgia: (1) the Late Miocene—Early Pliocene, (2) the Late Pliocene–Early Pleistocene, (3) the Middle Pleistocene, (4) the Late Pleistocene, and (5) the Holocene (Tutberidze 2012).

The Late Miocene–Early Pliocene. The powerful volcanic activity began at the end of the Miocene and ended in the Early Pliocene. The initial phase of volcanic activity involved explosive eruptions and ended with lava eruptions. The volcanites of this period are widely represented in the highlands of southern Georgia in the Erusheti highlands, Arsiani, Javakheti, and Samsari ranges. In geological literature sources, the volcanic formations of this period are known as the "Goderdzi suite". Some isolated outlets of small volcanites of the same period are observed at the northern periphery of Georgia's block and together form the Vanati-Tskhinvali-Kvasatali volcanoes' group.

The Late Pliocene–Early Pleistocene. After a break in the Middle Pliocene, volcanism was revived in the Late Pliocene and continued in the Early Pleistocene. Volcanism is widely seen in the Javakheti Upland. At the initial stage of the evolution of volcanism, there were strong areal eruptions (dolerite-basalts) that were subsequently replaced by central types of eruptions (andesite basalts, andesites, unimportant amounts of dacites). These rocks have built the Javakheti and Nialiskuri ranges (dolerites, andesite-basalts, and andesites), Akhalkalaki, Dmanisi, Tsalka, Gomareti, Kvemo Kartli plateaus (dolarist-basalt), and part of the Abul-Samsari Range (andesites, dacites, rhyolites, rhyolite dacites, and obsidians). The dolerites prevail among the erupted products. In the territory of Javakheti Upland, the volcanic and volcanic-sedimentary formations of this period are known as the "Akhalkalaki suite", which represents the alternation of dolerites, andesite-basalts, andesites, and located among them the lacustrine sediments. The volcanism of this period is revealed in the Borjom-Bakuriani region of the Lesser Caucasus folded system. This includes the basaltic lavas (with limited distribution), andesite-basalts, and andesite composition and their pyroclastolites, the eruption centers of which are maintained in the relief in the form of small size hillocks (Mukhera and Saghravi; Andesite). Volcanites include the basalt bodies of Perevisa, Didi Goradziri and Patara Goradziri in the Chiatura-Sachkhere region.

The Middle Pleistocene volcanism is mainly revealed in the Keli volcanic upland. This includes the Patara Nepiskalo volcanoes' group (Tskhradzma), represented by andesite-basalt, andesite (mainly) and dacite lavas and their pyroclastolites. To this stage of volcanism, evolution belongs the Mt. Pidarkhokhi (dacites and pyroclastolites).

The Late Pleistocene volcanism is weakly revealed in the volcanic upland of southern Georgia. Volcanic activity is expressed in effusive and explosive eruptions. To this age group belong the Mtkvari stream (dolerites), volcanic formations (andesites, dacites and their pyroclastolites) of Tavkvetili, Godorebi, and Shavnabada in the Abul-Samsari Range. The volcanic formations of this period are widespread in the Kazbegi volcanic region of the Greater Caucasus folded system. These include the lavas of andesite-basalts (with less distribution), andesite (mainly) and dacite composition and their pyroclastolites developed in the Gveleti, Tsdo, Chkheri, Arsha and Suatisi gorges. The lava streams and their pyroclastolites are related to the Kabarjina, Tsiteldziri, Miliona, Narvani, Sakokhe, Sadzele, and other volcanic centers. In the Keli volcanic upland and its surroundings, the northern Narvan-Khokhi, Pidarkhokhi, Char-Khokhi, southern Shadilkhokhi, Sirkhokhi (mostly the andesites and their pyroclastolites), and Gudauri-Mleta (andesite-basalt) stream are built of the Late Pleistocene formations.

The Holocene. The volcanic eruptions of the Holocene period in the Kazbegi volcanic region created the andesite and dacite lava flows of Shevardeni, Kichuttsveri, Tkarsheti, Khorisari, Goristsikhe and their pyroclastolites, and in the Keli volcanic upland—the Levinson-Lesing (andesites, dacites, radiodacites), Didi Nepiskalo (dacites, radiodacites, and liparites), Khodzi stream (andesites), Tsiteli Khati (andesite-basalt, andesite), and the Tsitelimta volcanic cone on Samsari (andesite, dacites).

References

Astakhov N, Javakhishvili A, Tsagareli A and Tsereteli D (1964) Geomorphological map of Georgia, in the Atlas of Georgian Soviet Socialist Republic, Tbilisi (Georgian/Russian), pp 45–46

Gamkrelidze IP (2012) Geology. Chapter in the National Atlas of Georgia. Publishing house "Cartograpia", Tbilisi (in Georgian), p 29

Gamkrelidze IP (1992) Geological map of Georgia. Chapter in the Atlas of Georgia. Publishing house "Cartograpia", Tbilisi (in Georgian), p 9

Gobejishvili R (2011) Relief of Georgia. Publishing house "Universali", Tbilisi (in Georgian)

Gobejishvili R, Lomidze N and Tielidze L (2011) Late Pleistocene (Wurmian) glaciations of the Caucasus. In: Ehlers J, Gibbard PL, Hughes PD (eds) Quaternary glaciations: extent and chronology. Elsevier, Amsterdam, pp 141–147. https://doi.org/10.1016/b978-0-444-53447-7.00012-X

Gobejishvili R, Tsereteli E (2012a) Quaternary sediments. National Atlas of Georgia. Publishing House "Cartograpia", Tbilisi (in Georgian), p 32

Gobejishvili R, Tsereteli E (2012b) Geomorphology. Chapter in the National Atlas of Georgia. Publishing House "Cartograpia", Tbilisi (in Georgian), p 45

Khazaradze R, Tvalchrelidze M, Mamatsashvili N (2000). Geological structure. Chapter in the Geography of Georgia. Publishing house "Metsniereba", Tbilisi (in Georgian)

Tielidze LG (2016) Glacier change over the last century, Caucasus Mountains, Georgia, observed from old topographical maps, Landsat and ASTER satellite imagery. The Cryosphere 10:713–725. https://doi.org/10.5194/tc-10-713-2016

Tsereteli D (1964) Quaternary sediments map of Georgia, in the Atlas of Georgian Soviet Socialist Republic, Tbilisi (Georgian/Russian), pp 33–34

Tutberidze B (2012) Neogene and Pleistocene volcanism, Chapter in the National Atlas of Georgia. Publishing house "Cartograpia", Tbilisi (in Georgian), p 40

Vekua A, Lortkipanidze D, Rightmire GP, Agusti J, Ferring R, Maisuradze G, Mouskhelishvili A, Nioradze M, Poncede Leon M, Tappen M, Tvalchrelidze M, Zollikofer C (2002) A new skull of early Homo from Dmanisi, Georgia. Science 297(5578):85–9. https://doi.org/10.1126/science.1072953

Levan Tielidze is a Ph.D. of Ivane Javakhishvili Tbilisi State University in the field of geomorphology-glaciology-cartography and a senior research scientist at Vakhushti Bagrationi Institute of Geography. He is also a researcher-scholarship holder of the Department of Earth Sciences of the Georgian National Academy of Sciences. The field of his research is mountain cartography and glacial-geomorphological study of the mountainous areas in the Quaternary (Late Pleistocene and Holocene). He is a member of several international organizations. In 2014/2015, he conducted a research stay at the Climate Change Institute of the University of Maine, USA, and in 2015/2016–2017 at the University of Northern British Columbia, Canada. He is the author of about 40 scientific papers, maps, and eight monographs.

Ramin Gobejishvili (1941–2014) was a Doctor of Geographical Sciences, a geomorphologist-glaciologist. Until his last days, he was Professor of the Faculty of Exact and Natural Sciences, Ivane Javakhishvili Tbilisi State University and the chair of Geomorphology and Cartography. During 1996–2008, he was the president of the Geographical Society of Georgia, and in 2005–2007—the director of the Institute of Geography. He was the author and coauthor of about 150 scientific works and ten monographs.

Bejan Tutberidze is a Head of the Geology Department of the Faculty of Exact and Natural Sciences at TSU, Professor, Doctor of Geology. He is Chairperson of the Academic Council, working at TSU since 1971. His research fields are petrologic-geochemical studies of Neogene-Anthropogenic volcanic and volcanogenic-sedimentary formations in Georgia and the establishment of the regularity of their distribution in time and space; volcanic equipment monitoring; studies of post-volcanic hydrothermal processes; and ore and non-ore mineralization processes.

Levan Maruashvili (1912–1992) was a prominent Georgian geographer. Doctor of Geographical Sciences (1954), Professor (1965). Honored Scientist of the Georgian Soviet Socialist Republic (1966). Honorary member of the Russian Geographical Society (1985). He produced over the 500

scientific and popular-scientific works in Georgian, Russian and English languages.

Nikoloz Astakhov (1911–1993) was a Doctor of Geographical Sciences, Professor (1965). He was researcher at the Department of Geomorphology of the Vakhushti Bagrationi Institute of Geography. His research area was structural geomorphology.

Roger Wheate is Associate Professor and founding member of the Geography program at the University of Northern British Columbia (UNBC), Canada. His teaching and research interests cover the application of cartographic output, remote sensing, and GIS across the spectrum of Natural Resource and Environmental Studies (NRES) faculty areas and the integration of geomatics technologies, including feature extraction and terrain visualization. Special interests include mountain cartography and glacier mapping and updating using remote sensing. He is the Canadian national delegate to the International Cartographic Association (ICA) and a member of the ICA Mountain Cartography Commission. He has been a member of the Canadian Cartographic Association since 1975 and was President in 1999.

2 Climate, Hydrography, and Soils of Georgia

Levan Tielidze, Vazha Trapaidze, Lia Matchavariani and Roger Wheate

Abstract

This chapter details the climate of Georgia; it describes the climate regions and areas such as the western and eastern Greater Caucasus, Kolkheti Lowland, Imereti Highland, Shida Kartli and Kvemo Kartli plains, Saguramo-Gombori middle highland, Iori Upland and Alazani Valley, mountainous Achara-Guria, Trialeti, Meskheti, and Javakheti Upland. Subchapters describe the air temperature and atmospheric precipitation. The fresh water resources of Georgia are also characterized: rivers, lakes, glaciers, marshes, water reservoirs, and underground waters. The last section is dedicated to the soils of Georgia.

Keywords

Climate of Georgia · Rivers of Georgia
Lakes of Georgia · Glaciers of Georgia
Marshes of Georgia · Water reservoirs of Georgia · Underground waters of Georgia
Soils of Georgia

2.1 Climate

Most global climate zones are represented in the relatively small territory of Georgia, ranging from the humid subtropical to eternal snow and glacier zones (Javakhishvili 1981). The diversity of the climate of Georgia is identified both by its location on the northern border of the subtropical zone among the Black and Caspian seas, and by the specific complexity of its landscape: the mountain ranges of different orientation and elevation play a role in creating the local climate. The Greater Caucasus mountains protect Georgia from the direct intrusion of cold air masses from the north, while the Black Sea moderates temperature fluctuations and contributes to variations in precipitation, especially in western Georgia.

L. Tielidze (✉)
Department of Geomorphology, Vakhushti Bagrationi Institute of Geography, Ivane Javakhishvili Tbilisi State University, Tbilisi, Georgia
e-mail: levan.tielidze@tsu.ge

V. Trapaidze · L. Matchavariani
Department of Geography, Faculty of Exact and Natural Sciences, Ivane Javakhishvili Tbilisi State University, Tbilisi, Georgia

R. Wheate
Natural Resources and Environmental Studies, University of Northern British Columbia (UNBC), Prince George, Canada

L. Tielidze
Department of Earth Sciences, Georgian National Academy of Sciences, Tbilisi, Georgia

L. Tielidze (ed.), *Geomorphology of Georgia*, Geography of the Physical Environment,
https://doi.org/10.1007/978-3-319-77764-1_2

Due to its relatively lower latitude location and moderate cloudiness, Georgia receives considerable direct solar radiation heat, with annual sunlight of 1350–2520 h. Total radiation is 115–153 kcal/cm^2 per year while the radiation balance is variable, from the maximum (52–53 kcal/cm^2) in the humid subtropical lowlands and the minimum (25 kcal/cm^2) in the high mountainous zone of the Greater Caucasus (Mumladze and Lomidze 2012a).

Atmospheric processes, from temperate and subtropical zones, participate in the generation of the climate. Zonal circulation is disturbed by the meridian circulation, during which cold air masses intrude from the northern latitudes, and warm air masses from the south. In such cases, the Greater Caucasus and the southern highland of Georgia are of particular importance. Air masses in Georgia mainly intrude from the west and east, but the weather often changes due to warm air masses intruded from the south. In the formation of climate the air masses of moderate latitudes, as well as both the marine and continental arctic and tropical air masses are all of great importance. Subtropical processes, through which air masses intrude into Transcaucasia and, in particular, in the territory of Georgia, are grouped into the following types: (1) western, (2) eastern, (3) bilateral, (4) anticyclonic condition, and (5) wave disturbance, developed in the southern regions of the Transcaucasia (Kordzakhia 1961).

2.2 Climate Zones and Regions

Georgia is in the subtropical belt according to the solar radiation regimes. By the nature of atmospheric circulation and related weather conditions related, the territory of Georgia is divided into the two circulation air regions and one sub-region; these are: (1) subtropical marine humid climate region, (2) transitional from subtropical continental climate to marine climate region; the latter includes the sub-region transitional from dry subtropical climate to moderate humid climate of the Asia Minor highland (Javakhishvili 1981; Mumladze and Lomidze 2012b) (Fig. 2.1).

The first zone covers the area of western Georgia and is characterized by the distinctive features of the humid subtropical maritime climate. The lower part is distinguished by mild winters and relatively cool summers, with moderate temperature amplitudes, abundant precipitation, and high humidity.

The second zone covers the territory of eastern Georgia and is characterized by moderately humid subtropical climate. Here, the winter is colder than in the first zone and precipitation is relatively small low. A subzone covers the central steppe part of the southern highland of Georgia. Here, the climate is more continental, the summer is hot, the winter is cold and the amount of atmospheric precipitation is less than in the other areas located at the same elevation in Georgia.

Significant fragmentation of the landscape transforms the general atmospheric circulation and causes great differences in numerical values of meteorological elements creating climatic regions in the territory of Georgia.

2.2.1 Western Greater Caucasus

The western Greater Caucasus is located in the marine subtropical humid climate region. Due to lower latitude and higher solar altitude, it receives a large amount of solar energy throughout the year. It is under the influence of humid winds blown from the west and Black Sea. The main feature of the climate is the altitudinal zonality. The average annual temperature (5–12 °C) in the lower parts becomes negative at a height of 2500 m asl. The average January air temperature, which is 1–2 °C at a height of 700–800 m, drops to −15 °C in the highland zone, decreasing to absolute minimum temperature of −25 to −40 °C. The temperature in July–August averages 6–22 °C in the western Greater Caucasus, and absolute maximum of 20–42 °C. Annual precipitation varies within 1800–3500 mm, except for landlocked lower areas

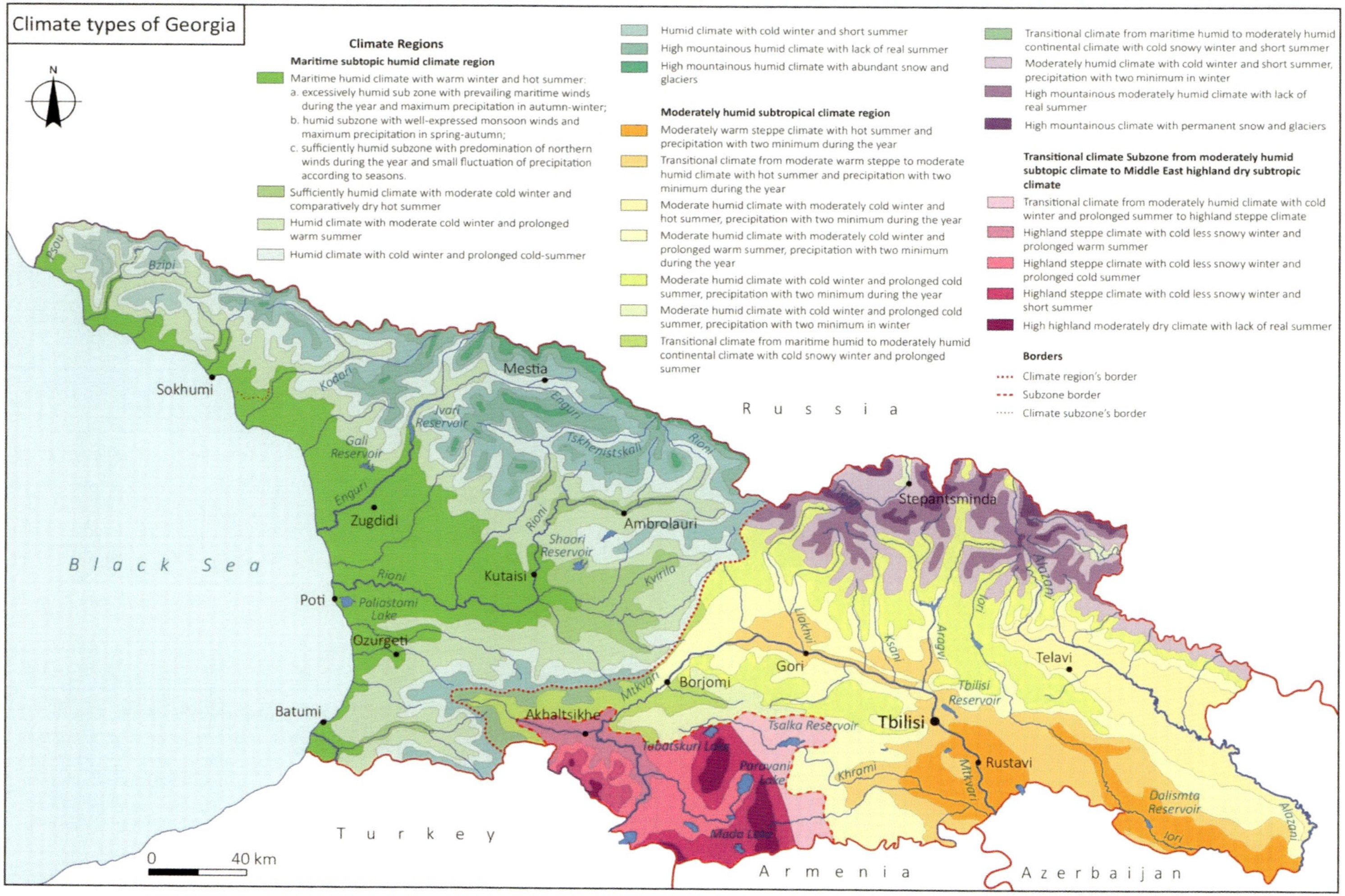

Fig. 2.1 Climate types of Georgia (Kordzakhia 1964; Mumladze and Lomidze 2012b)

(in Svaneti and Racha-Lechkhumi), where it ranges from 900–1200 mm. The windward slopes receive more precipitation.

2.2.2 Eastern Greater Caucasus

The eastern Greater Caucasus is located in the region transition from subtropical continental climate to the marine climate. The average annual temperature is 8–10 °C in the lower part, and −6 °C at 3500 m asl. The coldest months (January, February) temperature are −3 to −15 °C, with absolute minimum between −26 to −42 °C; the warmest months (July and August) have temperature of 2–18 °C, and the maximum is 16–40 °C. The precipitation increases with height and territorially varies within 800–1800 mm.

2.2.3 Kolkheti Lowland

The Kolkheti Lowland is characterized by an excessive humid subtropical climate. The region is mostly affected by the Black Sea, therefore, the winter is not cold and summer is relatively cool; the Greater Caucasus prevents the intrusion of cold air masses. Average annual temperature is 13–15 °C, January 2–7 °C, July–August 22–23 °C, absolute minimum temperature is −9 to −27 °C and the maximum 40–43 °C. In winter, eastern winds prevail, in summer—western winds, and breezes and foehns are frequent. Annual precipitation of 1400–3000 mm falls mainly in winter and autumn, but persistent snow cover is rare. With the influence of sea proximity and frequent western winds, cloudiness, and humidity are high all year round.

2.2.4 Imereti Highland

The impact of the sea on the Imereti highland is weakened, the climate is relatively dry, and the winter is noticeably colder than in the Kolkheti Lowland, but the signs of humid subtropical climate are still maintained. Average annual temperature is 5–13.5 °C, January average temperature in the lower part is 0 to −3 °C and in the high mountain region −5 to −6 °C; temperatures can drop to between −20 and −3 °C. The average temperature of the warmest month is 15–23 °C and the maximum temperature is 30–40 °C. Precipitation varies from 900–1800 mm; cloudiness is high (60–65%), as well as the relative humidity (75–80%). Mountain-gorge winds are frequent, and like the Kolkheti Lowland, eastern foehn winds prevail in winter and western winds in summer. Thunderstorm and hail are frequent in the region mainly in warm periods.

2.2.5 Shida Kartli and Kvemo Kartli Plain

The dry subtropical climate prevails in the Shida Kartli Plain, which is influenced by the surrounding high mountain ranges. Inversions develop during cold spells so that winter is colder than in the other areas of Georgia, at the same height. Average annual temperature varies between 9 and 11 °C; January −1 to −4 °C, and August 20.4–22.3 °C. The absolute minimum temperature is −26 to −31 °C and the maximum is 35–40 °C. Western and eastern winds prevail; western winds are cool and humid, and eastern winds are humid and cold in winter and hot in summer. Due to the frequency of windy days, favorable conditions are created for wind power stations. The first power plant was put in operation in October, 2016 (Fig. 2.2).

Annual precipitation is 500–800 mm in the Shida Kartli Plain, where droughts are frequent, and during warm spells, evaporation is much higher than precipitation. In 1929, to combat the drought, the Tiriponi channel was constructed, which starts on the left bank of the Didi Liakhvi River to the south of Tskhinvali town, goes to the southeast and crosses the rivers of the Patara Liakhvi, Adzura and Charebula and ends at the Kirbalula River near the Patara Khurvaleti Village. The length of the channel is ~42 km and the water discharge is 14 m^3/s. In Gori region 30.6 thousand hectares are irrigated by this channel (Apkhazava 1985).

Fig. 2.2 Wind power stations in Shida Kartli Plain (*photo by* L. Tielidze)

Air masses intrude into the Kvemo Kartli Plain from the east, and from the west through the Mtkvari (Kura) River gorge. The weather is especially affected by the wave disturbances in the south of Transcaucasia, and the abundant precipitation during the warm spell, as well as thunderstorm and hail (Javakhishvili 1977). The average annual air temperature is 12 °C, January is 0.2 °C; July and August are especially hot (23–25 °C). The absolute minimum temperature is −20 to −25 °C and the maximum 40–41 °C. The annual precipitation is 400–600 mm; the south is especially dry and droughty. The evaporation is much higher than precipitation, and snow rarely occurs.

2.2.6 Saguramo-Gombori Middle Highland, Iori Upland and Alazani Valley

The Saguramo-Gombori middle highland experiences a moderately humid climate, cold winters, and long cool summers. Average January temperature is −2 to −6 °C and July 15–20 °C. Precipitation does not increase significantly with elevation, rising from 800–850 mm in the mountain foothills to 1000 mm on the crests. Nevertheless, the air is quite saturated with water vapor, the average annual relative humidity is 75–80%. Fog is frequent along with the mountain-gorge winds.

Dry continental climate prevails in the Iori Upland, where the winter is cold and summer is hot and dry. The average annual air temperature is 10–11 °C, the average in January is 1–3 °C and July–August 22–24 °C; the absolute minimum is −24 to −32 °C and the maximum 40 °C. The annual amount of precipitation is 400–500 mm and snow cover is rare. Western winds dominate and especially strong winds blow in winter.

In the Alazani Valley a humid subtropical climate dominates with hot summers and moderately cold winters. The valley is well protected from the west and north. The air masses intrude freely only from the southeast, which affects the peculiarity of the climate of this region. The average annual temperature is 11–13 °C and the January average is 0 to −1 °C, but there are exceptional years, when temperatures fall below 0 °C. The average air temperature of the warmest month is 21–25 °C. The absolute minimum is −25 to −27 °C and maximum reaches 40 °C. In most parts of the region the annual amount of precipitation is 800–1300 mm. This region is distinguished from the other regions of Georgia not by the frequency of days with hail, but with the size of hail and the damage caused.

2.2.7 Mountainous Achara-Guria

The mountainous Achara-Guria region faces the west and the Black Sea. It is characterized by a humid climate and is especially hot in the lower part. The average annual air temperature varies between 2–12 °C. The average temperature of January is 0 °C (up to 1100 m asl), falling to −8 to −10 °C on the mountain the crests. The warmest month is August (10–20 °C). On the seaward slopes, throughout the year, the west and southwest winds dominate, bringing a large amount of moisture. This region is distinguished by abundant precipitation, although the middle part of the Acharistskali River gorge is dry, protected from the west by high mountain ranges. Snow may commence in November, remaining for 1–3 months at a height up to 1400 m a.s.l., and for 6–7 months—up to 2000 m. The maximum depth of the snow cover is 4–5 m.

2.2.8 Trialeti Range

Although the Achara-Imereti Range weakens the intrusion of air masses from the west, its influence is clearly visible to the neighborhoods of Borjomi-Bakuriani. In the eastern slopes of the Trialeti Range, air masses intruded from the east leave a visible trace. Northern and southern slopes of the Trialeti Range, the alongside air currents are formed here; the amount of precipitation is low and the climate is continental. Borjomi Gorge experiences warmer winters, but it is cold in the high mountainous parts. The average January temperature is −2 to −11 °C; the absolute minimum can decrease to −40 °C. The temperature of the warmest month is 9–20 °C, and the maximum reaches 37 °C. It is distinguished with mild cloudiness, long hours of sunshine and deep snow cover. Climatic conditions and the landforms promote conditions for winter sports events, and notably the mountain-ski resort Bakuriani (Fig. 2.3) on the northwest slope of the Trialeti Range.

2.2.9 Meskheti and Javakheti Highland

The Meskheti region is located in the depression and the effect of the intruded air masses is weakened, precipitation is negligible, climate is dry continental. Winters are cold because of strong inversions and the average January temperature is −2.5 to −9 °C. The absolute minimum is −30 to −38 °C; summer is moderately hot, average July–August temperature is 16–21 °C, maximum reaches 39 °C. Annual precipitation is 500–700 mm, and especially dry in summer. Snow commences in October, becoming persistent in December and remains until the end of March.

The continental climate prevails over the Javakheti highland, with very cold winters. The climate is transitional from moderately humid to highland dry climate. The average annual temperature is 4–6 °C, January −5 to −10 °C, July's 15–16 °C; absolute minimum is −34 to −41 °C and absolute maximum is 30–35 °C. Precipitation amounts to 600–750 mm per year. The frequency of frosty days, low cloud cover, and especially dry winters are related to the frequency of the anticyclones characteristic to this area. The snow cover appears in December and remains till March.

2.3 Air Temperature

In the territory of Georgia the highest average annual temperature (14–15 °C) is characteristic to the Black Sea coast and the Rioni River Gorge (up to Kutaisi). In eastern Georgia the warm climate is also typical in the middle part of the Mtkvari River gorge and part of the Alazani Valley, where the average annual air temperature reaches 13 °C in some locations, e.g., in Lagodekhi and Tsnori (Gagua and Mumladze 2012) (Fig. 2.4).

The specific peculiarity of the thermal nature of the ground surface of the territory of Georgia is the large contrasts in temperatures, which is

Fig. 2.3 Bakuriani mountain-ski resort located on the northwestern slope of the Trialeti Range (*photo by* R. Gobejishvili)

mainly caused by the vertical zonality. Together with the above cited warm areas, there are areas, where the average annual temperatures are below zero; for example, Mamisoni Pass −2.2 °C and Kazbegi (high mountainous) −5.8 °C.

The average winter temperatures are high in the Kolkheti Lowland, due to the physical and geographical location, as well as the frequency of the eastern rhumb winds of foehn nature. In most territories of western Georgia, the average temperature of January is above 0 °C becoming negative from 550–600 m asl. Batumi experiences the warmest January, where the average air temperature reaches 7 °C. The temperature gradually decreases from the sea to the east and by elevation. It is especially cold in this month in the higher zone of the Greater Caucasus (high mountainous Kazbegi, −14.7 °C; Mamisoni Pass, −11.6 °C), as well as Javakheti Upland (Efremovka, −10.6 °C; Paravnistskali, −8.2 °C), where along with the absolute height the landform also influences the decrease in temperature. For example, in Gudauri (2194 m a.s.l.) it is warmer by 3.2 °C than in Efremovka (2114 m a.s.l.), where the average temperature of January is −10.6 °C.

Javakheti Upland is the coldest region of Georgia, and may be referred to as the "Siberia" of Georgia. Thus this area is distinguished by low absolute minimum temperatures (Kartsakhi, −41 °C; Efremovka, −39 °C, etc.). On the northern slope of the Greater Caucasus (high mountainous Kazbegi), at the height of 3653 m a.s.l., the air temperature of −42 °C was documented in January 1983 (Tielidze and Wheate 2018).

If annual and January average temperatures decrease from the sea to the east, the average July temperature, the central month of summer, contrarily increases. It is very hot in July in the middle and downstreams of the Mtkvari River, particularly in the Kvemo Kartli steppe part (in Gardabani this indicator reaches 25.5 °C), where the average July temperature is 2–3 °C higher than in the coastal and other lowland regions of western Georgia. This climate indicator is also high in Kakheti, namely, in Tsnori 24.9 °C, in Lagodekhi 24.5 °C and in Gurjaani 24 °C. In western Georgia, the relatively hot July is characteristic of the marginal eastern central part of the Kolkheti Lowland (Samtredia 23.2 °C and Sakara 23.5 °C) and the Apkhazeti coast (Akhali Atoni 23.5 °C), where in summer the cloudiness and precipitation are half those in the Achara coast.

The entire intermountain lowland of Georgia with its foothill hillocks is distinguished by absolute temperature maxima. With increase in

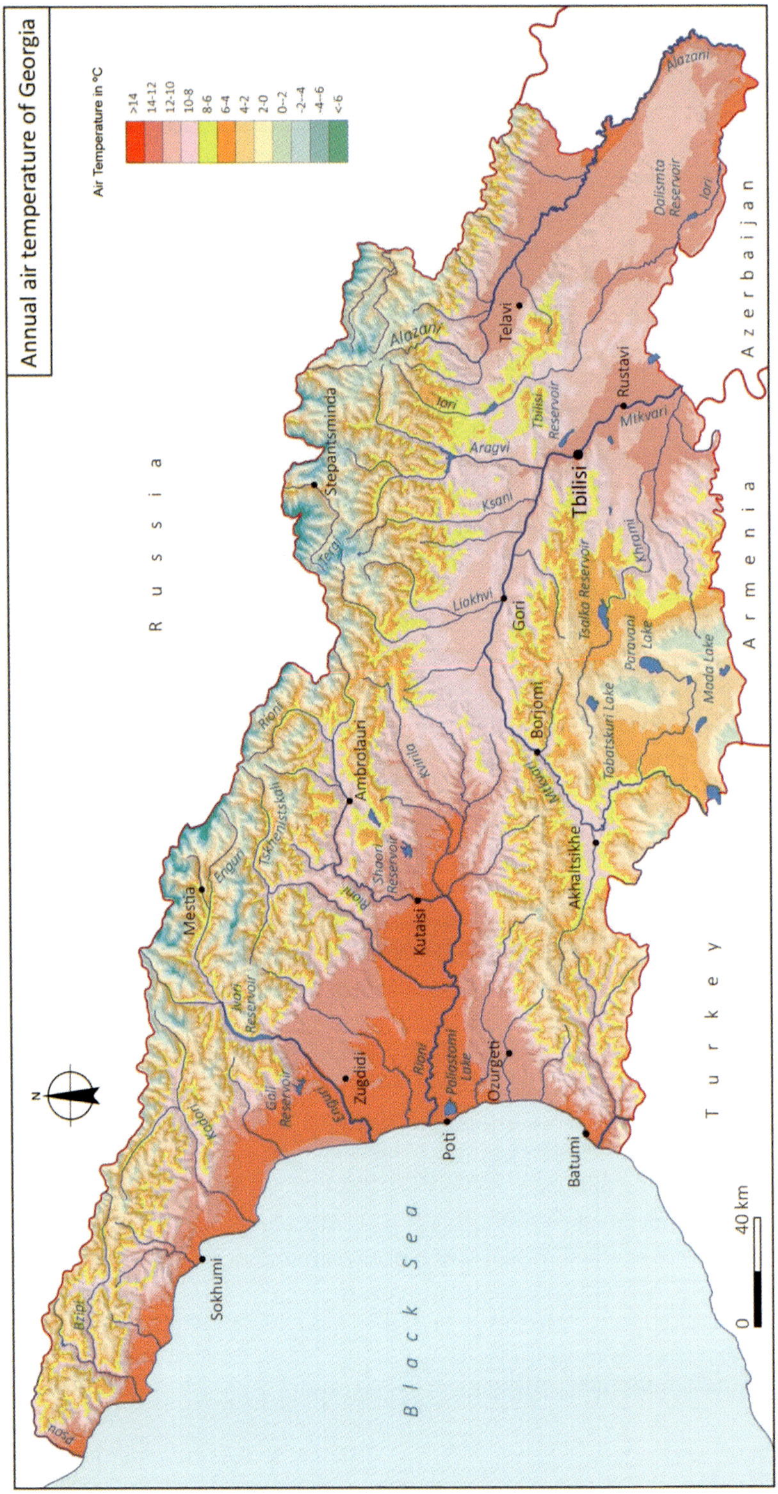

Fig. 2.4 Average annual air temperature in Georgia (Gagua and Mumladze 2012)

height, temperature decreases, although the landform and ground surface feature have a significant impact on the vertical zonal distribution of air temperature. For example, in Barisakho (located 390 m lower than Akhalkalaki) the absolute maximum temperature is 33 °C, and in Akhalkalaki, located at a height of 1716 m, it is 34 °C.

Autumn in Georgia is considerably warmer than spring. The difference between the October and April average temperatures is 1.5–2.5 °C on average. However, in the coastal area, for example, in the Akhali Atoni, it reaches 4.8 °C. This difference is reduced from the sea towards the east.

2.4 Atmospheric Precipitations

The complex orographic structure of Georgia and its interaction with the air masses, spread across the country, determines the large contrasts in the distribution of precipitation. From west and northwest of Europe the moist air masses are spread into Georgia across the Black Sea. In the cold seasons, the air masses from the Caspian Sea contain enough moisture, but in the warm seasons, the moisture content is significantly reduced. Air masses that come from the Caspian Sea to western Georgia do not produce precipitation. Warm, dry weather accompany air masses from the south.

Most precipitation Georgia (30–35%—in western Georgia and 20–30% in eastern Georgia) is related to the intrusion of air masses from the west. When air masses intrude from the east, 10–15% of the total annual precipitation falls only in the eastern part of the country. Abundant precipitation follows the bilateral intrusion of air masses simultaneously from the west and east. Wave disturbances and convection processes in the south of Georgia are associated with significant precipitation.

Average annual precipitation in Georgia is 1250 mm, but varies widely. The coastline of Achara is the highest in precipitation, exceeding 4500 mm at Mt. Mtirala near Batumi (Dolidze and Kartvelishvili 2012) (Fig. 2.5). In the north, along the sea coast, precipitation gradually decreases to 1400–1500 mm.

The seaward slopes of the Apkhazeti, Samegrelo and Racha (Tkibuli region) ranges are characterized by abundant precipitation. In the Kolkheti Lowland, the amount of precipitation decreases with distance from the sea. Despite proximity to the sea, precipitation is much less in the depressions of Shida Achara, Zemo Svaneti, Kvemo Svaneti and Racha and in Zemo Imereti, than in the other regions of western Georgia.

In comparison with the western part of the country, much less precipitation falls in eastern Georgia. Only mountainous southern slopes and especially the high mountainous regions of the Greater Caucasus are characterized by abundant precipitation, which amounts annually to 1200–1800 mm. In the agricultural areas such as Shida Kartli and Kvemo Kartli, Samtskhe-Javakheti, Shiraki and Eldari, the annual precipitation does not exceed 400–600 mm. Small amounts of precipitation fall on the northern slopes of the Georgia's Greater Caucasus (Kazbegi, Pirikita Khevsureti, Tusheti). The spatial distribution of precipitation during the warm and cold periods of the year is not significantly different from the annual distribution. However the intra annual distribution of precipitation in the eastern and western parts of Georgia substantially differs. The maximum precipitation in eastern Georgia is observed in May and June, with a weak secondary maximum in September–October. The minimum is in January and a secondary in August. Maximum precipitation in western Georgia is observed in most cases in October, the rest is distributed in November, December and January. The annual minimum is in April–May in the coastal zone and nearby areas and in July–August in the internal regions.

2.5 Fresh Water Resources

Fresh water resources are one of Georgia's main strengths. The mountainous nature, abundant precipitation, especially in the Black Sea basin, endow Georgia with an average height of the fresh water layer formed during a year, that is

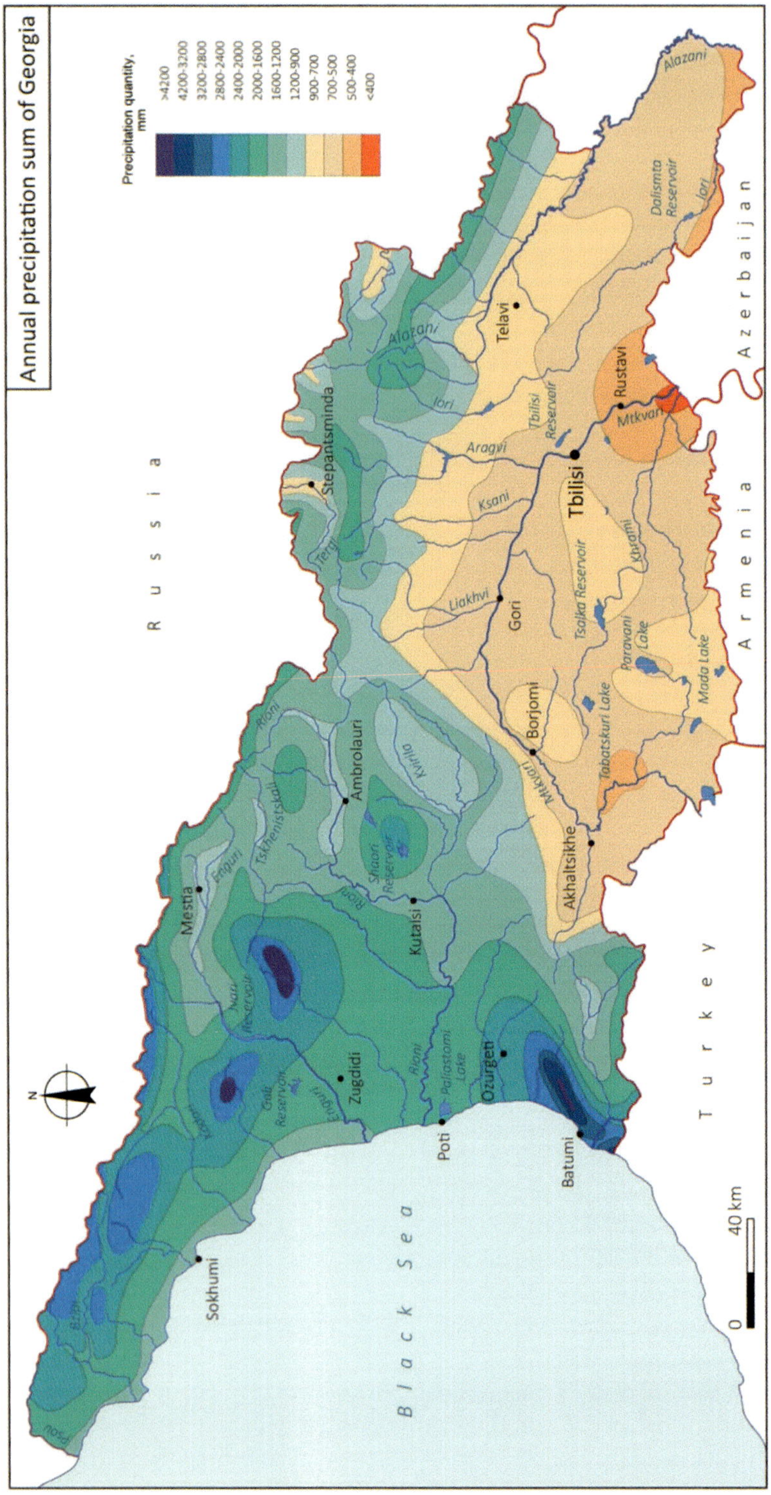

Fig. 2.5 Average annual precipitation in Georgia (Gogishvili et al. 2012)

only behind Norway, Switzerland and Austria among European countries. This is mainly due to the annual precipitation layer, which amounts to 93.3 km^3 in volume (Trapaidze 2012).

The average annual water amount per square kilometer is unevenly distributed; for western Georgia it amounts to 1.34 million m^3/km^2, and for eastern Georgia 0.37 million m^3/km^2. The average annual runoff, amounts to 57 km^3.

About 41% of the runoff belongs to underground waters, with the addition of 8.68 km^3 of the water flowing from neighboring countries annually, including 7.75 km^3 of transit runoff through the Mtkvari and Chorokhi rivers from Turkey. Freshwater water resources of the rivers of Georgia amount to 61.45 km^3, about 78% or 48 km^3 out of which flows into the Black Sea, and 13.45 km^3 flows into the territories of neighboring countries.

As a result of long-term observations, the water resources of Georgia are distributed by water bodies as follows: surface runoff 31%, transit runoff 9%, lakes 0.7%, underground runoff 22%, water reservoirs 3.3%, glaciers 30%, marshes 2% and others 2% (Trapaidze 2012).

2.5.1 Rivers

There are ~26,060 rivers in Georgia and their total length is approximately 60,000 km. 18,109 rivers belong to the Black Sea basin and 7951 to the Caspian Sea basin, comprising 30% of the total rivers of Georgia.

A large percentage of the rivers of Georgia belongs to the mountain river type. Due to the fragmentation of the surface of the relief, there are no rivers with great length and wide basins; the average length of the rivers is 2.3 km.

Rivers shorter than 25 km are the basis of the hydrographic network of Georgia, accounting for 99.4% of the total number of rivers, and 84.5% of the total length of the rivers. Rivers can also be tallied as small, medium and large very large rivers.

The average density of the river networks of Georgia, i.e., the area of the total length of the rivers in the total area of the territory, is 0.86 km/km^2, but they are not equally distributed. In general, with the decrease of precipitation from the west to the east, the network density is similarly reduced. The average density is 1.07 km/km^2 in western Georgia, while it is 0.68 km/km^2 in eastern Georgia. Against the background of this general trend of the distribution of the river network density, the results of the orographic influence are sharply highlighted. Specifically, on the western slope of the Chakvi Range, where the humid air masses bring 2500–4000 mm of precipitation annually, the density of the river network is 2.0–2.6 km/km^2 (the rivers of Cholokhi, Kintrishi, Chakvistskali, Partskhana, etc.). To the east of the range, with the reduction of precipitation to 1000–2000 mm, the network density is also reduced to 0.99 km/km^2 in the upper stream (Skhalta River) of the Acharistskali River.

Temperature conditions influence the density of the river networks. This is particularly evident in the high mountainous zone of the Greater Caucasus—higher than 2400–2600 m above sea level, where the river network density is reduced with air temperature. In the regions with average annual air temperature of −2 to 3 °C, the river network density is 0.3–0.4 km/km^2; in the colder regions, it is reduced to 0.06–0.08 km/km^2 and in the zone with average annual air temperature of −5 to −6 °C there are no rivers. Precipitation here is accumulated in the form of firn and ice and do not contribute to surface runoff.

In addition to climatic conditions, the geological structure plays an important role in distribution and density of the river networks. For example, the fissured mountain rocks of Javakheti Upland sharply reduce network density and increase groundwater reserves. River density in the Paravani River basin is ~0.35 km/km^2, and in some tributaries 0.7 km/km^2. A similar situation is found in the highly humid karst zone of western Georgia, where the density of the river networks is 2–2.5 times lower than in neighboring non-karst zones at 0.05–0.6 km/km^2. In this zone, the rivers—Iupshara, Amtkeli (the Kodori River basin), Shareula, Tkibula and Shabatghele (Rioni River basin) disappear, after which they again flow out on the surface.

The variation in density of the river networks according to elevation is complex. In southern Georgia upland and karst zone, the density decreases with increased elevation from 2.0 to 0.5 km/km^2 due to precipitation reduction and karst processes, and in high mountain regions (above 1800–2000 m a.s.l.) it decreases to 1.2–0.2 km/km^2 due to temperature. The upper limit of the presence of river network is 3200–3300 m in average, which in its turn, corresponds to the average height of the firn line.

The density of the river networks in the rest of Georgia ranges from 0.35 to 1.3 km/km^2. From the Black Sea coast the density of the river networks increases with elevation from 0.3 to 2.2 km/km^2 due to the reduction in swamps. It should be noted that in many regions, the actual density of the river network, considering irrigation systems, is much more than the natural density. The total length of only the channels of irrigation systems of the Zemo Kartli Upland is 16,500 km, which is 3 times higher than the total length of natural rivers of the region.

Although the rivers of Georgia are short in length and have small catchment basins, they are characterized by high falls, fast flow and potential energy reserves, especially the rivers of the Black Sea basin. They flow through narrow and deep gorges, where they often create canyons, sweeps, rapids, waterfalls, and other riverbed forms.

In western Georgia, the rivers of the southern slopes of the western part of the Greater Caucasus Range, such as Bzipi, Kodori, Enguri, Tskhenistskali, Rioni and its tributaries, originating from the eternal snow and glaciers, are distinguished by abundant water, along with the rivers of the western slope of the Chakvi Range: Chakvistskali, Kintrishi, etc. The rivers in the Kolkheti Lowland are typically short, originating in the hillock zone of the foothills and are characterized by large meanders and low speeds.

In eastern Georgia, the Mtkvari River and its first rate tributaries: Didi Liakhvi, Aragvi, Alazani, Iori, Paravani, Ktsia-Khrami, etc., contain abundant water. The first four rivers—Didi Liakhvi, Aragvi, Alazani, and Iori start from the southern slopes of the Greater Caucasus. The other rivers—Paravani, Ktsia-Khrami, etc., flow down from the uplands of the southern highland with lower flow.

In addition, dry ravines are mainly found in the Iori Upland, Kvemo Kartli Plain and the lowland near the Mtkvari River, where the water flows only during snow melt and heavy rains.

The highest flow river within Georgia is Rioni (Fig. 2.6) with average annual runoff 12.66 km^3, then the Mtkvari River 9.39 km^3. The rivers of Chorokhi (8.73 km^3), Enguri (6.04 km^3), Kodori (3.94 km^3) and others are also high flow levels. There is a water reserve of about 187 million m^3 in average in the riverbeds of Georgia, which is 0.30% of the total annual runoff of river waters (Trapaidze 2012).

2.5.2 Lakes

Georgia is a country of small lakes, numbering approximately 856 lakes, with more than half having an area less than 0.1 km^2. Thus the total area of Georgia's lakes is only 170 km^2, or 0.24% of the entire country.

The lakes range in depth from very shallow lakes 20–30 cm deep (saline lakes in the Iori Upland, etc.) to deep lakes (Ritsa—101 m, Patara Ritsa—76 m, Amtkeli—67 m, Keli—63 m, etc.). The most common lakes have a depth of 2–10 m. Some lakes in the drought zone of eastern Georgia are completely dried up during the summer period and filled with water only during snow melt and rainy periods.

Because of the small size of the lakes of Georgia, the volume of water is not large; only the Tabatskuri Lake volume exceeds 200 million m^3. Average water reserve of the lakes of Georgia is 723.24 million m^3, and part of this volume is involved in river runoff (Trapaidze 2012).

The lakes of Georgia are quite diverse in genesis as well. Lakes are formed as a result of both, endogenous and exogenous geological processes. Endogenous processes led to the formation of tectonic and volcanic depressions. The lakes of Kartsakhi, Bazaleti, etc. (Fig. 2.7) are of a tectonic origin; the lakes in the Javakheti Upland are mostly of volcanic origin.

Fig. 2.6 Rioni River valley (in Racha) (*photo by* L. Tielidze)

Fig. 2.7 Bazaleti Lake of tectonic origin (Google Earth imagery, 6/5/2015)

The floodplain lakes ("narionalebi") in the downstream of the rivers of Rioni, Khobi, Supsa and Alazani) and karst lakes (Satsurblia, Shavmoreva and Devi) belong to the exogenous lakes. Glacial lakes (Aduedaadzishi, Levani, Okrostskali and many nameless lakes) are found in the Bzipi and Kodori Ranges and Javakheti Upland; pond lakes (Ritsa, Keli, etc.) in the Bzipi, Ksani and other river basins; lagoon lakes (Paliastomi Lake in the Pichora River Basin, Anikhtsara and Inkiti in the Bichvinta Cape, Nurie in Batumi, etc.) in the Black Sea coastal zone; landslide lakes the Turtle Lake in the outskirts of Tbilisi, Udziro in the Gombori Range, Tsunda in the Mtkvari River gorge, etc.

2.5.3 Glaciers

In the territory of Georgia the glaciers are distributed both on the southern and northern slopes of the Greater Caucasus Range and on its sub-ranges (in the most upraised areas of the ranges of Bzipi, Ckhalta, Kodori, Egrisi, Lechkhumi, Racha, Khokhi, Kuro, Kidegani, and Atsunta). From west to east the glaciers are located in the following river basins—Bzipi, Kelasuri, Kodori, Enguri, Rioni, Didi Liakhvi, Aragvi, Asa, Tergi (Terek), Arghuni, and Pirikita Alazani (Tielidze 2016).

In total, there are 725 glaciers documented in Georgia with total area (2014) of 369.8 ± 8.7 km^2 (Tielidze and Wheate 2018), which is 0.53% of the whole territory of Georgia. Most of the glaciers are concentrated in western Georgia. The largest glacier of Georgia is Lekhziri with total area 23.76 ± 0.36 km^2 (Fig. 2.8).

Distribution of glaciers of Georgia by individual river basins is given in the Table 2.1 (Tielidze and Wheate 2018).

Water reserve trapped among the glaciers is an important characteristic for glacier resources. Measurements of thickness and volume (e.g., by

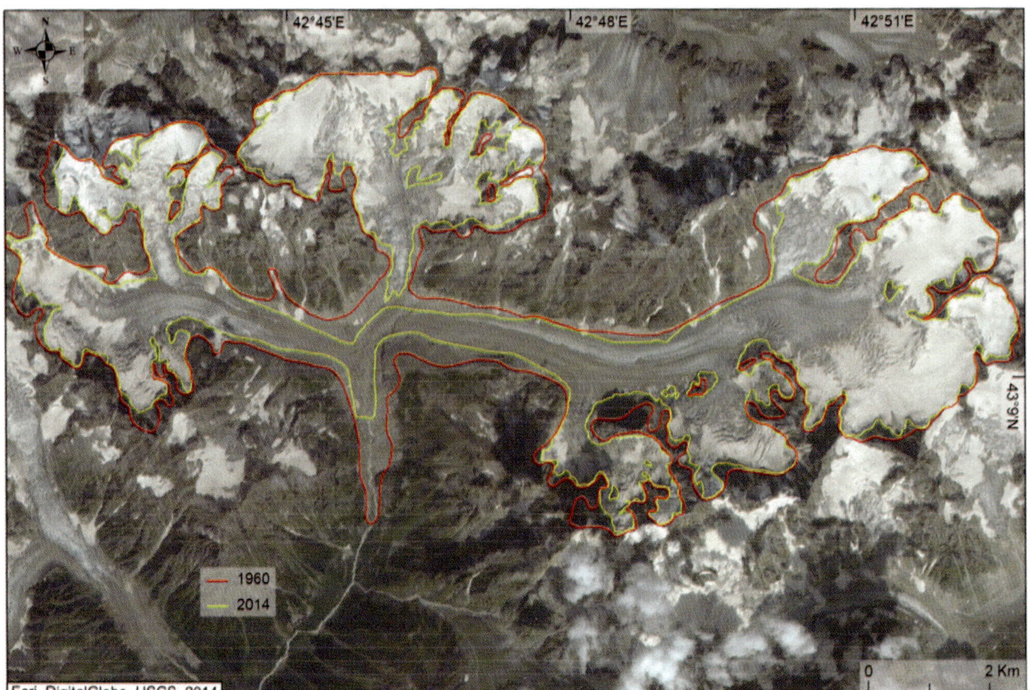

Fig. 2.8 Lekhziri, the largest glacier of Georgia

Table 2.1 Distribution of the glaciers of Georgia by river basins

№.	River basins	Number of glaciers	Glacier area, km^2
1	Bzipi	19	3.2 ± 0.4
2	Kelasuri	2	0.9 ± 0.06
3	Kodori	161	42.2 ± 2.8
4	Enguri	289	225.3 ± 9.2
5	Khobistkali	11	0.6 ± 0.08
6	Rioni	120	50.9 ± 2.6
7	Liakhvi	12	2.0 ± 0.16
8	Aragvi	1	0.3 ± 0.04
9	Tergi headwaters	74	40.2 ± 3.2
10	Asa	4	0.8 ± 0.1
11	Arghuni	12	0.8 ± 0.14
12	Pirikita Alazani	20	2.6 ± 0.2

radio echo sound methods) of the glaciers of Georgia have so far not been conducted. However some researchers (Ukleba 1997) assessed the glaciers' volume based on the empirical relations between glacier area and glacier volume. The amount of ice in the glaciers of Georgia assessed by this method is about 30.13 km^3. In the Enguri River basin alone, 22.462 million m^3 of ice is accumulated, exceeding by four times the average annual runoff of the Enguri River.

Glacial water runoff, in alignment with glacier areas and volumes is distributed between regions and basins. About 88% of the glacial water runoff belongs to western Georgia, where just the glaciers of the Enguri basin produce 56% of total runoff.

The average annual glacier runoff of the glaciers of Georgia is 1.50 km^3 in total, i.e., only 5% of the total water reserves of the glaciers of Georgia participates in water circulation annually. Most of the water reserves (95%) is trapped among the glaciers of Georgia (Trapaidze 2012).

2.5.4 Marshes

Marshes are the undrainable wetland areas in Georgia that are covered with peat layer about 30 cm thick. Peat structure dictates its ability to accumulate a large amount of water (more than 95% of its volume). In the recent past, marshes had a significant place in Georgia, especially in the Kolkheti Lowland while peat bogs were scarce (Fig. 2.9). The area of marshes together with wetlands was 256,000 ha, including 225,000 ha in western Georgia (Trapaidze 2012).

At present, part of the marshes and wetlands are drained. Marshes are found only in the western and coastal areas of Kolkheti Lowland and their total area is about 627 km^2. This area is characterized by low elevations, as well as by slight sloping of the surface and hydrographic networks that facilitate the accumulation of water masses.

The marshes influence on the climate, hydrological and hydrogeological regime represent an important factor in determining the ecological balance of the area. By genesis, the marshes of Georgia belong to the lowland marshes and are in the first stage of development. Smaller areas of highland marshes are represented in Javakheti Upland and in some other regions too.

In nature, according to the water cycle, the water in the wetlands is constantly moving in time and space, but this movement is very slow. The Kolkheti Lowland waters are mainly used for evaporation and water filtration in the river network; therefore, the course of the renewal of the reserves of these waters is long-term and they are being slowly restored.

Fig. 2.9 Kolkheti marshes (*photo by* G. Lominadze)

2.5.5 Water Reservoirs

An important part of the water resources of Georgia is accumulated in their reservoirs. Currently, 44 reservoirs with a volume of more than 0.5 mln. m^3 are operating. Their total volume is 3.32 m^3, and annually renewable useful volume is 2.27 km^3 (Trapaidze 2012).

Among the water reservoirs, eight are built in western Georgia and except for one irrigation reservoir, all are for energy purposes, with overall volume 0.85 km^3. Due to the relatively low humidity of eastern Georgia, most of the reservoirs there are for irrigation purposes.

The water reservoirs serve several types of water users. For example, the main purpose of the Zhinvali water reservoir is energy supply, irrigation and drinking water supply. However, in addition to the main purpose, all the reservoirs are used for fish farming, recreation and mitigation of natural disasters, such as floods.

Some water reservoirs in Georgia have been created in the depressions of former lakes, including Shaori, Tbilisi and other water reservoirs. Some artificial water reservoirs are natural lakes, but conditionally they are treated as water reservoirs, because their feeding is produced by artificial channels, e.g., the Kumisi Lake, near Tbilisi (Fig. 2.10).

Most water reservoirs of Georgia (Jvari, Tkibuli, Tbilisi, Zhinvali, etc.) were created for the purpose of seasonal regulation of the river runoff, while the Shaori, Sioni and Tsalka water reservoirs for a multi-year regulation. Gali, Lajanuri, Gumati, Vartsikhe and other water reservoirs have the ability of only daily regulation.

The water reservoirs are distinguished by fast mixing and quick recovery of their water masses. For example, in the Jvari water reservoir, a full renewal of water is done in 125 days, while the Sioni water reservoir in a year (Ukleba 1997).

Fig. 2.10 Kumisi Lake water reservoir (Google Earth imagery, 29/12/2017)

2.5.6 Underground Waters

Georgia is rich in underground waters, caused by the abundant precipitation (93.3 km^3 per year). These total 21.7 km^3 (311 mm), which is 43% of the surface runoff of the entire territory and 23% of annual precipitation (Trapaidze 2012). Most of these waters are of low mineralization and useful for drinking.

Two kinds of potable groundwaters are found. The hydro-carbonate underground waters belong to the first group, with mineralization degree 0.2–0.3 g/l. and reserve southern Georgia is 0.63 km^3. Potable underground waters with mineralization of 0.3–0.1 g/l belong to the second group, the reserve of which is 2.582 km^3 in the territory of Georgia (Khmaladze 2009).

Atmospheric precipitation and surface waters are the main source of underground water resource renewal. The peculiarities of geological-hydrogeological and physical-geographical conditions of the territory of Georgia identify the diversity of factors in the formation of underground water resources and the unequal distribution of these resources. The water content decreases from west to east, associated with decreasing atmospheric precipitation. In addition, there is an increase in the water content of rocks along with increasing elevation. This is a result of increase in precipitation by elevation, and the role of increasing condensation in feeding water content.

As a result, the water content complexes of high mountain regions are more abundant in water and have more springs than the same complexes in lower locations. This is especially true for the carbonate water horizons of the Paleogene-Upper Cretaceous, Lower Cretaceous and Upper Jurassic periods in the folded zone of the southern slope of the Greater Caucasus within the boundaries of western Georgia.

2.6 Soils

Georgia is characterized by a wide range of soil cover, due to the diversity of soil forming factors. Complex geological structure and heterogeneity of bed rocks, the relief peculiarity and contrasting climate conditions, specificity of vegetation cover and biodiversity, all influence the mosaics of soil cover of Georgia and peculiar geographical distribution. The genetic types characterized for humid subtropical lowland areas and arid regions are widespread in the foothills, mountain-forest and mountain-meadow zones (Matchavariani 2012).

The diversity of soil cover of Georgia is especially noticeable in the lowlands and regions with plain relief; in the mountains such differences are less evident. Each physical-geographical region in Georgia has a complex of characteristic soil types (Sabashvili 1965; Urushadze 1997).

Greater Caucasus region is characterized by the common structure of the vertical zonality of soil cover. However, according to local climatic and other conditions, the soil types' spectra are different within regions. In the southern slopes of the Greater Caucasus, the Brown Forest Soils (Cambisols Eutric and Cambisols Dystric) soils prevail. Above, the Mountain-Meadow subalpine and alpine soils are represented, reaching a high mountainous nival zone in the form of primitive soils.

In the **southern highland** the Mountain Chernozems dominate. At the higher massifs, ranges and slopes, the Mountain-Meadow soils (Leptosols Umbric) are widespread. Lesser Caucasus outer slopes are characterized by Brown Forest (Cambisols Dystric), mountain-forest Cinnamonic (Cambisol Chromic) and Mountain-Meadow (Leptosols Umbric) soils.

In the **Kolkheti Lowland** hydromorphic soils are common, such as Marsh Soils (Gleysols), Subtropical Podzols (Luvisols Albic) and Alluvial Soils (Fluvisols). In the upraised parts of the lowland, the Red Soils (Nitisols Ferralic) and Yellow Soils (Acrisols Haplic) can be found.

Mtkvari Lowland is characterized by dry subtropics soils—Grey and Meadow Grey Cinnamonic Soils (Cambisol Chromic), as well as Salt Soils (Solonetz Humic).

In Georgia, three soil-geographical regions of Western Georgia, Eastern Georgia and Southern Georgia can be distinguished with the corresponding sub-regions, zones and areas (Sabashvili 1965). Within the soil region of western Georgia three soil-geographical sub-regions can be distinguished: the Greater Caucasus, Intermountain lowlands and Transverse ranges of southern highlands; within the soil region of Eastern Georgia—the three sub-regions of the Greater Caucasus, Intermountain lowlands and uplands, and Transverse ranges of southern highlands; and within the soil region of Southern Georgia—the sub-regions of Akhaltsikhe intermountain depression and volcanic uplands of southern Georgia can be distinguished. Each soil-geographical sub-region includes the corresponding soil zones and areas.

2.6.1 Soil Region of Western Georgia

In the sub-region of intermountain lowlands of the soil region of Western Georgia, the Kolkheti Lowland soil zone is mainly represented by several soil-geographical areas, where the Marsh and Gley Podzolic Soils (Gleysols), Subtropical Podzols (Luvisols Albic), and Alluvial Soils (Fluvisols) are common. Within the sub-region of the Greater Caucasus the following zones can be distinguished: hilly foothill soil zone with Yellow Soils (Acrisols Haplic), Red Soils (Nitisols Ferralic) and Raw Humus Calcareous Soils (Leptosols Rendzic); mountain-forest soil zone with Raw Humus Calcareous (Leptosols Rendzic) and Brown Forest (Cambisols Eutric) soil varieties; mountain-meadow soil zone with Mountainous Meadow Turff, Turf-Peat, and Primitive (Leptosols Umbric soils. Within the sub-region of southern mountainous transverse range of the soil region of Western Georgia the hilly foothill soil zone is represented, where except for the red and yellow soils the varieties of Brown Forest (Cambisols Eutric & Cambisols Dystric), Yellow Brown Forest (Acrisols Haplic), Raw Humus Calcareous (Leptosols Rendzic) and Mountainous Meadow (Leptosols Umbric) soils can be found.

2.6.2 Soil Region of Eastern Georgia

In the sub-region of intermountain lowlands and uplands of the soil region of Eastern Georgia, two main zones are represented: the desert steppe and steppe soil zone with Grey Cinnamonic and Meadow Grey Cinnamonic (Cambisol Chromic), Black (Vertisols), and Salt (Solonetz Humic) soils; transitional forest-steppe and forest soil zone of valleys with Meadow Cinnamonic (Cambisol Chromic), Chernozem like, Saline and Alluvial (Fluvisols) soils. The sub-region of Greater Caucasus of the same region also includes three soil zones as follows: foothills transition forest-steppe and forest soil zone with Chernozem like, Cinnamonic Forest, Grey Cinnamonic, and Raw Humus Calcareous soils; mountain-forest soil zone with Cinnamonic Forest, Raw Humus Calcareous and Brown Forest soils; mountain-meadows soil zone with the varieties of Mountain-Meadows and Primitive soils.

2.6.3 Soil Region of Southern Georgia

The soil areas, united in the zones of valleys and mountain foothill transition forest-steppe and forest soils of the sub-region of Akhaltsikhe intermountain depression of the soil region of southern Georgia, include the Meadow Cinnamonic, Grey Cinnamonic and Cinnamonic Forest soils, and the mountain-forest soil zone—the Cinnamonic Forest and Brown Forest soils. The soil zone of the mountain steppe of the sub-region of volcanic uplands of the same region includes the soil-geographical areas of Mountain Chernozems soils; and the mountain-meadows zone—the Mountainous Meadow Chernozem like (Leptosols Mollic) soils.

Thus, tens of the names of soil nomenclatures are distinguished in the territory of Georgia. The main genetic types, which form the altitudinal belts of soil structure of Georgia, include as follows: Mountainous Meadow Soils (Leptosols Umbric), Brown Forest Soils (Cambisols Eutric & Cambisols Dystric), Mountain Chernozems, Black Soils (Vertisols), Raw Humus Calcareous Soils (Leptosols Rendzic), Cinnamonic and Meadow Cinnamonic Soils, Grey Cinnamonic and Meadow Grey Cinnamonic Soils (Cambisol Chromic), Salt Soils (Solonetz Humic), Red Soils (Nitisols Ferralic), Yellow Soils (Acrisols Haplic), Subtropical Podzols (Luvisols Albic), Gley Podzolic Soils (Gleysols), Alluvial Soils (Fluvisols), Marsh Soils (Gleysols) and others. The above-mentioned zonal soil types are divided into many subtypes, genus, and species according to the soil formation factors (Sabashvili 1965).

In the areas of the country, which are subject to strong industrial impact, the soils are strongly modified and "anthropogenic soils" are formed occupying 0.1% of area (>80 km^2) in Georgia (Matchavariani 2012).

References

Apkhazava I (1985) Georgian Soviet Encyclopedia, vol 9. Tbilisi (in Georgian), p 725

Dolidze J, Kartvelishvili L (2012) Atmospheric precipitations; chapter in the National Atlas of Georgia. Publishing house "Kartograpia", Tbilisi (in Georgian), p 67

Gagua G, Mumladze D (2012) Annual air temperature map of Georgia in the National Atlas of Georgia. Publishing hause "Kartograpia", Tbilisi (in Georgian) p 57

Gogishvili N, Gogichaishvili G, Dolidze J (2012) Average annual precipitation map in the National Atlas of Georgia. Publishing house "Kartograpia", Tbilisi (in Georgian), p 68

Javakhishvili Sh (1977) Georgian SSR climatography, Tbilisi (in Georgian)

Javakhishvili Sh (1981) Georgian Soviet encyclopedia. Georgian SSR, Tbilisi (in Georgian)

Khmaladze G (2009) Water resources of Georgia, Tbilisi (in Georgian)

Kordzakhia M (1961) Climate of Georgia. Publishing house "Metsniereba", Tbilisi (in Georgian), 249 pp
Kordzakhia M (1964) Climate types map of Georgia, in the Atlas of Georgian Soviet Socialist Republic, Tbilisi (Georgian/Russian), pp 97–98
Matchavariani L (2012) Soils, chapter in the National Atlas of Georgia. Publishing house "Kartograpia", Tbilisi (in Georgian), p 99
Mumladze D, Lomidze N (2012a) Air temperature, a text in the National Atlas of Georgia. Publishing house "Kartograpia", Tbilisi (in Georgian), p 56
Mumladze D, Lomidze N (2012b) Climate types map in the National Atlas of Georgia. Publishing house "Kartograpia", Tbilisi (in Georgian), p 47
Sabashvili M (1965) Soils of Georgia. Monograph, Publishing house "Metsniereba", Tbilisi (in Georgian), 552 pp
Tielidze LG (2016) Glacier change over the last century, Caucasus Mountains, Georgia, observed from old topographical maps, Landsat and ASTER satellite imagery. The Cryosphere 10:713–725. https://doi.org/10.5194/tc-10-713-2016
Tielidze LG, Wheate RD (2018) The Greater Caucasus glacier inventory (Russia, Georgia and Azerbaijan). The Cryosphere 12:81–94. https://doi.org/10.5194/tc-12-81-2018
Trapaidze V (2012) Water resources, training course. TSU publishing house, Tbilisi (in Georgian), 133 pp
Ukleba N (1997) The use of water resources of Georgia in the national economy, Tbilisi (in Georgian)
Urushadze (1997) Main soils of Georgia. Monograph. Publishing house "Metsniereba", Tbilisi (in Georgian), 267 pp

Levan Tielidze is a Ph.D. of Ivane Javakhishvili Tbilisi State University in the field of geomorphology-glaciology-cartography and a senior research scientist at Vakhushti Bagrationi Institute of Geography. He is also a researcher-scholarship holder of the Department of Earth Sciences of the Georgian National Academy of Sciences. The field of his research is mountain cartography and glacial-geomorphological study of the mountainous areas in the Quaternary (Late Pleistocene and Holocene). He is a member of several international organizations. In 2014/2015 he conducted a research stay at the Climate Change Institute of the University of Maine, USA, and in 2015/2016–2017 at the University of Northern British Columbia, Canada. He is the author of about 40 scientific papers, maps and eight monographs.

Vazha Trapaidze is a Ph.D. Hydrologyst, Associate Professor of Ivane Javakhishvili Tbilisi State University (TSU) Department of Hydrology, Oceanology and Meteorology, Faculty of Exact and Natural Sciences. His research areas are hydrology and water resources, hydraulics, climate change impact, coastal processes and oceanography. He is a co-author of about 90 scientific works, 3 textbooks, and 3 monographs.

Lia Matchavariani is Professor of Ivane Javakhishvili Tbilisi State University (TSU), Head of the Soil Geography Department, Director of the Institute of Applied Ecology at the Faculty of Exact and Natural Sciences of TSU; Doctor of Science in Geography (Geo-ecology) from TSU (2006), Ph. D. in Agrarian Sciences (Soil Science) from Georgian State Agrarian University (GSAU) (1989). She has over 180 publications (scientific papers, monographs, textbooks, etc.); Her research areas are soil science, micropedology, Earth and Environmental sciences.

Roger Wheate is Associate Professor and founding member, of the Geography program at the University of Northern British Columbia (UNBC), Canada. His teaching and research interests cover the application of cartographic output, remote sensing and GIS across the spectrum of Natural Resource and Environmental Studies (NRES) faculty areas and the integration of geomatics technologies, including feature extraction and terrain visualization. Special interests include mountain cartography and glacier mapping and updating using remote sensing. He is the Canadian national delegate to the International Cartographic Association (ICA) and a member of the ICA Mountain Cartography Commission. He has been a member of the Canadian Cartographic Association since 1975 and was President in 1999.

Part II

Landscape Types of Georgia

Coastal Landscape of Georgia

3

George Lominadze, Irakli Papashvili, George Kavlashvili, Levan Tielidze and Nino Chikhradze

Abstract

This chapter deals modern lithodynamics and morpho-dynamics of the Black Sea coast of Georgia, also the history of the development of the coastal zone of the Black Sea in Late Pleistocene and especially in Holocene are shortly described. Different sinking zones are distinguished and tectonics are discussed. The zones such as the main lithodynamic systems of the Kodori, Rioni (Poti), and Chorokhi rivers are distinguished. In each lithodynamic system, the conditions and rates of coastal erosion are highlighted; submarine relief and beaches are characterized; the amount of solid runoff and coastal deformation conditions are discussed; various local and anthropogenic factors of terrestrial decrease are also highlighted and the accumulative sections are distinguished, where the rate of terrestrial increase is observed.

Keywords

Coastal landscape of Georgia · Black sea Kodori lithodynamic system · Rioni lithodynamic system · Chorokhi lithodynamic system

3.1 Research Area

The length of the Black Sea coastline within Georgia is 310 km (9% of the total length). It is one of the most important natural formations for the country's general geographical location, resource potential, recreational zone, and the main artery of foreign relations. In addition, the Black Sea has a great influence on the formation of the climate of Georgia.

The eastern part of the Black Sea coastline is located almost entirely within Georgia and is mainly represented by the shores of accumulative genesis. The accumulative shores occupy almost 80% of the total length of the entire coast, while the remaining 20% cover abrasive and abrasive-accumulative shores (Fig. 3.1).

The submarine part of the Black Sea coast of Georgia is characterized by a complex morphological structure. It is mainly represented by a narrow shelf that is at times indistinct. The shelf is mainly determined by 80–150 m isobaths; the

G. Lominadze (✉) · G. Kavlashvili · L. Tielidze · N. Chikhradze
Department of Geomorphology and Geoecology, Vakhushti Bagrationi Institute of Geography, Ivane Javakhishvili State University, Tbilisi, Georgia
e-mail: g_lomin@hotmail.com

I. Papashvili
Gamma Consulting Ltd., Tbilisi, Georgia

L. Tielidze
Department of Earth Sciences, Georgian National Academy of Sciences, Tbilisi, Georgia

L. Tielidze (ed.), *Geomorphology of Georgia*, Geography of the Physical Environment,
https://doi.org/10.1007/978-3-319-77764-1_3

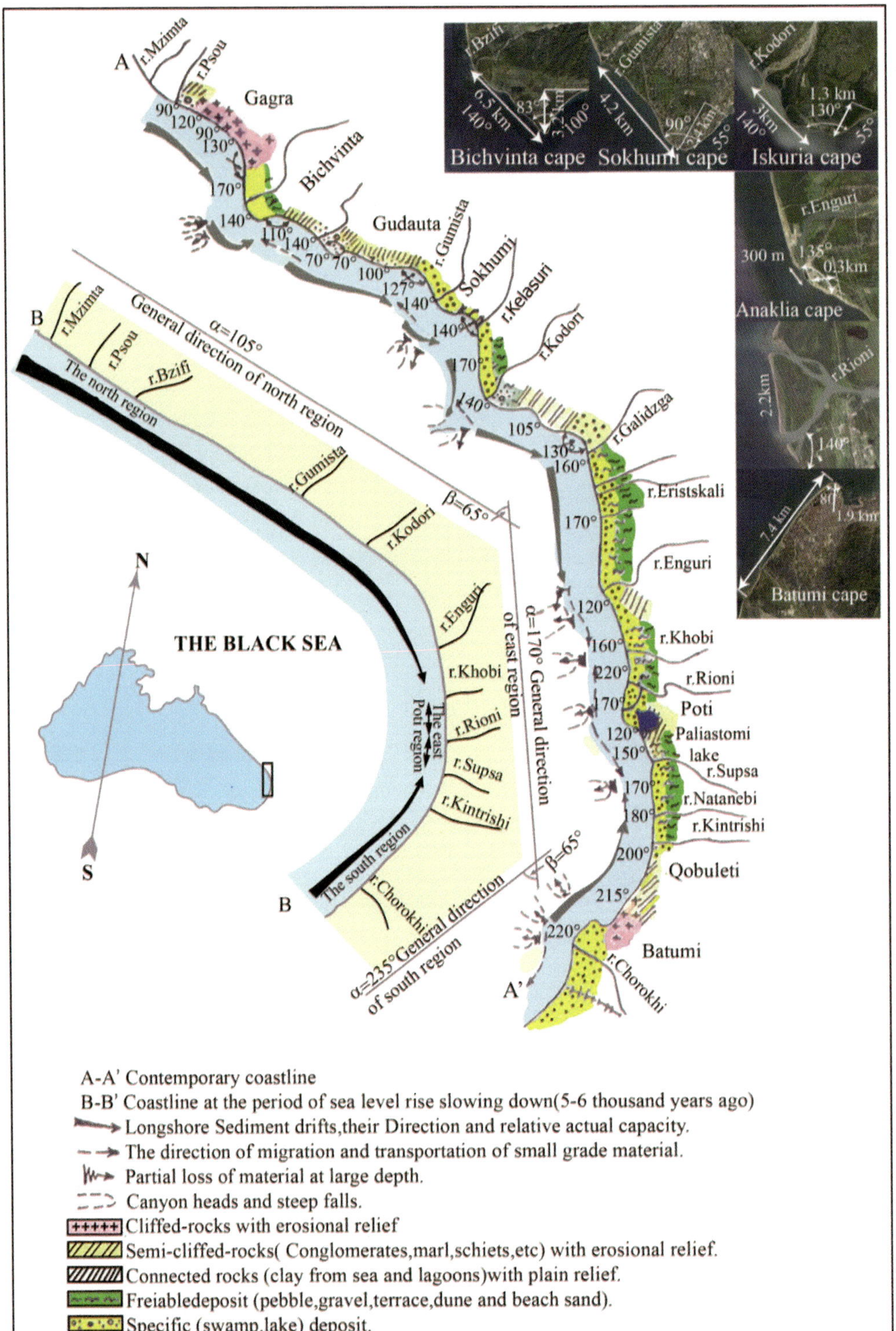

Fig. 3.1 Schematic map of morpho and lythodynamics of the Black Sea coastal zone in Georgia (Kiknadze 1991)

widest is in the sectors of Gudauta and Ochamchire. The shelf is dissected by the heads of the submarine canyons at the confluences of large rivers to the sea and extends almost to the coastal zone (Leontiev and Safyanov 1973; Kaplin et al. 1991).

3.1.1 Brief History of Coastal Zone Evolution

The analysis of geomorphological and stratigraphical materials confirms that the formation of the modern coast began mainly in the second half of Late Pleistocene and ended in the Holocene, about 5000–6000 years ago (Janelidze 1980). In the first half of the Holocene, the sea ingression within coastal terrain was underway. For this period, the configuration of the coastal strip differed significantly from today, and the coastal line was distinguished more by curves than abundant lagoons. From the second half of the Holocene, the shore obtained a more straightline shape. This process was expressed by the formation of powerful coastal embankments (dunes), as well as the occurrence of vast coastal marshes and small coastal lakes in the areas of large gulfs (Janelidze 1980; Alpenidze 2014). In the terrestrial part behind the embankments, the lower wetlands were formed, where lakes are found in some areas today; in particular, the Inkiti Lake and Skurcha Lake in the northern part of the coast, the Paliastomi Lake in the central part and the Nurie Geli Lake in the southern part within Batumi. Therefore, it can be said that during the second half of the Holocene there was not much change in the Black Sea coast of Georgia. The basis for such supposition is also the results of a complex analysis of the sediments under the Paliastomi Lake, according to which, in the second half of the Holocene era (over the last 6000 years), the Black Sea level variation amplitude did not exceed 2–3 m (Janelidze et al. 2015).

Eustatic and isostatic processes are important in coastal evolution. In the conditions of bottom uplifting, beaches are emerged and maintained depending on the development of the wave profile on the underwater slope. The beach forming deposits washed from the bottom are brought out towards the beach by waves. Such a process is observed in the Bzipi (Bzipi-Miusera), Akhali Kindghi and Ureki-Shekvetili district (Alpenidze 1985; Zenkovich 1962). This process is especially clearly expressed in the coastal section of the rivers of Supsa and Natanebi. Here, the land and the submarine slope adjacent to seashore, experience uplifting, and the sea level rising rate is far behind the positive tectonic movement rate. It seems that the process was underway in the entire second half of the Holocene. It contributed to the emergence of two parallel powerful embankments (dunes) near the coastline, the maximum heights of which reach 12–14 m. This process is underway today, which helps to feed the existing beach strip by deposits and its further sustainability.

Thus, we can discuss that the Black Sea coastal zone of Georgia has maintained stable conditions for the last 4000–5000 years, an indicator of which is the sustainability of the relic embankments in the accumulative shores. The embankments were not eroded during this period, and there was no sea water intrusion in the wetlands behind them.

3.1.2 Coastal Zone Tectonics

The transition zone of the Caucasus and the Black Sea is characterized by high tectonic activity, which has led to the existence of a narrow shelf, although there are also extensive structural formations such as the Gudauta, Ochamchire and Supsa-Natanebi watershallows. Their width in some cases reaches about 20 km. The delta-type accumulative bodies are formed in the confluences of large rivers in the shelf zone reaching great depths. In the underwater canyons, where the deposits of large volume flowed, erosion processes developed, leading to the advancement of the heads of the canyons towards the shore line. For instance, the erosive invasions in the heads of the underwater canyons of Enguri and Supsa are clearly depicted in bathymetric maps (Janelidze 1980; Lominadze and Papashvili

2013; Lominadze et al. 2014). Accumulative lowlands allocated along the coast are intensive accumulation zones of debris material, in particular, the Rioni intermountain sinking area is sinking from the Tertiary period until today. The sinking of the coastline of the lowland is still in progress, which is proved by results of repeated leveling. for example, the area of the Rioni River near Poti is sinking by 6.5 mm annually, while at the Chaladidi Village by 3.1 mm and in the north of the Rioni River confluence by 6.3 mm, etc. (Lilienberg and Matukova 1969).

Terrestrial tectonic sinking compensation was carried out at the expense of accumulation of debris material by the rivers from the Greater Caucasus and Achara-Trialeti Range. Accumulation was particularly strong in the coastline, where sediment accumulation was contributed by the hindering influence of sea level as an erosion basis on the drainage of surface waters, along with tectonic sinking of the land. The debris material formed as a result of the accumulation with permanently sinking due to the influence of tectonic movements. Horizons of this material originated in the past are allocated in different depths from the terrestrial surface (Janelidze 1980; Kiknadze 1991).

The results of the lithological and stratigraphical analysis of drilling materials indicate that in the coastline, up to the depth of 35–42 m from the terrestrial surface the deposit packs are developed, originated in the Holocene epoch. In some areas the horizon of fossil deposits are located beneath the Holocene deposits, the absolute age of which is 10,200–10,500 years using the radiocarbon method (Janelidze 1980; Janjghava 1979). The pack of the Holocene deposits is represented by the Holocene complex, characterized by a sharp facial variation both in terms of depth and horizontal distribution. This proves that the deposition of debris material and relief evolution in the coastal zone of the Kolkheti Lowland was underway against the background of frequent variation of accumulation conditions. In turn, this process was caused by tectonic sinking of the land and by influence of temporal variability of glacio-eustatic fluctuations of the sea with frequent migrations of river confluences and riverbeds.

3.2 The Coastal Zone Morphology and Lithodynamics

In the Black Sea shore of Georgia, the underwater canyons are often invaded into the gentle submarine slope and great depths are quite close to the coastline. Lateral type canyons are characterized by a particularly close allocation to the shore. Their heads are away from the river confluences with the sea, but are located closer to the submarine part of a beach, and by their location they often determine the coast line shape. Such canyons are in the southern part of the coast the Batumi and in the northern part of the coast the Inkiti submarine canyons (near the Bichvinta Cape) (Fig. 3.1). As a rule, the heads of such canyons form clearly visible accumulative capes and deeply intruded gulfs towards the alongside migration of deposits; they are the Bichvinta, Sokhumi, Kodori, and Batumi capes. Within these accumulative capes the shifting of deposits of considerable volume composing the alongside flow is observed at great depths of the submarine slope.

The issue of the formation of Georgia's coastline and its sustainability is largely dependent on the qualitative-quantitative characteristics of the alluvium brought to the sea by the rivers and the conditions of their redistribution along the shore.

The fine fracture part of the material brought out by the rivers to the confluence, shifts in the open drifting sea and is deposited there. The second part, beach forming for the particular region, joins the alongshore flow and participates in the process of formation and renovation of the beaches. This process depends on a number of factors: the amount and composition of coastal deposits, wave regime, inclination and peculiarities of the slope, the coastline shape, its orientation towards the wave field, etc. The listed factors are in close interidentified relationships and their combination determines the sustainability of the shores. In the conditions of stability of sea level and terrestrial tectonic regime, individual lithodynamic systems are formed, which are in a state of dynamic balance. They are characterized by the established regime of sediment movement, balanced shore contour and evolutionary development in time and space. In this case, short-term variations in the coastal zone are seasonal or cyclic and are related to the climatic conditions of the region.

The schematic lithodynamic system represents the combination of three main areas:

1. The area of the deposit feeding source of the system—the marine confluence of the large river in the conditions of Georgia, where the beach forming material is brought out in the coastal zone;
2. The transit area, where the deposit is shifted by the influence of waves in the form of an alongshore flow;
3. The area, where the deposit stream starts to dissipate, or in most cases, a diffusion with the flow of the neighbouring system takes place.

Obviously, within the system it is possible the small rivers are involved in the combined alongshore flow of the system, but in this case the main structure of the flow does not change, only its recovery takes place.

In case of anthropogenic interference in coastal processes, the dynamic system begins to move to a new regime, quantitatively and qualitatively different from the previously existing one. In fact, the earlier system is under degradation and in many cases under differentiation. Within it the subsystems with different budget and transportation regime of deposit are formed. These changes, as a rule, take place within the shorter, engineering time period and significantly differ from the natural, evolutionary development of the shores.

3.2.1 Kodori River Lithodynamic System

The lithodynamic system of the Kodori River begins from the shore allocated in the south from the sea mouth of the Kodori River. The delta of the Kodori River itself, is actually covered by a shelf and its lower seaside part is a combination of the heads of underwater canyons. In the south of the Kodori River, the shelf is gradually expanding. In the Skutcha gulf, up to 5 m depth, the bottom is covered by boulders and large stones and is a typical bench; in some areas they form about 100 m wide strips with the rocky remnant boulders of the abrasion. From the depth of 5 m and deeper, the sand distribution zone begins, and from the depth of 15 m the aleurolites and silt-sandy sediments are distributed. The relative height of the boulder mass on the bottom is 1–1.2 m. The underwater rocky reefs and mollusks clusters begin 600 m away from the coastline to the sea.

Near the modern confluence of the Kodori River, where the Kodori dynamic system begins, the Kodori (Iskuria) Cape has a soft contour and it does not seem to hinder the alongshore movement of the deposits; but in the east from the confluence, at a distance of 7–8 km, the obvious signs of shore erosion are observed. Despite the fact that the Kodori River brings a considerable amount of solid material [275,000 $m^{3/y}$ (Jaoshvili 1986)—362,000 m^3/y (Jaoshvili 2003)], its main part is lost in the underwater canyon. From the modern confluence of the Kodori River up to the Iskuria Cape, the azimut of the coastline is about 140–170°, while the section of the eastern side of the gulf up to Ochamchire equals 105° on average.

The Kodori (Iskuria) Cape creates a 5 km long gulf in the east. Due to the distance from the west of the cape, the gulf is protected from the westerly waves and therefore, the alongshore shifting of the deposits is still limited from this direction. It also plays a role in the bench-shaped bottom covered with wave hindering boulders. There is a change in the power of the alongshore flow between the Skurcha Gulf and the shore located in 7–8 km to the east, evident by the average width of the beaches, which is respectively 20–25 m and 5–7 m; in some places it is even less. In the same sections the beaches are bounded by 10–15 m high active cliffs, the abrasion intensity of which has increased significantly due to the reduction in beach width in recent years. The process is expressed by intense destruction of the cliff and creation of the clusters of eucalyptus trees growing over the cliff but now fallen down massively. Hence, it should not be surprising that in the Skurcha Gulf the beaches are much larger in comparison with the other sections, despite the closure of the gulf from the west. Due to the erosion of the cliffs in the eastern section and the protection by Iskuria Cape from the west, the existing material moves to the gulf from east to west. In addition, the Skurcha Gulf beaches are fed by the material brought from the bottom of the sea, which makes the coastline more sustainable.

More eastward at a distance of almost 15 km up to the Mokvi River confluence, the sea shore experiences strong erosive impact. The remnants of old buildings (near the confluence of the Tamishi River), including the large diameter sewage pipes and other objects of the 1970s located along the shore, are visible. The beach is almost entirely eroded, resulting in the outcropping of the brown-gray relict lagoon clays containing peat and other organic wastes. The eroded beach forms a 1–1.5 high cliff in some areas. At a distance of 20 km in the east from the Skurcha Gulf, when approaching the Mokvi River confluence, the character of the shore modifies and the beach becomes wider and the signs of erosion of the shore disappear. The reason for such change is the Ochamchire Port, which was built in 1936.

Ochamchire City is partly built-up in the southernmost periphery of the terraces of the Quaternary complex of the Kodori River. Two dams were built to protect the water area of the port from waves in the west and the east. Accumulation processes in the east of the Mokvi River is related to the construction of those dams. These constructions in the north of the river caused both the accumulation of its solid runoff and the alongshore deposit stream. In fact, in the second year since the construction of the port, significant changes began in the coast located in its south. It became obvious that the port facilities led to the disconnection of the existing alongshore stream. The negative process was gradually spread to the south so that in the 1970s it reached the confluence of the Ghalidzga River located 6 km away from the port, the erosion in the south of the river was observed even before. The most intensive retreat of the coastal line was directly from the port building at 1 km, and reached 5 m per year. As a result of the development process, the coast located in the south of the port retreated towards the land by150 meters and destroyed some buildings. At the same time, a bench composed of large boulders and stones on the sea bottom appeared in front of the eroded shore. In the south, the clay bottom is outcropped, and in some places the bottom is represented as a clay spots surface.

Within Ochamchire the beach had already disappeared in the 1950s. Subsequently they tried to reinforce the sea shore with a concrete wall, but it collapsed in a few years. Then they built a massive wall near the old wall, which also collapsed in a few years. At the beginning of the 1960s, a new wall and a line of groins were constructed; among their intervals the beach material was artificially disposed, brought from the confluence of the Mokvi River in the north of the port. The construction of the groins within the city was completed by 1974. By 1986 along the main groins (57 units) to the south up to the Ghalidzga River confluence, 600,000 m^3 material was artificially disposed along the 4 km coast. After that the sea shore became stable; the effect of artificial filling with the deposits was high and this condition remained for 20 years. The largest single filings were made in 1982, when 113,000 m^3 materials were brought in, and the largest amount of 190,000 m^3 materials were disposed by 1986. Surveys conducted during this period have shown that everywhere, except for the 1 km section directly near the port, the advancement of the shoreline towards the sea was observed. This phenomenon is conditioned by the lack of the deposit material in the existing section, i.e., volume of the alongshore deposits exceeded its capacity. The angle of the wave front at the shore line is 40–45°, i.e., it is highly effective for deposit shifting. Also, the negative deformations and unstable conditions of the coast were observed in the south from the port, at the 1–1.5 km section of the Ghalidzga River confluence, or a section of the shore, where the line of shore protective groins end. Obviously, the groins hinder the southward shifting of the deposits and therefore, the coast is unstable.

At present, the situation in the coastal zone is still complicated; no artificial inputs are being carried out, which resulted in the damage of a large number of groins within Ochamchire, and the volume of the beach material among them is sharply reduced. The sea bottom is also eroded and the clays are again outcropped.

3.2.1.1 Enguri River Confluence Section

In the south of Ochamchire City, the sea shore acquires the azimuth of 170° and maintains it up to the confluence of the Enguri River. The shore here is also typically accumulative (Fig. 3.1), which is actively eroded from the 1940–50s. The obvious proof of such a process is an eroded cliff which is observed almost everywhere along the shore up to the Enguri River confluence. Along with scientific research, local residents also confirm the retreat of the sea shore. Still several forest enterprises were operating in the 1950s, at the confluence of the Gagida River. Today, their location is covered by the sea and is far away from the modern coastline by 200–250 m. The sea shore, from the south of Ochamchire City up to the Khobi River confluence has been of great interest since the 1970s. Such attention was caused by the construction of a high dam on the Enguri River and its possible negative impact on the condition of the surrounding sea shore.

In 1978, the main part of the liquid and solid runoff of the Enguri River was overthrown into the Eristskali River bed, a new confluence of which is located in the north 23 km away from the old confluence. Accordingly, the liquid runoff in the old confluence decreased from 5.2 billion m^3/year to 1.24 million m^3/year and the solid runoff from 1.5 million m^3/year to 260,000 m^3/year (Jaoshvili 2003). By that time specialists evaluated differently the impact of the consequences of the construction of the Enguri high dam on the further sustainability of the coastal area. According to one of the forecasts (Lominadze et al. 2006) the sea shore would have been eroded at a distance of almost 40 km due to the dam construction, and the depth of the maximum sea intrusion into the land would have been 4.5 km. The further development of the processes did not confirm the magnitude of the predicted forecast and the changes turned to be unimportant. Initially, during the first 8–10 years of the construction of the dam, due to sharp reduction of deposits in the Enguri River confluence the rate of the coastal erosion significantly increased and reached the peak of 9–11 ha/year in the north of the confluence, and 4.4 ha/year in the south. It should be noted that the Enguri underwater canyon and adjacent coastline are characterized by complex morpho and lithodynamics; the rate of deformation of individual sections in the above-mentioned coastal zone was quite different. This was caused both by local peculiarities and the existence of additional anthropogenic factors. For example, there is the accumulative section in the north of the Enguri River confluence at a distance of 3–4 km, where the indicator of terrestrial increasing is 2–3 m per year and is still in relatively stable condition. In 1988–1989 the hydro-technical coastal protection measure was carried out in this section; for the compensation of the shore abrasion, 200 thousand m^3 beach forming materials were disposed in the coastal zone from the Enguri River dehydrated bed resulted in a positive deformation of the coast. Currently, the coastal erosion processes on the coast near the Enguri River confluence are still underway, but the rate of erosion is about two times lower than that in the previous 10–15 years (Zenkovich 1976).

3.2.2 Rioni River (Poti) Lithodynamic System

During the last 150 years, the Poti lithodynamic system (the central part of the Black Sea coast of Georgia) has been greatly modified. In natural conditions the Rioni River flowed into the sea with two bayous. One bayou always joined the sea in the vicinity of the underwater canyon, therefore, the most volume of its deposit got almost completely in the underwater canyon and did not participate in the feeding of the underwater slope. And the confluence of the second bayou always joined the sea away from the head of the canyon. The confluence of the river underwent natural migration, and in different periods joined the sea either in the north or south of the underwater canyon. The deposit, transported through bayou away from the underwater canyon, actively participated in the accumulation processes of the coastal zone. In the eighteenth and twenty centuries this bayou merged with the sea in the south of the head of the canyon, where intensive accumulation took place.

The underwater canyon (deep-water canyon of the Rioni River, the head of which is intruded near the coastal zone) periodically hindered advancement of the shoreline within the Poti lithodynamic system. Under natural conditions, the deposit loss in the great depths of the underwater canyon was 70–80% of the solid deposit discharge; therefore, despite the income of large amount of deposits, the terrestrial growth was of a temporary character. Accumulation or erosion depended on the allocation of the second bayou of the Rioni River towards the underwater canyon. The periodicity of this process is unknown, but it can be supposed that the migration period of the river bayou is recent two to three centuries. In spite of such changes in the coastal zone, Rioni lithodynamic system maintained its stability in total. It is proved by the existence of the alongshore relict embankments, which separate the Kolkheti marshes and lower terrestrial areas from the sea. If not taking into account the river sections near the confluences, the embankments have not been degraded for the past five-six thousand years. The sustainable condition of the embankments caused the formation of peculiar flora and fauna in the coastal areas of the Kolkheti Lowland.

The natural evolution of the Rioni lithodynamic system was underway until the 1890s (Papashvili et al. 2015). After the construction and reconstruction of Poti Port (1873–1888 1906), the unified lithodynamic system of the Rioni River was divided into two parts. The piers of the Poti Port did not allow the river deposits to pass along the seashore to the northern direction. After that, the deposit accumulated at the groin in the south of the port, near the underwater canyon. Occasionally, due to gravitational processes, a large amount of the accumulated deposit mass shifted into the great depths of the underwater canyons in the form of “Portions”. The mass destructed into the canyon moved the bottom along, causing the head of the canyon near the Rioni confluence to become active and advance towards the coastline. The previously unified Rioni lithodynamic system was divided into two parts by Poti Port, which resulted in the development of coastlines located in the north and south of the port independent from each other. Until 1939, the river confluence was located in the south of the port. It was a two-bayou delta, the confluence of the northern bayou which was located near the southern groin of the port, and the influence of the southern bayou near the current lighthouse of Poti. During the spring floods the city of Poti was constantly inundated, which caused damage to the city.

In the autumn of 1939 in order to prevent the floods in Poti, the Rioni River was shifted to the port in the north to the new confluence. In 1959, from the Rioni River confluence at the 7 km of the upstream, the watershed dam and a regulator were constructed, through which about 1/3 part of the river runoff was directed to the new confluence located in the south of the Poti Port through the city channel.

This activity has significantly changed the development of the lithodynamic system of the two divided parts Rioni. Due to the lack of deposition, the southern part of the system suffered intensive coastal erosion, which resulted in the sea swallowing 300 hectares of land area near Poti. The shore erosion is still in progress here, though its rate is significantly reduced.

In 1939, after the Rioni River confluence was shifted northward, formation of a new delta was started in the north of the port. Initially, numerous sandy islands occurred in the new river confluence of the sea. By 1941, a distinctly visible delta-shaped body occurred that was composed of many large and small islands. At the same time, deposit migrations in the north and south directions had started. In the 1950s, a clearly visible two-bayou delta was created, which continued to grow rapidly. By 2005 the islands were simultaneously formed in the confluences of both bayous of new delta to the sea, which still exist and their area constantly increases. Therefore, it can be said that the new delta of the Rioni River has moved from a two-bayou phase into a multi-bayou phase development. In the end, the new delta occupied a 13 km coastline of the sea from the port to the Khobi River influence in the south. In this beach a tendency of increasing of coastline is observed. In total, from 1939 until today, terrestrial growth within the delta was 1300 ha. At the same time, accumulated deposits volume amounted to almost 240 million m^3. Today, the shoreline advancement in the central part of the new delta is 12–14 meters per year; on the southern flank near the harbor 6–8 m/y and on the northern flank at the Khobi River confluence 2–3 m/y (here the rate of shore advancement has been determined before the construction of the Kulevi terminal at the Khobi River confluence).

In the north of the Poti Port, the river deposit brought from the central part of the new delta is equally distributed on the submarine slope of the southern and northern flanks. The deposits migrate annually to the south and north with the volume of about 0.7–0.7 million m^3. The remaining 2.0–2.5 million m^3 is distributed on the underwater slope and the central part of the delta. Excessive accumulation leads to the expansion of the central part of the delta and therefore, the confluences of both bayous are distanced from each other.

In total, after the northward shifting of the main runoff of the Rioni River, the land growth (in 1939–2007) made 1204 hectares in the coastal strip located in the north of the Poti Port. Before the construction of Gumati and Vartsikhe Hydro Electric Powers (HEPs), the terrestrial annual growth was 41.3 ha, and after the construction, the terrestrial growth rate decreased and made 9.8 ha/year, or less by 76%.

The volume of solid deposits of the Rioni River also decreased sharply after the construction of Gumati and Vartsikhe HEPs. The above factor reduced the intensity of accumulation processes of the coast in the north of the port by about 4 times, although the shoreline erosion was not observed. The reduction of deposits' deficit in the south of the port was not possible even after partial restoration-passing of the river discharge through the city channel. Approximately, for the year of 1959, when the watershed dam was activated on the 7th kilometer of the Rioni River, bringing riverine solid deposits was also considerably reduced due to the construction of Gumati and Vartsikhe HEPs. Accordingly, transport of the deposits into the sea was also reduced through the city channel. Hence, the process of seashore erosion could not be stopped.

3.2.3 Chorokhi River Lithodynamic System

The relief of the main part of the submarine slope of the Kakhaberi Plain has developed for a long period in the Quaternary sedimentary rocks of the Chorokhi River. The relief of the underwater slope at the present confluence of Chorokhi with the sea and at Batumi Cape becomes complex. The heads of Chorokhi and Batumi underwater canyons are located close to the shore line (Fig. 3.1), where the bottom inclination reaches 15°–30°. The largest part of the Chorokhi deposit here moves to great depths.

According to the map of 1834, the Chorokhi River flowed into the sea with three bayous. One bayou (Mejinistskali) joined the sea in the surroundings of the Adlia Village, from the Chorokhi confluence to the Batumi Cape at a distance of 3.5 km. The second bayou joined the sea at the present confluence of Chorokhi and the third bayou—in the south from the present confluence of Chorokhi in the direction of Gonio settlement at a distance of 1.5 km. The allocation of the bayous of those times resulted in a penetration of certain volume of the Chorokhi River deposits in the sea, as this section was separated from the heads of the underwater canyon. As a result, the deposits brought by the Gonio and Mejinistskali bayous were completely spent on the feeding of the coastline. During that period the coastline was shifted at 509 meters into the sea near the

Mejinistskali confluence compared to the present day. Also, the Gonio shore was shifted into the sea. The Batumi Cape was not as visible, and it was 520 meters away from the sea. The coast had a convex shape towards the sea and presumably it was in a stable condition.

According to researched cartographic data, the reconstruction of Kakhaberi Plain's seashore changes in the last 180 years has been possible. The oldest cartographic survey was carried out in 1834 by P. Mangarin. The map of 1880 was also retrieved. These maps were referenced to the map planned in 1926. These maps were referenced to the cartographic materials of 1977, 1981, and 2012 (Lominadze et al. 2013). All the data marked are depicted on the satellite image of 2009 (Fig. 3.2).

In the second half of the nineteenth century, the bayou of the Mejini River was blocked by a land dam, in order to protect the low-lying areas along it from flooding during the inundations on the Chorokhi River. After the construction of the dam, the Chorokhi deposit was redistributed on the two remaining in the south bayous, from which its greatest amount was headed to the head of the underwater canyon. This effectively ceased the supply of deposit for the central part of the shore of the Chorokhi-Batumi Cape. The sharp change in the movement of beach-forming deposit led to the changes in the shore line. By 1880 the Mejinistskali confluence retreated by 80 meters and at the same time, the present day confluence of Chorokhi advanced by 200–250 m in the sea. The Batumi Cape was developing very intensely, which increased by 280 meters.

In the second half of the nineteenth century, the construction of the Mejini dam caused great changes in the coastal zone, which was reflected in the straightening of the coast line in the Chorokhi-Batumi Cape section. This process was conducted on the expense of the advancement of present confluence, retreat of the Mejini River confluence and intensive accumulation near the Batumi Cape.

In the following years, when the Chorokhi confluence approached the underwater canyon, the advancement of its shore to the sea was stopped. Mejinistskali delivered several hundred thousand cubic meters of sand along with the gravel-pebble deposit. This sand were distributed by waves along the shore, including on the underwater slope. Since the Chorokhi confluence approached the head of the submarine canyon, the alongshore flow was mostly composed of gravel-pebble material, as the sand and the fine-grained fractions were disappeared in the underwater canyons. Thus, along the straightening of the shore, the fractional composition of the beach deposit was also gradually changed.

Due to proximity to the submarine canyon at the confluence of the Chorokhi River, most of the solid runoff got to great depths without compensation (Russo et al. 2007). The larger deposit was accumulated in the form of a debris cone because of its mass, from which it was shifted northward to the Batumi Cape (Papashvili et al. 2015). Due to the sharp decrease in the sandy fraction of deposit, the discharge of alongshore flow has decreased and presumably, its volume was approximately 150,000–200,000 m^3 per year at the end of the nineteenth century.

In 1906, the 170-meter concrete spur was built in the Batumi Cape, which helped to accumulate deposits at a distance of 1.5–2.0 km from of Batumi to the south. However, at the same time, due to the straightening of the shore the coast located directly to the north of the Chorokhi River, underwent erosion. The spur has played a positive role in the increasing and sustainability of existing beaches along the old boulevard of Batumi.

In 1975, the southern Gonio bayou of the Chorokhi River was blocked by the dam. With this, the whole runoff of the river got into one bed existing today, and the river confluence occurred near the heads of the underwater canyon. As mentioned above, in such conditions the most part of the Chorokhi River deposit disappeared in the great depths of the underwater canyon. Though, even in this case, approximately 100–120 thousand m^3 of gravel-pebble material (average weight diameter 36 mm) in the form of the alongshore flow, was shifting northward. Nevertheless, the Adlia coast, located directly in the north of the Chorokhi confluence, was retreating back to the land that was related

Fig. 3.2 Changes in the Chorokhi-Batumi coastal zone in 1834–2012

to the straightening of the Chorokhi-Batumi coastline.

According to various studies, the annual runoff of the Chorokhi River significantly varies from 5.3 million m^3 (Jaoshvili 2003) up to 8.8 million m^3 (Mandich 1967). If we consider that a real solid runoff is between the existing data, it becomes clear that only 5–10% of the solid runoff of the Chorokhi River is enough for compensatory accumulation. In fact, it is less, considering that the increase in the wetland surface of the plain is mainly stipulated by biological processes. There is no doubt that the solid runoff of the Chorokhi River was changing in the second half of the Holocene, though it is unlikely to be less than today. Therefore, it can be said

Fig. 3.3 Anthropogenic influence in the Black Sea coastal zone of Georgia, 25/09/2015 (*photo by* G. Lominadze)

Fig. 3.4 An example of coastal erosion (abrasion) near the Batumi, 23/02/2012 (*photo by* G. Lominadze)

that the most part (90-95%) of river deposits was lost in great depths of underwater canyons during the second half of the Holocene. Otherwise, the area of the Kakhaberi Plain would be much larger and the sea shore would continue to grow.

Today, the anthropogenic influence in the Black Sea coastal zone of Georgia gains increasing influence (Fig. 3.3). In the accumulative shores, the volume of beach sediment continuously decreases, which causes the real danger of eroding the old embankments (dunes) (Fig. 3.4). Due to the current situation, it is expected that in case of the continuation of the embankment-dune degradation, the sea ingression is expected to cover the lower parts of the land, which will result in a severe violation of ecological sustainability in the coastal zone. Presumably, the ongoing process will be speeded up if the modern rate of sea level rises in the future. Therefore, it is necessary to pay particular attention, in order that the negative impact of anthropogenic factor on the morphodynamic regime of coastal zone development in the background of intensive social and economic use of coastal areas, does not acquire irreversible character (Janelidze and Chikhradze 2012).

References

Alpenidze M (1985) Bottom feeding of the alongshore flow of sediments. Geomorphology, USSR Acad Sci (2):65–70 (in Russian)

Alpenidze M (2014) The relief of the Black Sea coast of the middle part of Apkhazeti. Geography and modern problems of environment. Mater Int Elect Conf in the Institute Tbilisi, SSU, 2014 №. 1, pp 17–27

Janelidze Z, Chikhradze N (2012) On anticipated transformation of the coastal relief in the conditions of raising average multi-annual sea level (Colchis Lowland case study). Collected Papers. New Series № 4 (83). TSU Vakhushti Bagrationi Institute of Geography, Tbilisi, Georgia, pp 84–88

Janelidze Z, Chikhradze N, Janelidze T (2015) The Black Sea eustasy role in the emergence and development of Paliastomi Lake. In: Proceedings of the International conference on "Modern problems of geography and anthropology". TSU Publishing house. Tbilisi, Georgia, pp 91–94

Janjghava KI (1979) Engineering geology of the shelf zone and the Black Sea coast within the Caucasus. "Metsniereba", Tbilisi (in Georgian)

Janelidze ChP (1980) Paleogeography of Georgia in the Holocene. "Metsniereba", Tbilisi, p 178 (in Russian)

Jaoshvili ShV (1986) River sediments and beach formation on the Black Sea coast of Georgia. "Sabchota Sakartvelo", Tbilisi (in Georgian)

Jaoshvili Sh (2003) The rivers of the Black Sea, Tbilisi, «Sea-Search & Arena» (in Georgian)

Kaplin PA, Leontiev OK, Lukyanova SA, Nikiforov LG (1991) Shores (monograph). "Misl", Moscow (in Russian)

Kiknadze AG (1991) Morphodynamics of the coastal zone and optimization of its use (by the example of the Black Sea coast of Georgia). Thesis for the degree of Doctor of Geographical Sciences in the form of a scientific report. Tbilisi

Leontiev OL, Safyanov GA (1973) Canyons under the sea. "Misl", Moscow, p 261 (in Russian)

Lilienberg DA, Matukova VA et al (1969) Map of modern vertical movements and morphostructures of the Caucasus. Problems of modern movements of the earth's crust. Moscow, pp 142–155 (in Russian)

Lominadze G, Papashvili I (2013) Enguri submarine canyon erosive development in the epoch of Pleistocene-Holocene. In: 2nd international symposium on Kazdaglar (mount Ida) and Edremit, human-environment interactions and ecology of mountain ecosystem proceedings and abstracts, pp 497–504

Lominadze G, Megreli N, Russo G (2006) Changes in the dynamics of the Black Sea coastal zone (central part of Colchis, 1976–2002) under the influence of technogenic factors. In: Proceedings of the international electronic conference on changes in the natural environment at the turn of the millennium, Tbilisi-Moscow, pp 133–139

Lominadze G, Papashvili I, Khorava S (2013) Modern development of the coastal zone of the Kakhaberi Plain. Ivane Javakhishvili Tbilisi State University, Institute of Geography of Vakhushti Bagrationi Geography. Collection of Works, New Series № 5 [84], Tbilisi, 39–45

Lominadze GJ, Papashvili IG, Khorava SG (2014) The Black Sea Kolkheti coastal zone submarine canyons erosive development in the epoch of Pleistocene-Holocene. Environ Ecol Mediterr Reg (Chap. 25) 295–305 (Cambridge scholars Publishing)

Mandich AF (1967) The size of the solid flow of the rivers of western Georgia, No 2, Geography series. Vestnik MSU, Moscow, pp 134–137

Papashvili I, Lominadze G, Khorava S (2015) The tendences of Modern Development of the Rioni-Supsa Interfluve Region of the Sea Shore. In: Materials of international conference "Applied ecology: problems and innovations". Proceedings ICAE, Batumi, Georgia, pp 42–53

Russo G, Bilashvili K, Megreli N, Savaneli Z (2007) Dynamics of the deltaic canyon area of the Rv. Chorokhi, Georgia. In: 3rd international symposium on submarine mass movements and their consequences. Springer

Zenkovich VP (1962) Fundamentals of the study of the development of the seashores. Publishing House of the USSR Academy of Sciences, Moscow (in Russian)
Zenkovich VP (1976) The expected changes in the seashore in the north of Colchis. Communications of the AN GSSR, vol. 83, No 3, "Metsniereba", Tbilisi (in Russian)

George Lominadze is Head of the Department of Geomorphology and Geoecology of the Vakhushti Bagrationi Institute of Geography, Ivane Javakhishvili Tbilisi State University. His research areas are coastal geomorphology, hazards, resorts and their interaction with the environment. He is a co-author of about 60 scientific works and four monographs.

Irakli Papashvili is Head of the marine group of "Gamma Consulting" Ltd. His research areas are oceanography, dynamics and morphology of sea shore. He is the author of about 30 scientific works and 20 scientific-research reports.

George Kavlashvili is a Geophysicist and research scientist at the Department of Geomorphology and Geoecology of the Vakhushti Bagrationi Institute of Geography, Ivane Javakhishvili Tbilisi State University. His research area is the dynamics and morphology of sea shores, and geographical information systems. He is the author of about 10 scientific works.

Levan Tielidze is a Ph.D. of Ivane Javakhishvili Tbilisi State University in the field of geomorphology-glaciology-cartography and a senior research scientist at Vakhushti Bagrationi Institute of Geography. He is also a researcher-scholarship holder of the Department of Earth Sciences of the Georgian National Academy of Sciences. The field of his research is mountain cartography and glacial-geomorphological study of the mountainous areas in the Quaternary (Late Pleistocene and Holocene). He is a member of several international organisations. In 2014/2015 he conducted a research stay at the Climate Change Institute of the University of Maine, USA, and in 2015/2016–2017 at the University of Northern British Columbia, Canada. He is the author of about 40 scientific papers, maps and eight monographs.

Nino Chikhradze is a Ph.D. student and research scientist at the Department of Geomorphology and Geoecology of the Vakhushti Bagrationi Institute of Geography, Ivane Javakhishvili Tbilisi State University. Her research area is the use of photogrammetric methods in the study of coastal zone dynamics. In 2013–2014 she completed her Ph.D. internship at the University of Mino, Braga, Portugal. She is a Vice-Delegate of the European Speleological Federation and has published about 40 scientific works.

4 Karst Landscape of Georgia

Zaza Lezhava, Lasha Asanidze, Kukuri Tsikarishvili
and Levan Tielidze

Abstract

This chapter presents the basic conditions for the creation of Georgia's karst terrain and the extent of the distribution. Hydrological and hydrogeological features of karst are given, such as water regime, karst surface, and underground runoff of karstification-water content horizons and complexes. Morphological-genetic types of karst are considered, such as the mountain outcropping and meadow, mountain forest, hidden, buried, and remnant karsts. The surface and underground karst forms, such as poljes, uvalas, depressions, sinkholes, expanded cracks, wells, shafts, abysses, various types of karst caves, and others are characterized. The subchapter includes appropriate maps and images.

Keywords

Georgian karst • Caves of Georgia • Krubera abyss

4.1 Research Area

Georgia is distinguished with a diversity of karst processes. The limestone rocks occupy a significant area (more than 10%) of the entire territory of Georgia, with a variety of surface and underground karst landscapes. The karst is particularly well developed in western Georgia, where it stretches for ~325 km with width between 2–3 km and 30–35 km from the Psou River to the areas of Ertso Lake (Asanidze et al. 2017a; Lezhava et al. 2017). It comprises the northern hilly slopes of the Kolkheti Lowland and the zones of southern slopes of the bordering western Greater Caucasus. The vertical distribution of karst starts from sea level (Gagra) and extends to absolute height of 2757.6 m (the Speleologists Peak in the Arabika massif) (Fig. 4.1). Today, the number of surveyed and partially studied karst caves within the karst strip of Georgia exceeds 1500 (Asanidze et al. 2017b).

Karst phenomena are also common over important areas outside the karst strip of western Georgia, such as Guria hill and Zemo Svaneti; the Truso gorge, Gombori Range, Iori Upland, and southern slope of the Kakheti Caucasus; and

Z. Lezhava · L. Asanidze (✉) · K. Tsikarishvili
L. Tielidze
Department of Geomorphology and Geoecology, Vakhushti Bagrationi Institute of Geography, Ivane Javakhishvili Tbilisi State University, Tbilisi, Georgia
e-mail: lasha.asanidze@tsu.ge

L. Tielidze
Department of Earth Sciences, Georgian National Academy of Sciences, Tbilisi, Georgia

L. Tielidze (ed.), *Geomorphology of Georgia*, Geography of the Physical Environment,
https://doi.org/10.1007/978-3-319-77764-1_4

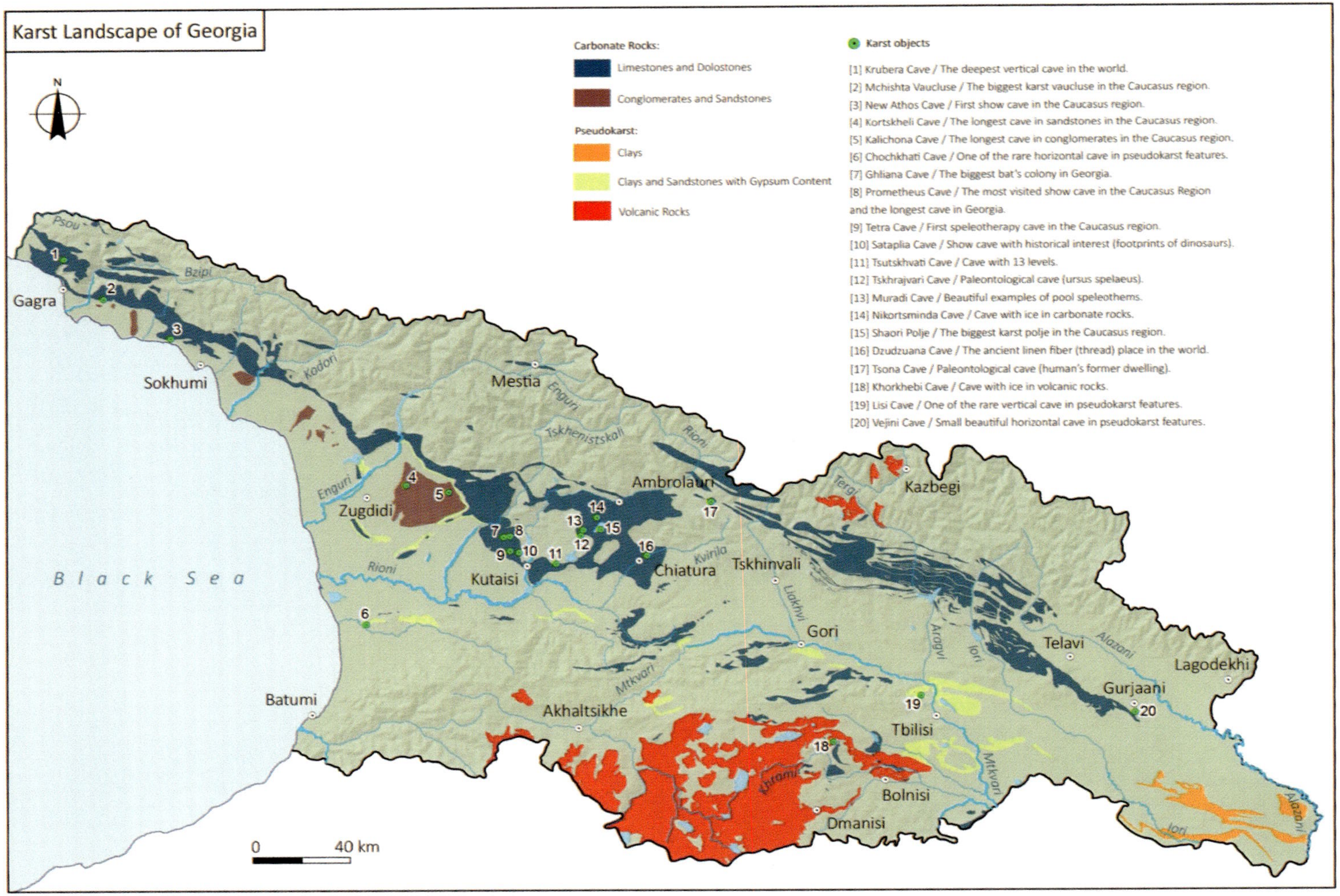

Fig. 4.1 The map of distribution of the karst relief of Georgia

the northern peripheries of the southern highland of Georgia. There are relatively unfavorable geological and climatic conditions for the development of karst in the mentioned regions that is reflected in the karst landscape and karst origin in general. In addition, the karst of these regions is still insufficiently studied.

The karst processes in Georgia are mainly developed in carbonate rocks, the age of which ranges from Upper Jurassic to Lower Pliocene (Asanidze et al. 2013a).

The limestones are of Upper Jurassic, Cretaceous, and Lower Paleogene age, and the fractured rocks (conglomerates, sandstones and breccias) with limestone cement represent the area of distribution of Neogene and Quaternary age clastokarst.

Capacities of limestone suites reach significant values (up to 2600 m in the Arabika massif). Lower Cretaceous (Urgonian) limestones are distinguished by pure chemical composition and mass, which determines the intensity of revelation of karst processes. The Upper Jurassic and Upper Cretaceous limestones are mostly thin layered and are karstified less intensely. In some areas, the karst conglomerates are represented by limestone fractures, and in some area by mixed particles of limestones and insoluble rocks (e.g., porphyrites), although this does not prevent the development of karst phenomena.

The limestone strip of Georgia is located between the folded system of the southern slope of the Greater Caucasus and the Block of Georgia according to the tectonic scheme of Georgia (Gamkrelidze 1964), as a result of which are presented both the orogenic and platform karst that are distinguished by different conditions and peculiarities of karstogenesis.

In the peripheries of the Greater Caucasus, the limestone suites are involved in the general high amplitude uplifting of the mountainous system of the Greater Caucasus, and the hillock strip bordered from the north of the Kolkheti Lowland experienced relatively weak movements (uplifting), in addition the tectonic tranquility was characteristic of some parts of the lower karst. The alignment of the karstified suites and sediments is almost horizontal in the Dzirula massif, Duripshi Plateau, Askhi massif, Maidan Plateau, and in some areas of the Block of Georgia, and the karst rocks are predominantly sloped by 20–30° in the rest of the area of the karst province.

4.2 Relief

The relation of karst formations to the southern slope of the Greater Caucasus and foothills has led to crossing the karst strip with numerous rivers, which flow down from the axial zone of the mountainous system and dissect the limestone strip into tens of isolated karst massifs that are separated by erosive gorges. Fragmentation into the small and large size karst massifs is a characteristic indicator of the karst strip of western Georgia. These massifs surround the horizontal alignment of karst formations (locked depressions, caves, etc.) of this area.

The western Georgian karst is characterized by a frequent and deep erosion dissection, which has a significant impact on the morphological, speleological, and hydrogeological peculiarities of the entire karst strip. Such intense dissection of the karst strip is caused by its location in the peripheral part of the sustainable upward movement zone, where the erosion processes took place and are still intensely underway. The cutting of the river networks and shifting of the groundwaters has taken place against the background of upward tectonic movements. In the Quaternary period alone, the Greater Caucasus is expected to be uplifted by 2–3 km (Tintilozov 1976; Maruashvili 1973).

In the modern stage of the study of Georgian speleology, those massifs are sharply distinguished by the density of karst cavities and the frequency of the openness of cave entrances, which are built of subhorizontally or slightly angled rocks and are characterized by more or less straight or inclined surfaces. It is noteworthy that most of the karst caves of Georgia are subhorizontal or inclined cavities that can be explained not by the great expansion of layered karst suites in comparison with massive limestone rocks, but mainly by the fact that the subhorizontal alignment of limestones creates favorable geomorphological and hydrogeological conditions for the formation of caves. Indeed, a significant part of the caves studied on the southern slope of the Greater Caucasus (Abrskili, Tkibula-Dzevrula, Uatapakhi, Kelasuri and others) are originated and developed in the layered limestone suites. More than 130 caves have been discovered in subhorizontal layered limestones in the Zemo Imereti Plateau alone (Lezhava 2015; Lezhava et al. 2016a).

4.3 Climate

Great diversity in the distribution of precipitation of the limestone massifs of Georgia has been created by significant vertical and horizontal fragmentation of the mountainous karst terrain of the southern slope of the Greater Caucasus, alternation of gorges and ranges of meridional and latitudinal directions and the variable orientation of slopes, their uneven distance from the Black Sea, along with the general circulation processes of atmosphere prevailing in different seasons.

The amount of precipitation in the karst strip of Georgia increases with increase in elevation. On the slopes of low-mountainous massifs, which are exposed to the sea, the increase in atmospheric precipitation reaches its maximum at a height of 300 m. The amount of precipitation at the higher elevations is somewhat reduced and again increases from 1000 to 2000 m. The critical increase in precipitation goes through the flattened crests of the limestone massifs (Bzipi, Arabika Gumishka-Psirtskha, etc.) within Apkhazeti at the absolute heights of 1800–2000 m (Chirakadze 1972). A similar picture is presented in the Okhachkue, Kvira, Migaria, Askhi, Racha, and Khvamli massifs, the southwestern slopes of which are under constant influence of (marine) humid air masses.

Thus, the karst strip of the southern slope of the Greater Caucasus is provided with abundant precipitation throughout the year, which leads to intensive and continuous karstification in the medium and lower limestone massifs. The limestone massifs (up to 1500–1600 m) are distinguished by the wide development of large karst sinkholes with a large number of underground forms. It should be noted that 75% of the caves studied in Georgia are located at 1100 m asl (Tintilozov 1976). There are somewhat different conditions in the high mountainous massifs (over 1600 m), where in the cold period (November–March), karstification processes either are suspended due to the sustainable snow cover (Kordzakhia 1961), or, at any rate, it has no substantial effect. Therefore, it is noteworthy that the protective influence of snow cover, along with the unfavorable hydrometeorological conditions, determines the abundance of undeveloped and weakly opened karst forms on the surfaces of high mountainous massifs.

4.4 Hydrological and Hydrogeological Peculiarities of Karst

The karst region of western Georgia is characterized by complex and diverse hydrogeological conditions, which are stipulated by the location of separate parts of the karst strip within the complex folds of the southern slope of the Greater Caucasus and Block of Georgia.

The water regime of karstified-water content horizons and complexes is directly dependent on the amount of atmospheric precipitation. The territory of Apkhazeti is distinguished by relatively high water content, where the average runoff is equal to 55 l/sec/km^2. Here, the average runoff from 30 l/s/km^2 (the Black Sea coastal zone) to 80–85 l/s/km^2 (above 2000 m) is in accordance with hypsometric levels. In the east, in the Enguri River basin, the average runoff is equal to 25–50 l/sec km^2. In the easternmost, in the Zemo Imereti Plateau, the runoff decreases to 20–25 l/s/km^2, and the most watered is the western section of the Racha limestone massif, where the runoff reaches 50–100 $l/s/km^2$ (Apkhazava 1971).

The high mountainous regions of outcropping karst development (Gagra and Bzipi ranges) are totally lacking the surface runoff due to the wide distribution of water absorbing forms, such as sinkholes, wells, and other surface karst forms. In these regions, along with rainwater, the snow melted waters have a significant role and often prevail in feeding underground basins with karst waters. The formation of large underground basins of karst waters within the karst strip of western Georgia, and karst genesis as a whole, takes place in somewhat different conditions in the sections of southern slope of the Greater Caucasus and the Block of Georgia.

The existence of interim water-stable horizons within the Arabika massif defines the level allocation of underground waters, and the plicate (folded) structures define the movement of these waters toward the mould inclination of synclines. And in the coastal part, the monoclinal structure of the water content horizons and the inclination of underneath water-stable layers toward the sea control the movement of underground waters and their discharge in the coastline. Therein, where the limestone and beneath water-stable suites sink below sea level, submarine discharging of underground karst waters takes place.

In the limestone massif of Bzipi developed in the same zone, two underground basins of karst waters are distinguished—Dzirkhva and Mchishta, which are mainly used to discharge the karst waters of the Bzipi massif. Mchishta is the largest underground karst river of the Caucasus. Its average discharge is 9.5 and maximum discharge is up to 200 m^3/s. The difference between its theoretical and actual runoff module is quite large (Kiknadze 1976) and as identified by studies, it collects water from a large area. By identification of the actual catchment basin of this giant underground karst vaucluse, the area (160 km^2) of its underground basin was defined (Gigineishvili and Tabidze 1970). Similarly to the Mchishta vaucluse, the areas of underground and surface water content basins of karst waters of considerable debit formed in different massifs often do not coincide with each other.

Within these massifs (Arabika and Bzipi), in the vertical descending circulation area, the world's deepest abysses (Krubera 2197 m, Sarma 1830 m, Illyuziya-Snezhnaya-Mezennogo 1753 m, etc.) are developed. The depth of karst waters underground circulation exceeds 2500 m and extends beyond sea level (submarine discharge takes place in the Gagra–Gantiadi region).

The medium and low-mountainous karst massifs located on various hypsometric heights, namely, from 950 m (the Gaucha massif) to 2156 m (the Okhachkue massif), are characterized by a two-storied structure of underground basins of karst waters and a predominant development of horizontally angled circulation of hydrodynamic zones. Synclinal basins are mainly represented here, as well as the monoclonal ones. In the construction of low-mountainous massifs, the covering water-stable horizon is actively involved in, which sometimes sinks below sea level.

The subhorizontal alignment of the rocks within the Kudaro and Racha limestone massifs leads mainly to the horizontal circulation of karst waters and therefore the formation of extensive subhorizontal caves. Underground rivers (Shareula, Kheori, Tsivtskala, etc.) as large karst sources flow out in the daylight at different absolute markers, which is conditioned by their connection with the upper or lower karstified-water content horizons (Fig. 4.2). Isolated underground flows are also common here.

The basins of karst massifs of Georgia (Askhi, Urta-Unagira, Tskaltubo-Kutaisi, Okriba-Argveti, and Zemo Imereti Plateau) are characterized mainly by horizontal circulation of karst waters and the caves allocated in levels.

The carbonate rocks within the Block of Georgia are deeply sunk to the south and below sea level and are covered by the powerful water-stable covering deposits thus creating artesian conditions—hydrodynamic and fixed regime zones of depth circulation. It is estimated that the water resources of the Bzipi artesian

Fig. 4.2 Shareula karst stream (the Racha limestone massif) (*photo by* L. Asanidze)

basin are equal to 24.3 m^3/sec, and that of Tskaltubo—15.0 m^3/sec (Kiknadze 1976).

In some parts of the limestone cover of the Block of Georgia, unlike the geosynclinal type of karst, the common level of underground waters is observed. Namely, in the Zemo Imereti Plateau, within the Chiatura Block structure, the presence of common level of underground waters has been identified by conducting experiments (Lezhava 2015; Lezhava et al. 2015a, b).

It is interesting to note that within the individual massifs of Georgia, both hanging sources (streams) of the aeration zone and sources of full saturation and depth circulation zones simultaneously operate. Exits of hanging streams, which are represented by spectacular episodic and permanent waterfalls (Fig. 4.3) are often related to the aeration zones, especially to their lower parts.

Particularly interesting region for the development of hanging streams functioning throughout the year is the Askhi limestone massif and in particular, the Abasha River basin, where they are discharged at the absolute heights of 600–800 m in the form of waterfalls. Permanent underground water flows have also been observed in the hearth of limestone massifs during the joint studies conducted by Georgian and foreign speleologists (Migaria, Askhi, Eki Mountain, etc.) (Lezhava et al. 2015a, b) (Fig. 4.4).

4.5 Morphological and Genetic Types of Karst

Within the karst strip of Georgia morphological-genetic types of mountain outcropping and meadow, mountain forest, hidden, buried, and remnant karst are represented.

The mountain outcropping and meadow limestone karst occupies a large area on the high surfaces of the Arabika, Bzipi, Askhi, and other massifs. In the Arabika massive corrie surfaces occupy the area of 6 hectares (Tintilozov and Maruashvili 1971) only in the serum gorge (Odtabalagani region). In the other areas, outcropping karst is mainly found in separate islands and in the vicinity of the meadow karst (Fig. 4.5).

Fig. 4.3 Kinchkha waterfall (the Askhi limestone massif) (*photo by* K. Tsikarishvili)

In the subalpine and alpine aerials of Georgia's mountainous limestone massifs, the strongly karstified surfaces are covered with thin (10–45 cm) soil-meadow cover, which does not prevent the area seeping of atmospheric precipitation under the ground. Tight links are observed between the outcropping and meadow karstified sections. The karst sinkholes, rarely wells (including snow-icy ones), shafts and locked depressions are widely developed there. It should be also noted that the morphological features of outcropping and meadow karsts in the foothill zone, and especially, in low-mountainous karst, are considerably different from the similar areas of higher karst, and it is quite logical, as far as the karst areas of mountain-meadow higher zone are characterized by complex terrain, harsh climatic conditions, with active participation of the snow melted waters in karst development and nival processes, to a certain extent.

Fig. 4.4 Saadamio Cave (the Eki limestone massif) (*photo by* K. Tsikarishvili)

A relatively wider limestone strip, covered with mountain forests, nearly continuously stretches to the south of the outcropping and meadow karst surfaces. Due to its characteristic soil-vegetation cover and climatic regime, this zone creates the individual genetic type in the territory of Georgia; it is a special version of a meadow karst—a karst of the mountain-forest strip (Maksimovich 1963; Gvozdetsky 1954).

The forest plays an important role in holding atmospheric precipitation on the surface of the earth and their gradual (limiting) spending.

Fig. 4.5 Outcropping karst (Godogani Plateau) (*photo by* K. Tsikarishvili)

Appropriately abundant precipitation, along with favorable temperature conditions, contribute to the feeding of numerous karst sources, which are discharged in the peripheral sections. In the case of absence of the forest massifs, most of them would have stopped functioning a long time ago. Under the forest cover, the limestone strip, apart from the strongly inclined slopes of individual ranges, is distinguished by much intense karstification rather than outcropping and meadow karst regions. In the case of favorable topographical conditions (smoothed or slightly angled surfaces), the straight cone-shaped sinkholes are often formed in the forest strip, which are much wider and deep (sometimes 150–200 m wide and 50–60 m deep) than in crestal outcrops. Significantly processed giant sized sinkholes are widely represented in the areas of Mamdzishkha (the Gagra Range) in the limestone massifs of Bzipi, Askhi, Racha, etc.

In the high mountainous limestone massifs, the fragments of typical covered karst are associated with the accumulation areas of the Late Pleistocene moraines and fluvio-glacial sediments (Tielidze 2017). Such "Islands" include the area near Gelgeluka (Arabika) covered with moraine deposits and the lower part of the Bgia dead gorge (the Bzipi massif). The hidden karst is particularly widespread in the foothills of Georgia, where the carbonate rocks are covered by unsoluble rocks, which are not genetically related to karst rocks. The typical areas of distribution of hidden limestone karst are the Duripshi, Jali, and Zemo Imereti Plateaus, as well as the Otkhari sloped Lowland and central Samegrelo (Odishi). The latter is one of the largest clastokarst regions in Georgia. Here, the Neogene age conglomerates and sandstones are karstified. The clastokarst region of the Otkhari foothill Lowland looks like the central Odishi by its feature, which is basically represented by a weakly cemented deluvium of the limestone breccias. It is noteworthy that while the Duripshi, Jali, and Zemo Imereti Plateaus are full of dense net of sinkholes, karst forms are not found on the plain surfaces of the central Odishi and Otkhari. Here, complex systems of caves are present, characterized by significant length, rarely branched corridors, and sharply expressed erosion morphology. Thus, these two regions (central Odishi and Otkhari) can be considered as a special version of hidden karst, as they significantly differ from the abovementioned regions (Duripshi, Jali and Zemo Imereti).

In addition to this list, in the kart strip of Georgia, the buried and remnant type of karst are distinguished, which are relatively less distributed and at the same time insufficiently studied. The well-expressed buried karst forms are known in the Chiatura structural Plateau. Here, the karst sinkholes are filled up with Oligocene sandstones and manganese ore, and in certain areas, the Cretaceous limestones are deposited on Upper Jurassic corrie formations. The relatively young buried karst formations are revealed in the higher areas of the Bzipi Range, Askhi massif, etc. (Tintilozov 1976).

In the karst strip of Georgia, the remnant karst is preserved only as a relic form. It is represented in the Ertso-Tsona depression (in the upper stream of the Kvirila River) by individual denudative rocky remnants of almost similar heights. The rocky remnants of old relief of Ertso-Tsona are distinguished with unimportant relative altitudes (70–80 m) and karstified surfaces. Similar relief is well-expressed Kudaro massif in the vicinity of the Ertso Lake. In both regions, the Kimmeridgian and Tithonian pure limestones go through karstification.

4.6 Surface and Underground Karst Landforms

The karst relief of Georgia is represented by various surface and underground karst forms. The first includes the various types of locked negative forms —poljes, uvalas, depressions, sinkholes, expanded cracks, corries or serums, and their accompanying karst forms: karst canyons and cliffs, bridges, dead gorges, and others. The underground forms include the wells, shafts, abysses, numerous and various types of karst caves, including the widely developed clastokarst caves.

The large karst hollows, represented by poljes, depressions, and uvalas, are formed both in higher and lower karsts. The Shaori, Tsebelda, Kubcha-Yasht, and Turchu poljes, or the wide hollows are characterized with flat bottom and constant rivers. The first of these has been modified regarding the construction of the hydroelectric power station and now there is an artificial lake there. The Shaori depression was formed by blocking the surface river gorges due to an overthrust developed there that is now the largest polje in Georgia.

On the contrary to poljes, the karst cavities are relatively widespread, characterized by the uneven bottom and lack of important constant water streams. They were formed as a result of movement of the syncline moulds building material due to erosive-denudational processes. Similar cavities are formed in the limestone massifs of Askhi, Migaria (the Tsipuria and Okunvare cavities), Okhachkue, and others. The bottom and slopes of all these cavities are complicated by numerous karst sinkholes. Karst cavities are also known in other high mountainous (Arabika, Bzipi Range, Racha, Rikhva, etc.) and low-mountainous (Eki Mountain, Urta, etc.) karst massifs.

The karst sinkholes are distinguished by a great variety of shapes and sizes. According to the shape (form), the rectangular, asymmetric, flat-bottomed, packed, and other types of sinkholes are common. The density of sinkholes spread over the limestone masses is different. Thus, their number in the Arabika massif, in the "Didvake", is about 80–90 per 1 km^2 on average. There are 210–220 sinkholes documented per 1 km^2 on the crest of the Bzipi Range. The Racha limestone massive is also distinguished by intensive development of sinkholes, the number of which reaches 40 per 1 km^2 in the western section (Asanidze et al. 2017c). In the mentioned massif, most of the karst sinkholes are filled with rain or snow melted water almost all year round (Fig. 4.6).

In some massifs, karst sinkholes merge and cross each other in such a way that they do not leave any free space. In the massive limestones, the straight cone-shaped sinkholes are mainly formed (Askhi, the southern cuesta of Khvamli, the south slope of the Bzipi Range, etc.). The asymmetric sinkholes are mainly the characteristic of the thin layered inclined limestones (some areas of Arabika, Bzipi Range, Okhachkue, etc.). Flat-bottomed sinkholes (with loose materials presented in the bottom and often used for agricultural crops) are common in the strip of the lower karst. This type is especially widespread in the Zemo Imereti Plateau, in the area of Kutaisi-Tskaltubo, Ertso-Tsona and in other areas (Tsikarishvili et al. 2015). There are also complex, packed (secondary) sinkholes, where young sinkholes are formed on the bottom of old sinkholes.

Corrie (serum) surfaces are less common in the karst province of Georgia than in the Crimea, Balkan Peninsula, and other districts of outcropping karsts, although they are still found in many areas. They are formed both in the higher karst massifs and in the lower karst strip, and the forms are completely absent only in the clastokarst areas. According to the capacity and angle of slope of limestone layers, as well as their chemical composition, the corrie formations of Georgia are of different nature. The deeply cut corrie fields are represented in the Arabika massif, in the southeast of the site of Gelegeluki. The surface of the limestone suites is sloped by 20° at a large area (6 ha) and the surface released after retreating the Würm glaciers is dissected by 1–2 m deep (8–11 m in some areas) corrie cracks. Similar corrie surfaces are found in the areas built of massive and chemically pure limestones of the Bzipi Range, in the massifs of Askhi, Kvira, Migaria, and Racha.

Within the karst strip of Georgia, the wells, shafts, and abysses are formed both in high and medium and low karts; but the listed forms dominate in the Arabika and Bzipi higher massifs which were influenced by the Quaternary glaciation. The Okhachkue, Migaria, Racha, and Askhi massifs are also distinguished by abundance of wells, shafts, and abysses.

The medium and especially high mountainous massifs of karst strips of Georgia are classic areas of distribution of deep abysses. It is noteworthy that four out of the first ten of the world's deepest karst abysses are located in Georgia (Asanidze et al. 2013b, 2017d). Today, in the world, the 1st,

Fig. 4.6 Karst sinkhole full of water (the Racha limestone massif) (*photo by* L. Asanidze)

2nd and 3rd places are distributed between the Krubera (2197 m), Sarma (1830 m) and Illyuziya-Snezhnaya-Mezhennogo (1753 m) abysses found in the high mountainous limestone massifs of Georgia. Thus, the Krubera abyss is the deepest in the world (Asanidze et al. 2017e), where the world's record of depth penetration was established by the leadership of Speleologic Federations of Ukraine. However, it should be said that the first entry into the Krubera abyss at a depth of 150 m was carried out by the Georgian speleologists in 1961. At the same time, it should be noted that in the discovery and study of karst caves in Georgia, together with Georgian specialists, significant contribution has been made by the professional expedition groups of Ukraine, Russia, France, Czech Republic, Poland, Slovenia, and other European countries.

Within the karst strip of Georgia, the caves are distinguished by abundance and diversity of caves. The speleothems forms in the caves are also diverse (Lezhava et al. 2016b) (Fig. 4.7).

Most of the caves are formed in limestones, limestone conglomerates, breccias, and sandstones. Currently, the total number of traced and partially studied caves in Georgia is over 1500 (Asanidze et al. 2017e). The total length of the documented caves is 275 km, and the depth is over 80 km (Tatashidze et al. 2009). Most of the researched cavities are gathered at

Fig. 4.7 Spherical example of pool calcite speleothems, deposited subaqueously in cave pools at multiple places of the Muradi Cave (Racha limestone massif) (*photo by* L Asanidze)

plateau-shaped and inclined surfaces or are opened on the slopes of the river gorges and in the limestone massifs up to the absolute height of 1100 m.

Lower and foothill karsts, particularly the plateau-shaped karst regions (Zemo Imereti, Duripshi and Central Odishi), are distinguished by significant density of subhorizontal and horizontal caves. Thus, by the last years' data, up to 130 caves are documented in the Zemo Imereti Plateau (Lezhava 2015; Lezhava et al. 2016b). Certain parts of the caves are well organized and are involved in the speleo-industry. Among them, the Akhali Atoni, Prometheus and Sataplia caves are notable.

Medium and high mountainous limestone massifs of karst strips of Georgia are distinguished by the abundance of snow-icy caves. In this respect, the upper section of the Illyuziya-Snezhnaya-Mezhennogo Cave system should be noted, the bottom of which is occupied by firn-glacial cone of many centuries at a depth of 200 m. It is the only abyss in the earth, in which snow, firn and the underground glacier extend at such a depth from the surface (Tatashidze et al. 2009).

It is also worth noting the Vakhushti Bagrationi, and Marteli abysses located in the Arabika massif, the Boga freezer in the Khvamli massif, and the Skhvava and Nikortsminda freezers in the Racha limestone massif, and a wide distribution of icy wells in the Gagra and Bzipi limestone massifs. The icy caves are also distributed in the southern highland of Georgia, in the cavities of volcanic and lava rocks (the Khorkhebi and Khizabavra freezers), on the southern slopes of the Greater Caucasus, in the volcanic boulder valley, and many other places.

With a significant distribution in the karst strip of Georgia, clastokarst caves are distinguished. The main areas of their distribution are the central Odishi (Samegrelo) and Duripsh plateaus. In addition, they are found in other places in Apkhazeti and Samegrelo. Namely, the Nazodelavo Cave (length 700 m) is the longest (clastokarst) cave in the Caucasus formed in the conglomerates. The morphological characteristics of Odishi conglomerate caves are broad entrances, hall expansions, hallways, small canyons cut in the bottom, erosive terraces, and weakly developed cave shapes.

The interesting and beautiful forms formed as a result of karst processes also include travertines (limestone tuff). These accumulative forms are widely distributed in Georgia within the Greater Caucasus and in the limestone part of intermontane. Travertines are found in the Greater Caucasus, both in the areas of distribution of limestone and non-limestone rocks. In the latter, the carbonate solution comes out from significant depths (from the limestone suites beneath the slates), which is caused by tectonic faults (in the vicinities of the Juta Village, the Kolotani River in the northern Khevsureti). The age of travertine formations varies from modern times and Holocene up to Middle Pleistocene and may be older.

References

Apkhazava IS (1971) The Black Sea and a terrestrial water network. In the book: "Geomorphology of Georgia". P/h "Metsniereba", Tbilisi (in Georgian)

Asanidze L, Tsikarishvili K, Bolashvili N (2013a) Cave tourism potential in Georgia. In: The 2nd international symposium on Kaz Mountains (Mount Ida) and Edremit. Proceedings & Abstracts, Edremit, 3–5 May 2013, pp 243–247

Asanidze L, Tsikarishvili K, Bolashvili N (2013b) Speleology of Georgia. In: 16th International Congress of Speleology, Brno, 21–28 July 2013, 1, pp 29–32

Asanidze L, Chikhradze N, Lezhava Z, Tsikarishvili K, Polk J, Chartolani G (2017a) Sedimentological study of caves in the Zemo Imereti Plateau, Georgia, Caucasus region. Open J Geol 7:465–477

Asanidze L, Lezhava Z, Tsikarishvili K, Chikhradze N, Polk J (2017b) Karst morphological processes and evolution of the limestone massif of Georgia from depositional, sedimentary, and structural investigations in Muradi Cave. In: Proceedings of 17th international congress of speleology, at Sydney, Australia

Asanidze L, Lezhava Z, Chikhradze N (2017c) Speleological investigation of the largest limestone Massif in Georgia (Caucasus). Open J Geol 7:1530–1537. https://doi.org/10.4236/ojg.2017.710102

Asanidze L, Avkopashvili G, Tsikarishvili K, Lezhava Z, Chikhradze N, Avkopashvili M, Samkharadze Z, Chartolani G (2017d) Geoecological monitoring of Karst water in the Racha Limestone Massif (Country of Georgia). Open J Geol 7:822–829

Asanidze L, Chikhradze N, Lezhava Z, Tsikarishvili K, Polk SJ, Lominadze G, Bolashvili N (2017e) Complex speleogenetic processes and mineral deposition in the Caucasus region of Georgia. J Environ Biol 1107–1113. https://doi.org/10.22438/jeb/38/5(si)/gm-30

Chirakadze GI (1972) Climatic parameters of the natural moisture content of Transcaucasia. Papers of Transcaucasus scientific researches of the Institute of Hydrometeorolog, issue 50(56), "Gidrometeoizdat" (in Russian)

Gamkrelidze P (1964) Tectonics. In the book: Geology of the USSR, Vol. X. The Georgian SSR. "Nedra", M (in Russian)

Gvozdetsky NA (1954) Karst (2nd edn) (Geografgiz M) (in Russian)

Gigineishvili G, Tabidze D (1970) About underground water intakes of the Bzib karst massif (Southern slope of the Greater Caucasus). Bull of AS GSSR, 60, No. 1, Tbilisi (in Georgian)

Kordzakhia M (1961) Climate of Georgia. Tbilisi (in Georgian)

Kiknadze TZ (1976) To the hydrogeology of the karst of western Georgia. Caves of Georgia, Speleological Collection, No. 6, P/h "Metsniereba", Tbilisi, pp 100–106 (in Georgian)

Lezhava Z (2015) Karst of Zemo Imereti Plateau and surrounding regions. Publishing house "Universali", Tbilisi (in Georgian)

Lezhava Z, Tsikarishili K, Bolashvili N, Asanidze L, Chikhradze N (2015a) Preliminary outcomes of the Eki Mountain Karst-Hydrological and speleological study. Proceedings of ICEA 2015. In: International conference applied ecology: problems, innovations, Tbilisi-Batumi, pp 89–95

Lezhava Z, Bolashvili N, Tsikarishvili K, Asanidze L, Chikhradze N (2015b) Hydrological and hydrogeological characteristics of the platform karst (Zemo Imereti plateau, Georgia). In: Proceedings of the fourteenth multidisciplinary conference: Sinkhole and the engineering and environmental impacts of karst, Rochester, Minnesota, pp 93–100

Lezhava Z, Tsikarishvili K, Dvalashvili G, Tushabramishvili D (2016a) Chiatura municipality karst caves. Tbilisi (in Georgian)

Lezhava Z, Asanidze L, Tsikarishvili K, Lominadze G, Chartolani G, Julakidze I and Naskidashvili A (2016b) Cave Murada-the unique speleological object on the Nakerala range. Sci Technol. Sci Reviewed Mag 2:43–50 (in Georgian)

Lezhava Z, Tsikarishvili K, Asanidze L, Bolashvili N, Chikhradze N, Chartolani G (2017) Ecological investigation of Karst water in the central part of Georgia. Lap Lambert Academic Publishing, Saarbrücken. ISBN 978-620-2-02210-1

Maksimovich GA (1963) Fundamentals of Karstology, Vol. I, Perm (in Russian)

Maruashvili L (1973) The fundamentals of speleology. Tbilisi (in Georgian)

Tatashidze Z, Tsikarishvili K, Jishkariani J (2009) The Cadastre of Karst caves of Georgia. Collected Papers, New Series 3(82) (Tbilisi) (in Georgian)

Tielidze L (2017) Late pleistocene and holocene glaciation. Chapter in Glaciers of Georgia, Springer, Geography of the Physical Environment. https://doi.org/10.1007/978-3-319-50571-8-6

Tintilozov ZK (1976) Karst caves of Georgia (Morphological analysis). Tbilisi (in Russian)

Tintilozov ZK, Maruashvili LI (1971) Karst and pseudo-karst relief. In the book: Geomorphology of Georgia, Tbilisi (in Russian)

Tsikarishvili K, Lezhava Z, Asanidze L, Bolashvili N, Chikhradze N, Chartolani G (2015) Geomorphologic Features of Sataplia-Tskaltubo Limestone Massif. Earth Sciences, Science Publishing Group. 4(5-1):108–112

Zaza Lezhava is a Ph.D. senior research scientist at the Department of Geomorphology and Geoecology of the Vakhushti Bagrationi Institute of Geography, Ivane Javakhishvili Tbilisi State University. He is a coauthor of about 60 scientific works and several monographs. His research focuses on speleology and karst waters.

Lasha Asanidze is a Ph.D. research scientist at the Department of Geomorphology and Geoecology of the Vakhushti Bagrationi Institute of Geography, Ivane Javakhishvili Tbilisi State University. His research areas are speleology, karstology, and geomorphology. During 2014–2016, he completed an internship at several US universities. He is an official delegate of the European Speleological Federation, and has published about 20 scientific works.

Kukuri Tsikarishvili is a Ph.D. main research scientist at the Department of Geomorphology and Geoecology of the Vakhushti Bagrationi Institute of Geography, Ivane Javakhishvili Tbilisi State University. His research focuses on speleo-geomorphology.

Levan Tielidze is a Ph.D. of Ivane Javakhishvili Tbilisi State University in the field of geomorphology–glaciology–cartography and a senior research scientist at Vakhushti Bagrationi Institute of Geography. He is also a researcher-scholarship holder of the Department of Earth Sciences of the Georgian National Academy of Sciences. The field of his research is mountain cartography and glacial-geomorphological study of the mountainous areas in the Quaternary (Late Pleistocene and Holocene). He is a member of several international organizations. In 2014/2015, he conducted a research stay at the Climate Change Institute of the University of Maine, USA, and in 2015/2016–2017 at the University of Northern British Columbia, Canada. He is the author of about 40 scientific papers, maps, and eight monographs.

Glacial and Postglacial Landscape of Georgia

5

Levan Tielidze

Abstract
In this chapter, the glacial landscape of Georgia is considered according to the individual river basins. Only those basins (Bzipi, Kodori, Enguri, Rioni and Tergi) are distinguished, where the glacial forms are mostly preserved such as trough gorges, corries (cirques) stadial moraines, sheep rocks, ledges, and others. Their distribution extent is given. We do not provide the description of the modern glaciers in this chapter because a separate monograph directly about the glaciers of Georgia has been published.

Keywords
Trough gorge · Corrie (cirque) · Stadial moraine

5.1 Research Area

Landforms created due to the action of Pleistocene, Holocene, Little Ice Age (LIA), and modern glaciers are well represented in the high mountainous zone of the Greater Caucasus, especially in central Greater Caucasus (Gobejishvili et al. 2011), as well as in Khokhi, Kuro, Shavana, Kidegani, Khevsureti, and Pirikita subranges. Complex tectonic-geomorphological structure and different climatic conditions determine the variation of the capacities/borders of old and modern glaciations (Fig. 5.1).

Excaration formations of relief including trough gorges are as follows: corries (cirques) hanging valleys, ledges, nunataks, trough saddles, sheep rocks, glacial scratches, etc. And the accumulation formations include the moraine deposits of various types and erratic boulders, which are mainly built of magmatic rocks (granites, diabases, porphyrites, andesites, etc).

A relief transformed by glaciers is common in those mountain ranges (Gagra, Bzipi, Egrisi, Racha, etc.), where there are no glaciers today at all.

In the Lesser Caucasus (within Georgia), the visible traces of glaciation are preserved in the higher parts of Achara-Trialeti, Shavsheti, Arsiani and Samsari ranges, and Erusheti highland.

L. Tielidze (✉)
Department of Geomorphology and Geoecology, Vakhushti Bagrationi Institute of Geography, Ivane Javakhishvili Tbilisi State University, Tbilisi, Georgia
e-mail: levan.tielidze@tsu.ge

L. Tielidze
Department of Earth Sciences, Georgian National Academy of Sciences, Tbilisi, Georgia

L. Tielidze (ed.), *Geomorphology of Georgia*, Geography of the Physical Environment,
https://doi.org/10.1007/978-3-319-77764-1_5

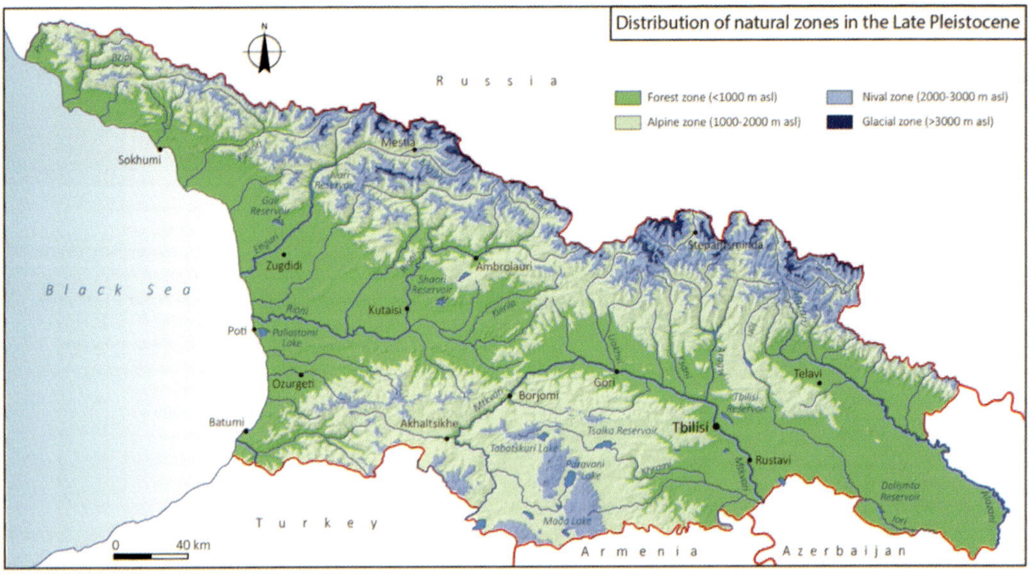

Fig. 5.1 Distribution of natural zones in the Late Pleistocene

5.2 Bzipi River Basin

Detection of glacial relief is very difficult in Bzipi River basin due to deep forest and fragmented relief. Nevertheless, corries, trough fragments, and sheep rocks free of glaciers are widely represented on the slopes above the forest strip. On the northern slope of Bzipi Range, in the heads of Psikva, Khimsa, and Mtsra rivers, the old glacial corries are located at an elevation of 2150–2450 m asl, the bottoms of most of which are occupied by lakes. The gorges of the left tributaries of Bzipi River—Bashkatsari, Psikva, Khimsa, and Mtsra rivers, the length of which does not exceed 4–5 km—maintain the trough form to the end.

In Bzipi basin, on the southern slope of the Greater Caucasus, the corries are located at the height of 2600–3000 m asl, most of which are occupied by glaciers; in some areas, there are forms free of glaciers (Gobejishvili 1995). The gorges of the tributaries of Bzipi River keep their trough forms almost until the confluence (only Psishi and Gribzi gorges are distinguished by erosive nature in their lower parts). In the heads of Bzipi River's right tributary—Baviu River, where there are no glaciers today, the old corries are widespread at the height of 2200–2500 m asl, most of which are occupied by small lakes. It is observable that in the heads of Baviu River, there should not have been a powerful glaciation in Late Pleistocene as the gorge lacks the trough shape.

Depending on distribution of signs of Pleistocene glaciation, Lashipse River basin included in Bzipi River basin is worth the attention, which keeps its trough form before joining the right tributary—Avadhara River. Old glacial corries are located in the heads of the gorge on the southern slope of Greater Caucasus and in Atsetuka Range at 2100–2450 m asl.

Glacial deposits (stadial moraines) are poorly preserved in Bzipi basin. Morphologically observable lateral or terminal stadial moraines in this gorge have not been observed, if we do not take into account the young stadial moraines remaining in the heads of the river at heights of 1800 m and 2100–2300 m asl. Below, in Bzipi gorge, the absence of stadial moraines should be caused by the great inclination and dissection of the slopes.

Boulders similar to the moraine materials are found at the mouth of Psishi River at 1350 m asl, which can be found on the right bank of Psishi River up to the shepherds' shelter. The remains of stadial moraines are located in Psishi River gorge at a height of 1580 m. Relatively young moraines are common at the heights of 2150 and 2300 m.

Boulders similar to the moraine materials are scattered in Khimsa River gorge above the mouth at a height of 1550 m. Relatively, young stadial moraines are found in the upper part of Khimsa at the heights of 1750, 2050 and 2250 m. The remains of the stadial moraines are preserved in Psikva River gorge at a height of 1750 m.

Glacial deposits are less distributed in the gorges of the rivers Lashipse, Mzimni, and Avadhara. Here, the young moraines are common along the sections of distribution of corries at the heights of 1950, 2100 and 2200 m. There are also the fragments of two stadial moraines in Avadhara River gorge at the height of 1650–1700 m. Powerful stadial moraines can be found at the confluence of Avadhara and Lashipse rivers at the height of 1470 m.

5.3 Kodori River Basin

Glacial landforms are preserved in Kodori River Basin, in the gorges of its tributary rivers Klichi, Ghvandra, Sakeni, and Chkhalta. In Klichi River gorge, which originates from Klukhori Pass, modern glaciers are quite widespread; most parts of corries are occupied by glaciers. Klichi River gorge maintains the trough form up to the confluence of its left tributary—the Nakhari River. More below, the trough form is destroyed by the actions of debris cones and river erosion, but in spite of this, the Pleistocene trough fragments are observed on the slopes of the gorge before the confluence of the right tributary river Achapara. Achapara River itself keeps its well-formed trough shape before joining the Klichi River, below which, the gorge is of erosive character.

Moraine deposits are widespread in Klichi gorge, and below the head of the river, they are presented at different absolute altitudes (1100–2450 m).

Glacial landforms (cirques, troughs, etc.) are widespread in the Ghvandra River gorge. Glacial corries are richly represented on the slopes of Ghvaghva and Klichi, most of which are occupied by lakes. Ghvandra gorge itself maintains the trough form up to the height of 1550 m (site of "Mindora"), and then it is erosive. In Ghvandra gorge, there are also a number of glacial deposits preserved in the form of stadial moraines and individual boulders. The youngest stadial moraines are located in the heads of Ghvandra River at 2150–2300 m asl, which lack vegetative cover (Khazaradze 2004). Below in the Ghvandra gorge, the remnants of the stadial moraines are represented beginning from the site of "Mindora" within 1090–1600 m.

An exceptional abundance of glacial forms is in Sakeni gorge. At present, modern glaciers are still widespread in the upper part of the gorge, which is bordered by Kodori and Klichi ranges. Sakeni River itself, which at the same time is considered a head of Kodori River, flows out from the great glacier of this gorge—Ghvandra. Most corries are occupied by glaciers that can be explained by the abundance of modern glaciers. Sakeni River gorge is trough shaped up to the height of 1500 m asl.

Small rows of young stadial terminal moraines are located in Sakeni River gorge at a height of 2100–2400 m. In this section, Sakeni gorge is represented by ledges, the surfaces of which are smoothed by exaration scratches. Below, at a height of 2050 m asl, there is a relatively powerful stadial moraine located transversally to the gorge, which is directly formed on the top of the ledge. Sakeni gorge is represented by a vast floodplain at a height of 1950 m asl, which is horizontal so that the river forms distinctive meanders. At the end of the floodplain, the gorge is blocked by a powerful stadial moraine composed of about 10 m size fragmented material of the rocks composing the crystalline heart of the Greater Caucasus.

Glacial deposits are particularly poorly preserved in Chkhalta River gorge, from the south it is bordered by a range of the same name. The main part of the gorge (especially the lower section) is erosive and lacks the traces of glacial activity. Only in its one section are the remnants of stadial moraines observed at a height of 650–700 m asl near the confluence of southern Prishi—the left tributary. The rocks of the Greater

Caucasus crystalline heart prevail in the composition of stadial moraines.

Two km below the Chkhalta River confluence, Zima River conflows Kodori River from the right side. In the area of confluence, the granite boulders are observed, which are considered to be brought by Pleistocene glaciers from the heads of Kodori (Reinhardt 1936). Existing granite material is not distinguished with great size and is well processed by river. Similar boulders are widespread in Zimi River gorge from its head to the confluence.

Glacial deposits are relatively poorly preserved in the gorges of Kodori River left tributaries. Only in the heads of Bramba River gorge are the remnants of young stadial terminal moraines observed. Below, traces of glacier activity are erased in the relief and only individual processed porphyritic boulders can indicate the existence of glaciers in the past.

5.4 Enguri River Basin

Landforms modeled by glaciers are widespread in the Enguri River basin. Forms (troughs, corries, exaration scratches and other microforms) produced by the Late Pleistocene glaciers are especially visible, creating typical alpine relief in the middle and high hypsometric zones of the basin. These forms are unevenly distributed and in some areas are transformed (Tielidze 2017a). Glacial forms in crystalline rocks are more resistant to physical weathering and are well preserved. In the area of distribution of clay slates, the weathering is intense and the glacial forms are modified. As for the forms developed by the older glaciation (Mindel, Riss), they are completely erased in relief due to a long erosive action and physical weathering, or are preserved in the form of individual fragments.

The glacial landforms are well preserved in the upper part of Darchi (Darchi-Ormeleti) River gorge—the right tributary of Nenskra River. The glacial corries are located on the slopes of Likhnimi Range at a height of 2200–2600 m. The gorge itself is trough shape along 7.5 km.

There are three medium-sized cirques, separated from each other by low hills, on the southeastern slope of Kharikhra Range in the regions of Khida Pass and Mt. Kharikhra, in the head of Lakhami—the right tributary of Nenskra. Continuation of the hills creates weakly developed troughs. The dividing hills are well processed/cut by glaciers, where numerous small lakes are formed. Lakhami gorge is mainly erosive and there are only weakly developed trough gorges, which are filled up with moraine material brought by mudflows from cirques. Similar short (1.0–1.5 km) troughs are in the upper parts of the gorges of right tributaries Devra, Tetnasheri and Kharami.

Okrili River trough gorge conflows Nenskra gorge at a height of 1250 m asl. At the confluence with Nenskra, the gorge is narrowed and forms the 100–150 m high step. The granite-gneiss rocks outcropped on the slopes are processed by the Late Pleistocene glacier. Modern glaciers are still represented in the upper part of Okrili trough. Kharikhra glacier is the largest glacier in the gorge. The rest of the glaciers are represented in the old corries as unimportant remnants.

One more tributary—Tskhvandiri gorge conflows the Nenskra gorge at 1350 m asl, which maintains a well-distinctive trough form at approximately 5.5 km distance from the head. Below, the gorge is narrowed and it continues until the confluence. In this section, on the slopes, the traces of Late Pleistocene glacier activity are observed (smoothed forms and scratches), which creates 250–300 m high step at the junction of Nenskra gorge. In the upper part at a height of 1875 m asl, Tskhvandiri trough is divided into three sub-troughs and is developed in the form of short (1.5 km) troughs. In this section, on the slopes of Kharikhra Range, within 2400–2900 m asl, several cirques free of glaciers are observed. Glacial forms are observed here and there in the form of remnants of corries lengthwise in the east of Tskhvandiri, caused by steep slopes and enhanced erosion of numerous water flows.

Dalari River gorge rich in glacial forms joins the Nenskra gorge from the right side at a height of 1620 m asl. There are numerous glaciers in the upper part and on the slopes of the gorge, as well as corries—the forms free of glaciers. Dalari gorge itself is a well-formed trough at about 5.5 km distance, connected with numerous short hanging trough gorges from both sides. There are three small corries with lakes in their bottoms next to each other on the northern slope of Tskhvandiri Range, which borders Dalari trough from the south in the easternmost part at a height of 2450 m.

From the north Dalari gorge is bordered directly by the southern slope of Greater Caucasus watershed range, which is characterized by steep slopes in this section. Therefore, the distribution of tributary troughs is limited here. Modern glaciers are mostly located along the crestal part of Greater Caucasus Range. The smoothed surfaces of the base rocks outcropped on the slopes indicate that the glaciers were represented there in the form of extensive, hanging tongues.

Above Dalari confluence the Nenskra gorge stretches at a distance of 15 km and leans against Chiperazau Pass. The gorge is an ideal trough in this section.

In the east, Nenskra gorge is bordered by Shdavleri Range, the northern part of which is mainly occupied by modern glaciers, and the southern part is free of snow-ice cover and instead corries free of glaciers are represented at a height of 2800.

At a height of 1200 m Nenskra gorge is joined from the left side by Tita gorge, which maintains the trough form up to 1650 m asl, and then it is erosive. The trough is asymmetric because the watershed of Tita and Marghi rivers is much lower than the range stretched northward of Tita gorge. Nenskra gorge with the length reaching 40 km is trough shaped from head to the Okrili River confluence at a distance of 25 km. This section of the gorge is mainly developed in the crystalline complex and is well preserved, and below, it is partly transformed. The trough bottom is covered by debris cones. At the junction of the Enguri gorge, on the left slope of the Nenskra gorge, at a height of 100–150 m above the bottom, the Late Pleistocene trough side is preserved, on which the village Tobari is built up. A relatively small similar plain is represented on the right slope of the gorge. The modern morphology of Nenskra basin, the abundance of glacial forms, and modern glaciers indicate that during the Late Pleistocene glaciation, there were two powerful glaciers located in the trough.

Abundance of moraine deposits that are spread both in Nenskra gorge itself and in the tributary gorges indicates the powerful glaciation of the Nenskra River basin. In the upper part of the Nenskra gorge, within 1950–2350 m, the relatively young terminal moraine heaps are represented. The powerful heap of the terminal moraine is located in the confluence of Nenskra and Dalari at a height of 1620 m asl. Three terminal moraine heaps are preserved in the Dalari gore itself within 1920–2320 m. Below in the Nenskra gorge, at the confluence of Tskhvandiri River, at a height of 1620 m, a powerful stadial moraine covered with forest is developed. There is a powerful suite of lateral moraines preserved on the left slope, in the same place. Remnants of terminal moraine washed by the river are located at a height of 1620 m. There is a powerful stadial moraine covered with dense forest at the confluence of Nenskra and Okrili left by the glacier. Powerful moraine accumulation is located on the right side of the river, above 0.5 km from the village Tita.

The next strong moraine, composed with large boulders of granites and gneisses, is presented on the right side of Nenskra at a height of 1000 m at the village Sgurishi. On the opposite side of the village Kari, on the left side of the Nenskra gorge, elongated vast plain area is located at a height of 320 m from the bottom of the gorge, where the village Zemo Marghi is built up. There is numerous boulder material scattered on the territory of the village. The gorge side of the plain area is cut and the 8–10 m thick section is outcropped, where the weathered uncategorized granite-gneiss boulders are observed. The plain area stretches along the slope as a narrow strip at a distance of 2.0–2.5 km and at the same time it is a Nenskra and Marghi rivers' watershed with cut crest. Granite and gneiss boulders are scattered in its surface.

In the lower part of Nenskra gorge, at the confluence of Lakhami River, a fragment of stadial moraine is preserved at a height of 800 m asl, on the right side of the river, on the 25–30 m high terrace. Even more below, at the confluence of Nenskra River (750 m), on the left slope of the gorge (at 100–150 m from the river), in the territory of the village Tobari, numerous granite and gneiss erratic boulders are scattered in the porphyritic boulders, which can be brought from the heads of Nenskra.

In Nakra basin, the glacial forms are mainly developed on the eastern slope of Shdavleri Range. Glacial corries are represented in stages, and in some areas several corries are formed in one large cirque. Namely, on the slope (2600–2650 m) between the peaks Gvergisheli and Gvergildashi several corries are represented that are separated from each other by hills of lesser capacity. Morphologically, a large cirque is stretched in the relief here, on the slopes of which the abovementioned corries are located in steps. The area along the slope of Shdavleri Range is characterized by having below almost all of the existing corries the hanging troughs are developed, the end parts of which are expanded in Nakra gorge.

A similar situation is presented on the western slope of Tsalgmili Range, which borders Nakra gorge from the east, but there is only one difference—here, in the heads of all left tributaries of Nakra River, the modern ice cover is developed. Late Pleistocene corries are located at a height of 2700–2800 m, the gorges are shorter but maintain trough feature to the end. Nakra gorge itself maintains trough character at a distance of ~16 km from the head up to the confluence of Utviri, its left tributary (1250 m), and then it is erosive.

Glacial deposits are widespread in Nakra gorge, which are preserved mainly in the upper part of the gorge as stadial moraines. A young powerful moraine hill is formed at a height of 2100 m asl, on the left side of a summer farm, the surface of which is covered with various herbs. More below, at a height of 2050 m, there is a stadial moraine of 25–30 m capacity preserved, which dammed a little lake. A powerful (30–40 m) moraine is preserved at a height of 1650 m, which is covered by a dense forest.

At a height of 1140 m asl, Enguri gorge is conflowed from the right side by Dolra gorge, in the heads of which the modern glaciers are widespread even today. Almost all cirques and corries there are covered by glaciers (Fig. 5.2). The gorge itself, the length of which is about 18 km is characterized by a trough shape (Fig. 5.3). It is joined by several trough gorges from both the sides in the upper part. At a height of 2120 m asl, the trough is joined by Kvishi trough from the right side, the length of which is about 3.5 km and there is a Kvishi Glacier in the head. From the end of the glacier, ~2 km below, on the bottom of the gorge the sheep rocks are outcropped, on the surface of which the scratches are visible. At a height of 1980 m asl, Dolra gorge is joined by Ushba trough from the left side, in the head of which Ushba Glacier is located. About 500 meters below from the end of the glacier tongue the sheep rocks are outcropped. Below, at the village Bagvdanari, Dolra gorge is conflowed by Guli River (Gulichala) gorge, in the upper part of which a Guli Glacier is stretched. On the left slope of the glacier, there are several corries free from the glacier remained within 2950–3100 m. Guli gorge maintains its trough form in the upper part, and below, it is partially modified by erosion.

Dolra trough is characterized by different forms in the separate sections. In the upstream, at a distance of 4–5 km, it descends with a wide bottom and gentle slope. Above 1.5 km before joining Ushba trough, the gorge is narrowed and the bottom inclination increases as well. From the place of joining, the trough takes meridional direction and gradually expands, and reaches the maximum in the surroundings of the village Mazeri (0.8–1.0 km). The trough bottom is overlapped by debris cones in some areas.

Fig. 5.2 The confluence of Northern and Southern Ushba glaciers in the Dolra River headwater (*photo by* L. Tielidze)

Fig. 5.3 Dolra River trough (*photo by* L. Tielidze)

Stadial moraines are widespread in Dolra gorge. Abundance and character of the moraines indicate the capacity of Late Pleistocene glaciation. Comparatively young moraines are represented near the modern glaciers. A powerful heap of stadial moraines covered with dense forest is represented near the junction of Dolra and Ushba at 1900 m asl. The moraine is mainly composed of large non-categorized materials. There are individual boulders with the capacity measured in tens of cubic meters.

Lateral moraines are richly represented on the western slope of Bali Range (the left slope of Dolra gorge), which starts from the Ushkhvanari Village and stretches to the confluence of Dolra. There are rows of lateral moraines at different altitudes, which are covered by forest. A powerful lateral moraine (1.5–2.0 km long) stretches along the slope from the gorge's bottom at a distance of 250–300 m, which later passes the crest along with the reduction in height of the range.

On the right slope of Dolra gorge, lateral moraines and moraine materials are poorly preserved that should be caused by the great inclination of the slope. Only in the final part of Detsili Range can be observed the large amount of moraine materials, over which the villages are built up.

Moraines of Würm glaciation are widespread in Mestiachala River gorge. Stadial moraine of 20–30 m capacity is represented at 1450 m asl at former touristic base on the left side of the river in the surroundings of the town Mestia. Particularly numerous glacial deposits are observed on the western slope of Gvalda Range bordering Mestiachala River gorge in the east. Here are powerful moraine hills represented at different altitudes of the gorge, which stretches at a distance of 3.0–3.5 km from the town Mestia. In the lower part, the moraine material covers the crest of Gvalda Range. The highest moraine is preserved at 420–450 m above the bottom of the gorge.

Mestiachala River gorge is free of glacial deposits along about 7 km above the aerodrome; only at a height of 1720 m asl, there is a large stadial moraine covered with forest at the confluence of the rivers Chalaati and Lekhziri. Above the confluence, in Lekhziri gorge, there are relatively young stadial moraines developed within 1750–2050 m asl. Numerous moraine materials are scattered in Chalaati gorge. Younger moraine rows are developed near the glacial tongues (Fig. 5.4).

No stadial moraines are observed below the town Mestia before the confluence of the rivers Mulkhura and Enguri. And at the junction, at 1350–1400 m asl (the Latali community), a large stadial moraine is represented. It should be assumed that the old (Late Pleistocene) glacier was in the stable condition for a long time in this section.

Mulkhura gorge is characterized by asymmetric trough shape before joining Tviberi and Tsaneri gorges that is caused due to the large debris cones developed along the right slope. Tviberi and Tsaneri gorges themselves maintain their well-expressed trough features at a short distance (2.0–2.5 km). And below, both gorges are erosive, though the glacial fragments preserved on the slopes indicate that these gorges should have been filled up with powerful glaciers in Pleistocene age. In the bottom and on the slopes of Tviberi gorge, there are lots of scattered moraine materials. Stadial moraines have a relatively limited distribution in Tsaneri gorge. Particular abundance of moraine material is observed in the crest of Ughviri Range and on its northern slope, which borders Mulkhura gorge from the south. There is a true moraine landscape formed on the saddle of Ughviri Pass, and the old former lakes are also observed.

At the village Ipari, Adishchala River conflows Enguri River from the right side; in the head of Adishchala, one of the most beautiful glaciers in the Caucasus—Adishi Glacier is located with its icefall (Fig. 5.5) (Tielidze 2017b). The gorge itself maintains its trough shape at a distance of 6.5–7.0 km, and below—it is erosive, though in some areas the traces of old (Pleistocene) glacier action are still observed on the slopes. Old glacial corries are located here at a height of 2650–2700 m asl. Glacial deposits are widespread in Adishi gorge in the form of the moraine rows of different periods.

Above Ipari, in about 12 km, another tributary —Khaldechala River conflows Enguri River from the right side, which maintains trough shape at about 4.0–4.5 km from the head, and then the river begins cutting the trough bottom and flows through the canyon bed. There are several corries free of glaciers represented on the left slope of Khalde trough (2600–2700 m asl) covered with

Fig. 5.4 Chalaati Glacier and gorge (*photo by* L. Tielidze)

herbal vegetation. In Khaldechala River gorge, the Pleistocene glaciation deposits are well preserved. Their abundance is especially noticeable in the upper part of the gorge.

Enguri maintains the trough shape only along the ~12 km from its head up to the village Murkmeli; furthermore it is deep erosive. In the heads of Enguri River, above Ushguli community, within heights of 2150–2250 m, three stadial moraines are developed, that indicates periodic suspensions of Shkhara Glacier in this section. In the territory of Ushguli community, there are plenty of boulder materials scattered on the slopes.

Glacial deposits are rare in the left tributary gorges of Enguri River. In some gorges, they do not appear at all or if observed, they are very modified and maintained in small fragments at different altitudes. Hence, the morphologically expressed stadial moraines were not been

Fig. 5.5 Adishi Glacier (*photo by* L. Tielidze)

observed by us in the gorges of the rivers Mushuristskali, Ghuristskali, Zeda Korsha and Kveda Korsha, Lasili, etc. Here, fragmented material is mainly found, which is the product of secondary fragmenting of rocks.

Late Pleistocene glaciation moraines are relatively well preserved in Lailchala River gorge within 1450–1700 m asl. On the right slope, at the confluence, on the 170 m high smoothed step, individual large-sized boulders remain, the remnants of stadial moraines.

5.5 Rioni and Tskhenistskali River Basins

In Rioni and Tskhenistskali basin, glacial landforms are relatively well preserved at the precincts, which are built from rocks (granites, gneisses, crystalline shales, porphyrites, and other metamorphic rocks) resistant to erosion. That is why the excess of the forms left by glaciers in the relief is observed in those parts of the basin that are directly leaning against the southern slope of Greater Caucasus Range. As for subranges (Svaneti, Lechkhumi, Racha, etc.), the glacial forms are preserved only in those precincts, where Bajocian suites (porphyrites) are outcropped in the form of islands. The main part of lateral ranges is built from the rocks (laser slabs, sandstones, etc.) unresistant to erosion and it is difficult to detect the old glacial relief.

In Svaneti Range, the southern slope of which directly enters the research area, glacial landforms are poorly preserved. This can be explained as the gorges, developed on its slopes through which the right tributaries of Tskhenistskali River flow, are cut into the Lias shales almost all of them and mostly are erosive.

In the area of Zagari Pass, where Svaneti Range descends to 2640 m, the trace of glaciers presence is almost erased. There are several corrie fragments only in the crestal part.

From Letperi Pass (2825 m) up to Mt. Lasili (3495 m), the southern slope of Svaneti Range is free of modern ice cover. It is difficult to find

Fig. 5.6 Zeskho Glacier (*photo by* L. Tielidze)

traces of old glaciers. Small depressions with lakes are observed in some places directly on the crest. It is likely that these depressions are fragments of old corries.

Laskadura River is one of the large tributaries of Tskhenistskali River, which takes its origin on the southern slope of Mt. Lahili (4009 m). Here are just several small-sized corrie type glaciers. Laskadura gorge maintains its trough shape up to 1800 m asl, and then it is erosive to the end.

Kheledula gorge, the largest tributary of Tskhenistskali, originating from Svaneti and Samegrelo ranges, is mainly erosive. Glacial landforms are only preserved in the heads of the river. Particularly large number of glaciers is observed in the part of Samegrelo Range, which directly enters the Kheledula basin. This should be caused by Bajocian porphiritic suite participating in the geological structure of this part of Samegrelo Range.

Glacial forms are much better preserved in the gorges of Koruldashi, Zeskho, Tskhenistskali River heads, and Ghobishura River. There are many corries located in the heads of Koruldashi River (southern slope of Greater Caucasus Main Range), which are still covered by modern glaciers. The gorge itself is a well-shaped trough up to 1800 m asl and erosive—relatively below, though the fragments of old trough are observed on both the slopes before joining Zeskho River.

The same situation is in Zeskho River gorge. Most parts of corries are still occupied by modern glaciers (Fig. 5.6), and the gorge itself maintains its trough shape up to the village Zeskho (1750 m) and it is erosive below up to the confluence.

Fig. 5.7 Chveshura River trough (*photo by* L. Tielidze)

There is a lack of glaciers in the heads of Tskhenistskali River. This results from shales prevailing in the structure of this section of Greater Caucasus Range, and corries free from glaciers are modified. Tskhenistskali itself maintains its trough form up to 1870 m asl.

The trace of old glaciation is well preserved in the form of glacial relief in the heads of Rioni River. Most part of troughs and corries are occupied by modern glaciers. Rioni gorge itself maintains the trough form until the confluence of Sasvanostskali River—its right tributary (1800 m asl).

Exceptional abundance of glacial forms is observed in the heads of Zopkhitura River—the left tributary of Rioni River, where the number of series of corries is represented on the southern slopes of the peaks Geze and Laboda, most of which are occupied by modern glaciers. The gorge itself is trough shaped up to 1700 m asl. In our opinion, in Late Pleistocene from Zopkhitura gorge, the larger glacier descended more than from the heads of Rioni gorge.

The abundance of glacial forms is observed in Chveshura River gorge—the left tributary of Rioni River, in the head of which Kirtisho—the largest glacier of Rioni basin is located. Numerous tributaries join Chveshura River from both the sides, in the heads of which the small-sized glaciers are located. Numerous corries are represented there (2500–3000 m asl). Chveshura gorge is a trough shaped up to the village Gona (Fig. 5.7). In some places, the trough fragments still remain on the slopes at a lower altitude.

One of the largest tributaries of Rioni River is Chanchakhi River, almost all tributaries of which flow from modern glaciers. Corries free of glaciers are richly represented in Chanchakhi gorge as well. The largest valley glaciers are Boko and Buba (Fig. 5.8). The glacial deposits preserved there prove that there were powerful glaciers in

Fig. 5.8 Buba Glacier (*photo by* L. Tielidze)

the heads (rivers Bokostskali and Bubistskali) of Chanchakhi River in Late Pleistocene.

5.6 Tergi River Basin

Numerous small tributaries join Tergi River from both sides, from the head up to the section of Kobi Village. Among the right tributaries, Tsotsoltastskali, Desistskali, and Esikomi are worth mentioning, which take their heads from the northern slopes of Keli volcanic upland. Smaller glaciers are preserved in the heads of the abovementioned rivers. Gorges are trough shaped before joining Tergi. Moraines are poorly preserved and are found in the heads only in the form of small-sized lateral moraines. In Late Pleistocene, these gorges were thoroughly occupied by glaciers. A similar picture is observed in this section in the gorges of the left tributaries (Tepistskali, Jimarastskali, Suatisistskali, and Mnaisistskali) of Tergi, in the heads of which the powerful glaciers are located, formed on the slopes of Mkinvartsveri (Kazbegi). These gorges are longer (5–6 km) than the right tributaries and are still maintaining the trough shape. Moraines are poorly maintained and are preserved only in the heads.

The main gorge of Tergi itself has a well-expressed trough shape up to the Desi Village (2200 m asl). Further down, the gorge is expanding and loses its glacial appearance due to debris cones intruded from both sides. Stadial moraines are quite modified in this section.

Below the Kobi Village, Tergi River flows through the vast gorge and there is no trace of glaciers activity. Snostskali River, its right tributary, is the largest in this section; many small tributaries join the river. Old glacial forms located in the heads of the tributaries prove that glaciation was relatively weak there and was only a local character. The exception is Juta River gorge, all the tributaries of which are characterized by trough shape. Juta gorge itself is trough shaped before joining Kora River, and the glacial material is also richly represented.

References

Gobejishvili RG (1995) Saqartvelos tanamedrove mkinvarebi da Evraziis mtebshi gamkinvarebis evolucia gvian Pleistocensa da Holocenshi (Present day glaciers of Georgia and evolution of glaciation in the mountains of Eurasia in late Pleistocene and Holocene), sadoqtoro disertacia, Tbilisi (in Georgian), 320 pp

Gobejishvili R, Lomidze N, Tielidze L (2011) Late Pleistocene (Wurmian) glaciations of the Caucasus. In: Ehlers J, Gibbard PL, Hughes PD (eds) Quaternary glaciations: extent and chronology. Elsevier, Amsterdam, pp 141–147. https://doi.org/10.1016/b978-0-444-53447-7.00012-x

Khazaradze R (2004) Kavkasiis Pleistocenuri Gamkinvareba (Pleistocene Glaciation in the Caucasus). Publ. Metsniereba, Tbilisi (in Georgian), p 319

Reinhardt AL (1936) Lednikovii period kavkaza i ego otnoshenie k oledeneniu alp i altai (Ice Age of the Caucasus and its relation to the glaciation of the Alps and the Altai). tr. II. mejd. konf. po izuchenie chetvertichnogo perioda evropi. L-M, bip. 2 (in Russian)

Tielidze L (2017a) Late Pleistocene and Holocene Glaciation. Chapter in Glaciers of Georgia, Springer, Geography of the Physical Environment. https://doi.org/10.1007/978-3-319-50571-8-6

Tielidze L (2017b) The New Inventory and General Description of the Modern Glaciers. Chapter in Glaciers of Georgia, Springer, Geography of the Physical Environment. https://doi.org/10.1007/978-3-319-50571-8-3

Levan Tielidze is a Ph.D. of Ivane Javakhishvili Tbilisi State University in the field of geomorphology-glaciology-cartography and a senior research scientist at Vakhushti Bagrationi Institute of Geography. He is also a researcher-scholarship holder of the Department of Earth Sciences of the Georgian National Academy of Sciences. The field of his research is mountain cartography and glacial-geomorphological study of the mountainous areas in the Quaternary (Late Pleistocene and Holocene). He is a member of several international organisations. In 2014/2015 he conducted a research stay at the Climate Change Institute of the University of Maine, USA, and in 2015/2016–2017 at the University of Northern British Columbia, Canada. He is the author of about 40 scientific papers, maps and eight monographs.

Urban Landscape of Georgia (On the Example of Large Cities)

Lela Gadrani, Mariam Tsitsagi and Levan Tielidze

Abstract

This chapter presents the general geographical location and issues of built-up areas of the major cities of Georgia (Tbilisi, Kutaisi, Rustavi and Batumi). The change in land use and dynamics of population over the last 30 years is discussed. In the case of Tbilisi, there are allocated areas such as bare soil area, green area, built-up area, and the agricultural area; also the main areas of land use transformation are highlighted. The reasons for the change in urban areas and population over the last 30 years are considered in the example of all four cities.

Keywords

Urban landscape • Tbilisi • Kutaisi Rustavi • Batumi

6.1 Introduction

Urban geomorphology is defined geographically confined to areas of concentrated urbanization, where the natural environment is anthropogenically modified and where natural processes modify anthropogenic structures (Thornbush 2015). One of the demonstrative examples of anthropogenic impact on the relief is a city. The city is an anthropogene (human created) landscape and constitutes a suitable field locality for study by the urban geomorphologist who must assess the type and magnitude of changes to lands and waters of the earth's surface in all physical settings (Coates 1984). Urbanization period of cities includes the most long-term stage, therefore, within geographical environment only the areas of large cities undergo anthropogenic impact. The growth of cities is the engine of the world economy and generates enormous social benefits by concentrating human creativity. Most global environmental problems of the late twentieth century can be linked directly to the massive transfers of resources and

L. Gadrani (✉) · L. Tielidze
Department of Geomorphology and Geoecology, Vakhushti Bagrationi Institute of Geography, Ivane Javakhishvili Tbilisi State University, Tbilisi, Georgia
e-mail: lizigadrani@gmail.com

M. Tsitsagi
Department of Hydrology and Climatology, Vakhushti Bagrationi Institute of Geography, Ivane Javakhishvili Tbilisi State University, Tbilisi, Georgia

L. Tielidze
Department of Earth Sciences, Georgian National Academy of Sciences, Tbilisi, Georgia

L. Tielidze (ed.), *Geomorphology of Georgia*, Geography of the Physical Environment,
https://doi.org/10.1007/978-3-319-77764-1_6

waste products required by large cities (Bathrellos 2007). Depending on the natural conditions, one of the main factors that determine the validity of the area for the development of the city is a relief of a site; especially for Georgia—a country with complex terrain.

Today, in the development of modern sustainable cities, these are significant factors: healthy climate, water supply, prevention of adverse effects of natural disasters, the possibility of territorial expansion, the urban organization of, and improvement of the service level, i.e., all the factors that more or less influence human and the whole functioning of the city. And all of the above lead to changes in the natural environment. Indeed, one of the main reasons for speeding up urbanization is a rapid growth of the city population, though this is not the only reason.

Below, we will consider the impact of urban processes on the relief on the example of the four main cities of Georgia: Tbilisi, Kutaisi, Batumi, and Rustavi (Fig. 6.1).

6.2 Tbilisi

Tbilisi is located in the south Caucasus, at 41° 41′ north latitude and 44° 47′ east longitude. The capital is situated on both the banks of the Mtkvari River in eastern Georgia (Fig. 6.2). The section of the Mtkvari River gorge (Tbilisi depression) from Zemo Avchala to Ponichala is characterized by a complex terrain formed during geologic periods. The height of the bottom of the depression varies in the range of 350–450 m asl, and its slopes reach 1500 m. In the north, it is bordered by the Saguramo Range, in the east and southeast by the Iori Upland, and in the south and west by the subranges of the Trialeti Range (Tatashidze 2000). The city stretches at ~33 km along the Mtkvari River and occupies an area of 503 km^2.

The river divides the capital into two parts. The left side of Tbilisi exceeds the right side both in area and number of population. The part of the city on the left bank of the Mtkvari River is stretched from the Avchala district to the Lochini River. The relief of the right bank is represented by the subranges of the Trialeti Range, which are steeped toward the Mtkvari gorge. That is why the mountains are an important obstacle for the city to grow on the right bank. Nevertheless, in such a geographical environment, the most densely populated areas are detected while other areas of the city remain unoccupied.

Tbilisi is located in complex geological and geomorphological areas, with a complex hydrogeology, characterized by aggressive underground waters. Landslides, mudflows, erosion, and floods are notable among the exogenous processes developed here. Landslide-gravitational phenomena and mudflows are the main processes that damage the infrastructure of the city (Gadrani and Gaprindashvili 2016). In addition, the relief in the most tense areas is extremely transformed by anthropogenic influence. With the increase of urbanization, new territories are also involved in this process (Fig. 6.3).

In the plain and mountain-hillock part of Tbilisi vicinities, the landscapes characteristic to the steppes of eastern Georgia are formed, and in the western and northern mountainous parts—the mountain-forest landscapes, which create the altitudinal zones. (Elizbarashvili et al. 2017). Today, the primary natural landscape of Tbilisi and its surroundings has been extremely violated by human activities and an anthropogenic landscape is formed instead (Bondirev et al. 2008), which is mostly represented by artificial terraces.

The rapid urban growth of Tbilisi began in the early nineteenth century, when in 1801 the Kartl-Kakheti Kingdom was united in the Russian Empire. The city became the center of the Gubernya of Georgia and new buildings of European style were built in the entire territory. The roads were built and the railway was constructed in order to connect Tbilisi with the cities of Russia and the Transcaucasia such as Batumi, Poti, Baku, and Yerevan. In the 1850s, Tbilisi was once again demonstrated as an important political, trade, and cultural center in the south Caucasus.

From the beginning of the twentieth century to 1960, the city was growing along both banks of the Mtkvari River, which contributed to the

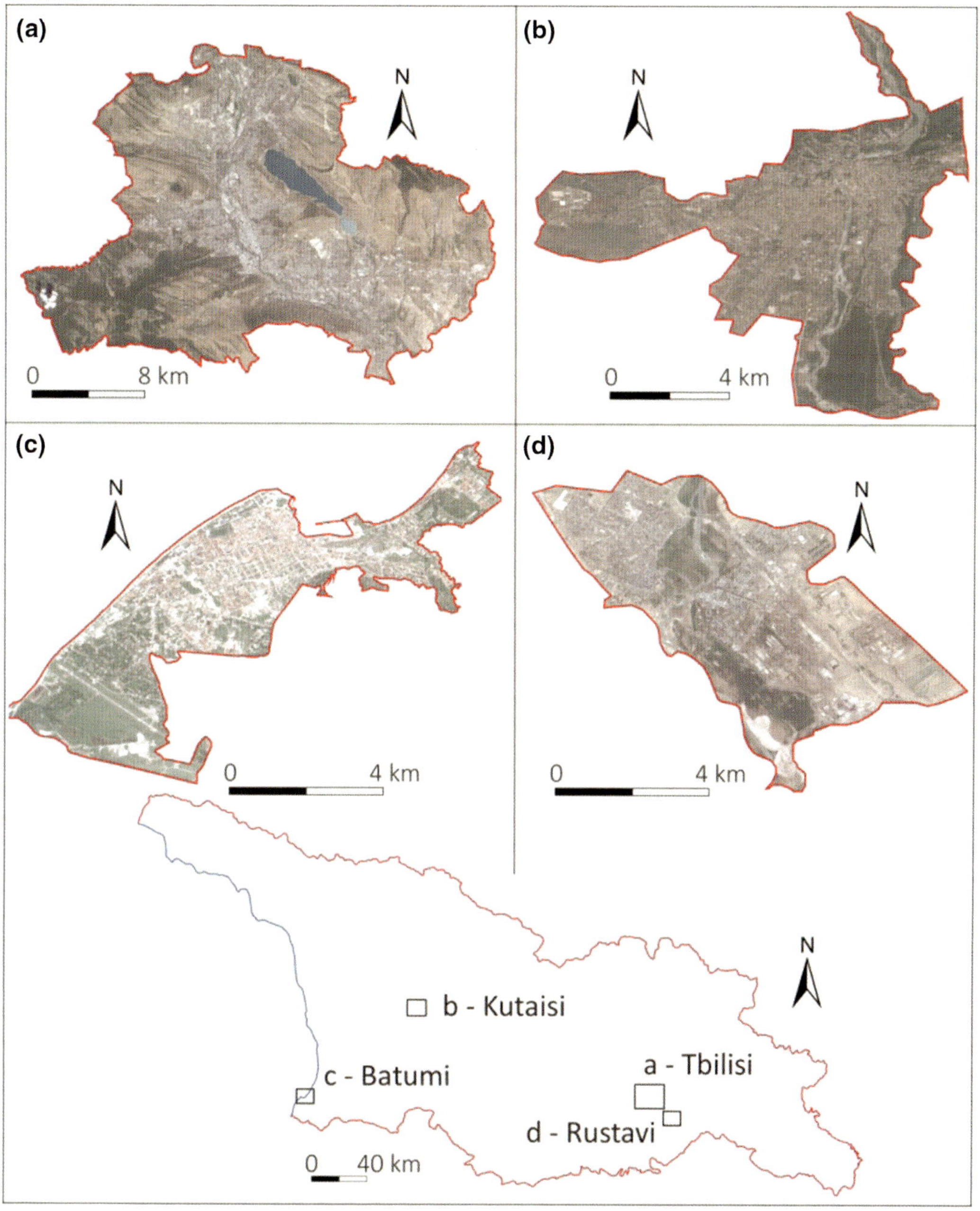

Fig. 6.1 The study area; **a** Tbilisi (Sentinel 2, 21.07.2017); **b** Kutaisi (Sentinel 2, 13.08.2017); **c** Batumi (Sentinel 2, 27.07.2017); **d** Rustavi (Sentinel 2, 21.07.2017)

prolonged structural planned development of the city. In 2007, the area expanded significantly in the southeast, southwest and northwest. The newly joined districts consist of agricultural, forest-covered, recreational and residential belts. In 2009, Tbilisi City Hall approved the new general plan of the capital—Prospective Development Plan of the Capital. Accordingly, the agricultural zones located in newly joined areas have acquired the status of the residential districts.

Fig. 6.2 Tbilisi—capital of Georgia (*photo by* L. Tielidze)

Fig. 6.3 An example of the transformed relief by anthropogenic influence (*photo by* L. Gadrani)

Although the forests provide numerous environmental benefits to urban/suburban settings, including reduction of urban heat island effects, enhanced water and air quality, regulation of storm water drainage, and runoff and provision of wildlife habitat (Nichol 2009; Giner and Rogan 2012), due to the unregulated constructions that lack strategic planning, even the green areas of the capital have become a construction zone. According to the 2009 General Plan of the City, the green area per capita is 5.6 m^2, which is both significantly lower than the same indicator for the 1980s (12 m^2 per capita) and the norm envisaged by European law (25 m^2 per capita) Patarkalashvili 2017). That is why it is necessary to pay attention to the improvement of the ecological condition of the expanded borders of the city while planning, and maintain creation of green zones (Elizbarashvili et al. 2017).

During the last century, the population of Tbilisi grew steadily and according to official statistical data, it reached maximum (1,267,000 people) by 1992. In the following years, when there was a political destabilization, the armed conflict in Apkhazeti and Tskhinvali region and economic regression caused a decrease in the population of Tbilisi. Due to the unbearable socioeconomic conditions, overall unemployment, political instability, and ethnic tensions, the population has moved to other countries. From 1992 to 2004 including, the number of population of the capital decreased by 15%. According to the 2014 census, 29% of total amount of population of Georgia live in Tbilisi (1,108,717 people). For comparison, in 1970, the population of Tbilisi was 19% of the total amount of population of Georgia (Department of Statistics of Georgia 2016).

6.2.1 Land Use Transformation

The process of development in the territory of Tbilisi initially included the narrowest part of the Mtkvari gorge and this is explained by defensive purposes. In the Soviet period, against the active industrialization background, the development of plain suburb areas was started, which has resulted in the replacement of some farm areas by a built-up area (Fig. 6.4).

According to the 1987 data, the most territory of the study area is occupied by bare soils, grassland, and other (36.54%) followed by green area (29.12%), then comes built-up area (23.63%) and the agricultural area (8.26%); the latter is located only in the suburbs of the city. The process of development of the suburbs has already begun in the period analyzed. After 29 years, by 2016, there is a sharp change in the situation, which implies that the built-up area (37.53%) has increased sharply, and as Fig. 6.4 shows, it is mainly at the expense of bare soils and green area. The reason for this is mainly the growth of the city population (Fig. 6.5), which in its turn is conditioned by internal migration. It implies both migration from village to town and

Fig. 6.4 Land use transformation in the area of Tbilisi, 1987–2016

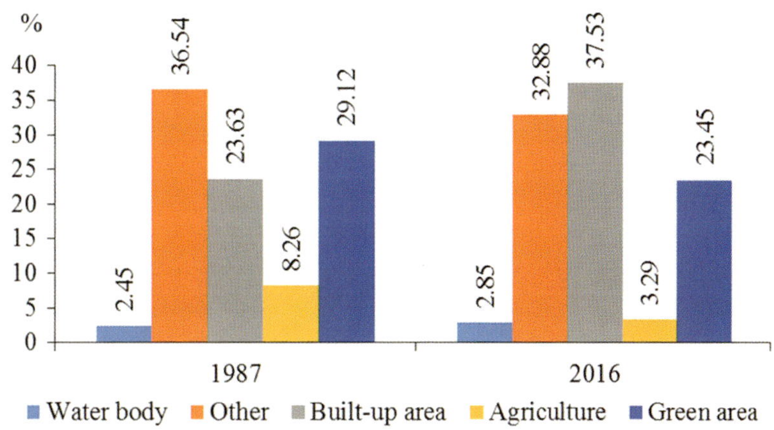

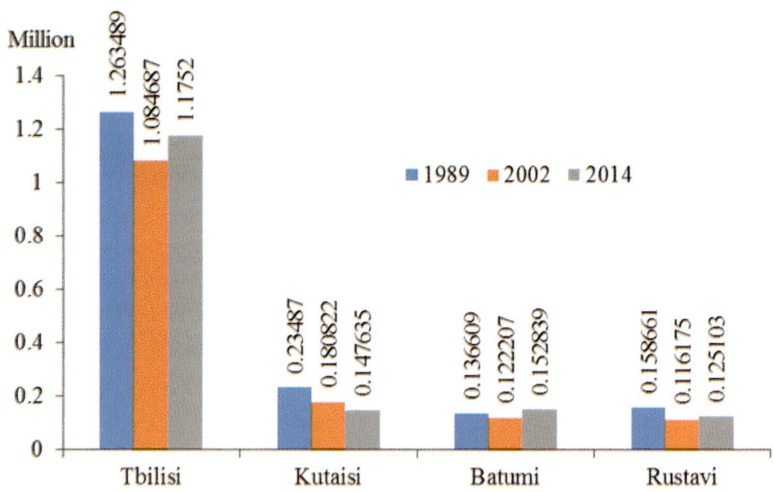

Fig. 6.5 Number of population according to the National Census of 1989, 2002 and 2014

the internally displaced persons as a result of armed conflicts. The second reason is the increase in city population's demand on private property, which in turn causes the building boom. In Tbilisi, construction became the fastest growing economic sector in the last 10–15 years. However, unfortunately in many cases, the expansion of the construction sector is associated with the destruction of green cover. The expansion of the city is obvious in the northeast (territory of Didi Dighomi) (Fig. 6.6). The built-up conversion of these agricultural lands was started

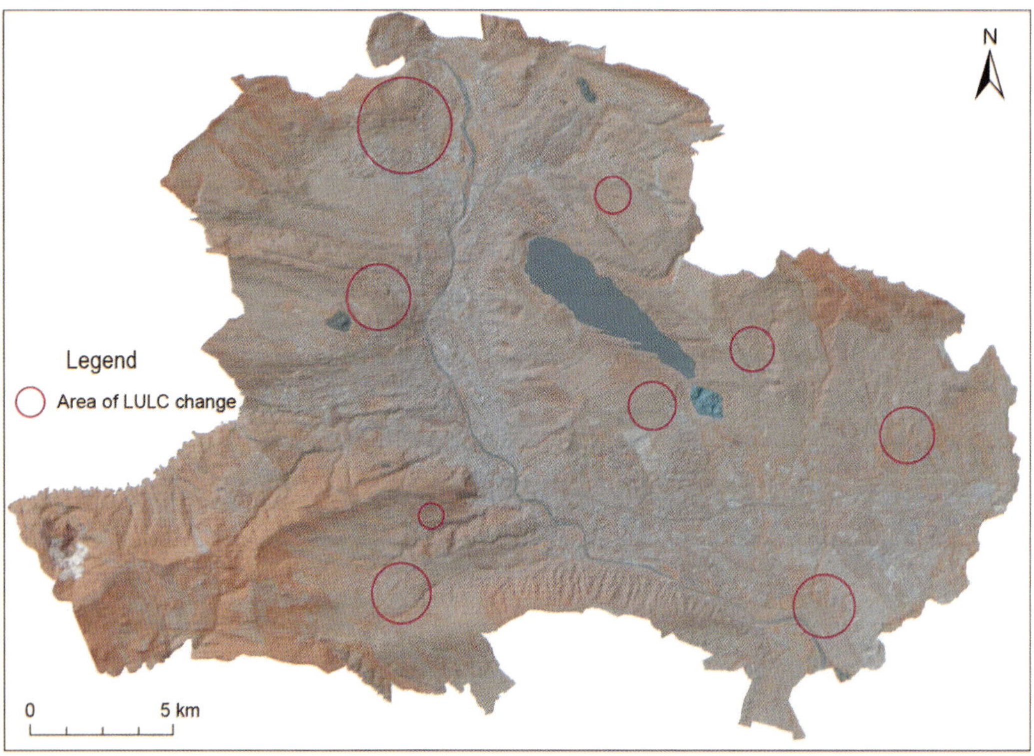

Fig. 6.6 Main areas of land use transformation in 1987–2016 in Tbilisi (Sentinel 2, 26.08. 2016)

at the end of the 1980s and is still underway. Also the important epicenters of transformations are the surroundings of the Tbilisi Sea and the Lisi Lake.

6.3 Kutaisi

Kutaisi belongs to the oldest cities of the world located in the central part of western Georgia. It is built in the area where the Rioni River flows from its narrow and deep gorge to the wide plain. Until recent period, the river divided the city built-up area into two equal east and west parts, but because of the large industrial construction in the post-World War II period, the right part of the city occupies much more area than the districts located on the left bank of the river. Kutaisi and its relief are characterized by a great diversity of natural conditions. Plain-lowlands and mountain-hillocks (Jaoshvili 1989) are bordering each other in the city and in the surrounding area.

On the right and left sides of the river, the most part of the territory of Kutaisi is occupied by residential blocks, but much more area is occupied by the industrial enterprises on the right bank. While the built-up area on the left bank is not growing anymore and its boundaries are quite clear, the right bank's built-up area is growing relatively rapidly and its border is marked by a change every year. The city has already joined the nearby villages and gradually occupies their territories. Both parts of Kutaisi are directly bordered by arable land and often the cultivated territories are intruded in the built-up districts.

Since 1921, with the establishment of Soviet power in Georgia, a new stage has been started in the urban building and architectural development of Kutaisi. The abolition of private property on the land, the municipalization of the residential fund allowed the city planning process the organizational approach. Kutaisi was conditionally divided into 11 districts; among them, two industrial and two semi-industrial. Residential areas were located mainly in the northern part of the city. For this period, the city planning structure maintained the character of the prerevolutionary city and its accompanying negative

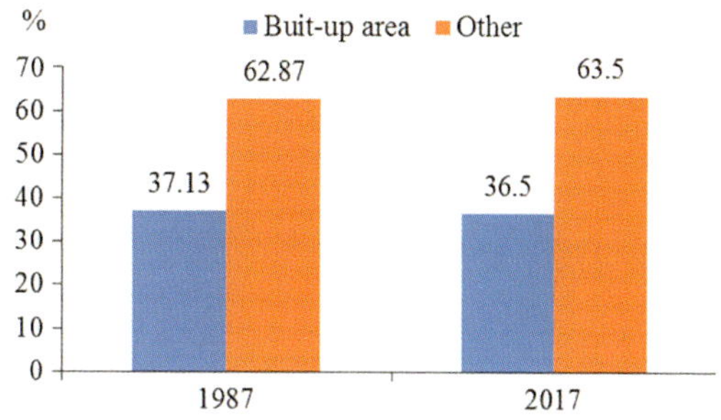

Fig. 6.7 Change in Kutaisi built-up area 1987–2017

moments. With the new construction the multi-storey housing appeared in Kutaisi built-up area, which changed the character of Kutaisi built-up area, especially in its central part.

In the case of Kutaisi, the built-up area is not significantly reduced in percentage (Fig. 6.7). Considering the deficiencies in the classification of remote sensed imagery, we can boldly say that the situation in Kutaisi has not changed much after the Soviet period regarding built-up area (Fig. 6.8).

It can be explained, first of all, as the statistical data indicate the decrease in the number of city population (Fig. 6.5), and on the other hand, the city has almost lost its entire industrial function after the Soviet period.

6.4 Rustavi

The administrative center of Kvemo Kartli is the Rustavi City. It is located on both banks of the Mtkvari River in Kvemo Kartli Plain, from the capital of Georgia to the southeast, at about 350 m asl. Rustavi occupies 6060 ha of inferred tile steppes. From the west, it is bordered by the Iaghluja and Chatmi mountains, while eastward the Gardabani and Ponichala valleys. The Mtkvari River divides the city into the left and right banks. The so-called "Old Rustavi" settlement is located on the left bank of the city, while the so-called "New Rustavi" settlement on the right one.

The city has undergone a great transformation over 60 years since its foundation, namely, from a small town it grew into a powerful industrial and cultural center. Among the geographical

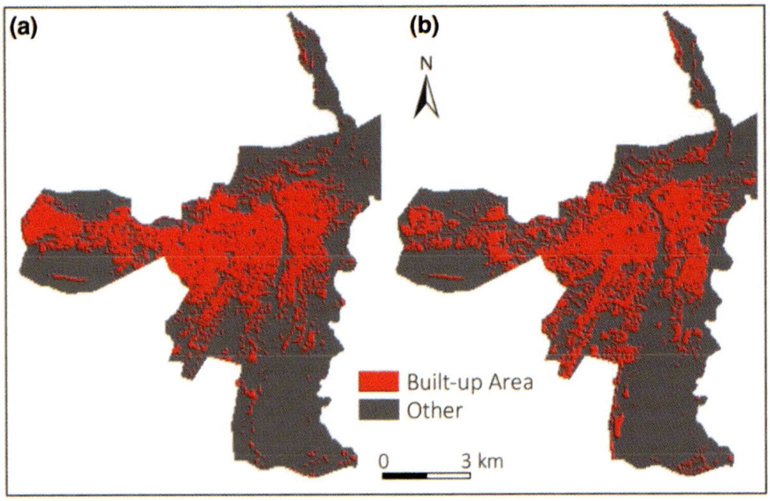

Fig. 6.8 Kutaisi built-up area, **a** 1987 and **b** 2017

features of Rustavi City the following are worth mentioning: the proximity to the capital of the country, high industrial potential, and frontier location (high flow rate of transport flows).

Rustavi is a typical industrial city. Heavy industry enterprises, railway station, administrative part, and the first residential districts were built on the left bank of the Mtkvari River. Subsequently, the housing construction was expanded mainly on the right bank of the Mtkvari River. In the formation of the city look, a large role has played the vast industrial area, linearly built-up residential blocks and a beautiful park created by the Mtkvari floodplain.

Social and economic crisis of independent Georgia turned out for Rustavi to be an obstacle in its development as a city. However, if we look at the data of the census, there is a slight increase in the number of population in the recent census (Fig. 6.5), explaining that the proximity to the capital has led to a strong wave of build-up has shifted from the west of Tbilisi to Rustavi. Because of the territorial proximity, scientists talked about a joint megapolis of Tbilisi-Rustavi as one of the means of discharging the capital. Even a slight change in built-up area (Fig. 6.9) indicates that in the last 30 years, there have been no significant changes in the direction of built-up area (Fig. 6.10), but it should also be noted that recently a new wave of build-up begins in Rustavi the reason of which is again and again the proximity to the capital and not the city's functional change.

6.5 Batumi

Batumi is the capital of the Autonomous Republic of Achara. It is located on the Black Sea coast in the Kakhaberi Lowland at 2–3 m asl.

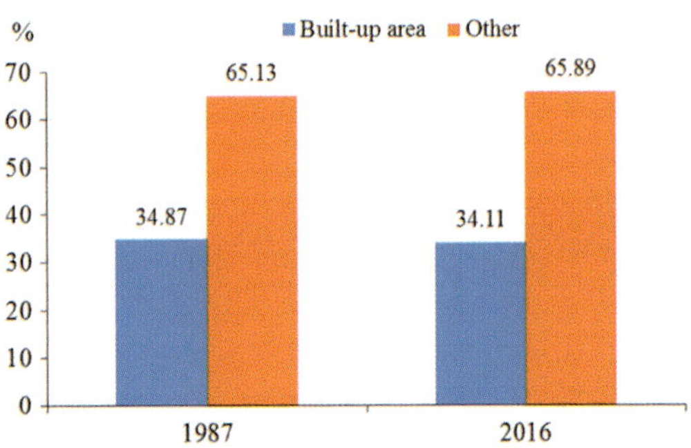

Fig. 6.9 Changing in the built-up area of Rustavi, 1987, 2016

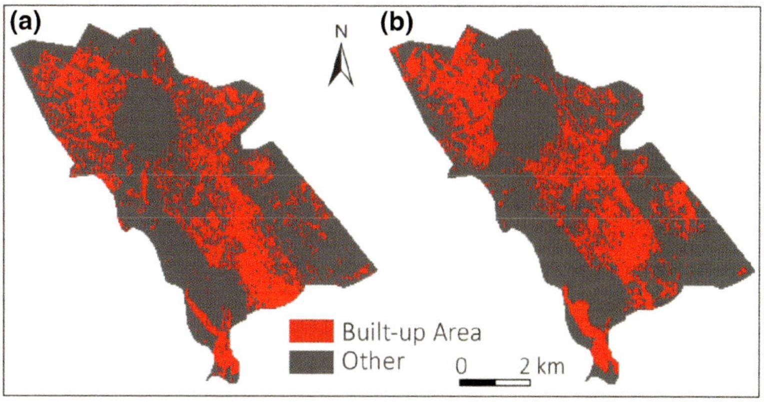

Fig. 6.10 Rustavi built-up area, **a** 1987 and **b** 2016

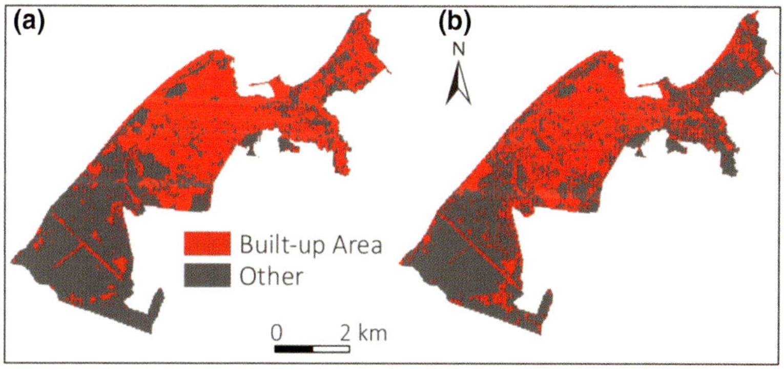

Fig. 6.11 Batumi built-up area, **a** 1987 and **b** 2017

The city is stretched from the northeast to the southwest at about 7 km. Batumi is a land and marine gateway with its maritime port, railway center, and industry. Currently, the territory of Batumi is 6494.31 ha.

The city is a complex urban formation. Its genesis takes its origin in the distant past. The city acquired a new political and trading function in the second half of the nineteenth century. It embarked on the way to be formed as a prospective port city in the Black Sea basin, which caused the accumulation of not only the job seekers from the interior regions of Georgia next to local population, but also the migrations outside the country, which was both of individual and group character and often forced. Gradually Batumi became a multiethnic city.

It should be noted that Batumi has clearly got ahead of the capital of the country by the development rate. In fact, Batumi is an unofficial summer capital. Batumi development is indeed important both in economic and social terms, however, there is a danger that in the unequal conditions of urban development, we will eventually get one developed center—Batumi, and a greatly backward periphery around it.

Nowadays Batumi preserves all the main functions that it has in the Soviet times, and the function of the tourist resort city was added to it, which promoted the rapid development of this sector of economy. Of course, the demand on private property increased, followed by a new wave of built-up. The built-up area also increased accordingly; if it was 13.39 km^2 by 1987, it increased to 13.50 km^2 (Fig. 6.11) by 2017. These numbers are not impressive, but if we take into account that the old lower buildings are replaced by multistorey buildings due to the lack of land in the city, then it is clear that similar to Tbilisi the emphasis is made on vertical built-up of the city.

References

Bathrellos GD (2007) An overview in urban geology and urban geomorphology. Bulletin of the Geological Society of Greece, Athens, pp 1354–1364

Bondirev I, Seperteladze Z, Salukvadze E, Chaladze T, Janelidze Z (2008) Transformation to natural ambience of the Tbilisi city and its vicinities. Poligraph, Tbilisi (in Georgian)

Coates DR (1984) Urban geomorphology. In: Finkl CW (ed) Applied geology. Springer, pp 598–601

Department of Statistics of Georgia (2016) http://geostat.ge

Elizbarashvili N, Meladze G, Salukvadze E, Svanadze D (2017) Methodology of landscape planning of urban agglomerations (on the example of Tbilisi-Rustavi). Dani, Tbilisi (in Georgian)

Gadrani L, Gaprindashvili G (2016) Modern geo-ecological conditions and terrain transformation of Tbilisi (Georgia). Landscape Archit Reg Plann 2(1):36–42

Giner NM, Rogan J (2012) A comparison of Landsat ETM+ and high-resolution aerial orthophotos to map urban/suburban forest cover in Massachusetts, USA. Remote Sens Lett 3(8):667–676

Jaoshvili (1989) Kutaisi (Monograph). Publishing House "Metsniereba", Tbilisi (in Georgian)

Nichol J (2009) Remote sensing of urban areas. In: Warner TA, Nellis MD, Foody GM (eds) The SAGE handbook of remote sensing. Sage Publications, London, pp 423–436

Patarkalashvili TK (2017) Urban forests and green spaces of Tbilisi and ecological problems. Ann Agrarian Sci, 1–5 (in Georgian)

Tatashidze Z (2000) Tbilisi—chapter in geography of Georgia, Part 1. Metsniereba, Tbilisi, pp 200–207 (in Georgian)

Thornbush M (2015) Geography, urban geomorphology and sustainability. Area 47(4):350–353

Lela Gadrani is a Ph.D. student and research scientist at the Department of Geomorphology and Geoecology of the Vakhushti Bagrationi Institute of Geography, Ivane Javakhishvili Tbilisi State University. Her research area is Urban Geomorphology and Urban Planning. In 2016, she conducted a research stay at the University of Politecnico di Torino, Italy.

Mariam Tsitsagi is a Ph.D. senior research scientist at the Department of Hydrology and Climatology of the Vakhushti Bagrationi Institute of Geography, Ivane Javakhishvili Tbilisi State University with some teaching experience (Georgian-American University, GAU). She is actively involved in the international projects and has several science publications. Her research focuses on soil erosion, remote sensing, and GIS.

Levan Tielidze is a Ph.D. of Ivane Javakhishvili Tbilisi State University in the field of geomorphology–glaciology–cartography and a senior research scientist at Vakhushti Bagrationi Institute of Geography. He is also a researcher-scholarship holder of the Department of Earth Sciences of the Georgian National Academy of Sciences. The field of his research is mountain cartography and glacial-geomorphological study of the mountainous areas in the Quaternary (Late Pleistocene and Holocene). He is a member of several international organizations. In 2014/2015, he conducted a research stay at the Climate Change Institute of the University of Maine, USA, and in 2015/2016–2017 at the University of Northern British Columbia, Canada. He is the author of about 40 scientific papers, maps and eight monographs.

Part III

Main Geomophological Units of Georgia

7 Morphological Division of the Landscape of Georgia

Alexander Javakhishvili, Levan Maruashvili, Ramin Gobejishvili and Levan Tielidze

Abstract

This chapter presents the morphological division of the landscape of Georgia, where the main morphostructural units are considered, such as the Greater Caucasus watershed range, the intermountain lowland of Georgia, and the highland of southern Georgia. The geological structure and tectonic conditions are discussed together with their distribution. The main orographic units are divided into separate sections by their morphological and morphometric features. In the Greater Caucasus zone the western, central and eastern sections are distinguished. The intermountain lowland is divided into the two subzones—the Kolkheti Lowland (western Georgia) and the Iveria Plain (eastern Georgia). There are two subzones distinguished in the highland of southern Georgia: Achara-Trialeti-Loki mountainous relief and volcanic highland of Georgia. The geodynamic processes, distributed across the study region, are also presented.

Keywords

Georgian Greater Caucasus • Intermountain lowland of Georgia • Highland of southern Georgia

A. Javakhishvili · L. Maruashvili · R. Gobejishvili
L. Tielidze (✉)
Department of Geomorphology and Geoecology, Vakhushti Bagrationi Institute of Geography, Ivane Javakhishvili Tbilisi State University, Tbilisi, Georgia
e-mail: levan.tielidze@tsu.ge

A. Javakhishvili · R. Gobejishvili
Department of Geography, Faculty of Exact and Natural Sciences, Ivane Javakhishvili Tbilisi State University, Tbilisi, Georgia

A. Javakhishvili · L. Tielidze
Department of Earth Sciences, Georgian National Academy of Sciences, Tbilisi, Georgia

7.1 Main Geomorphological Units

The landscape of Georgia is characterized by a diverse and complex morphology, morphometry, and morphography. The complexity of the landscape depends on its geotectonic features, and the diversity—on the altitudinal zonality of the relief and peculiarities of geodynamic processes. The highest point of Georgia is the Mt. Eastern Shkhara (5203 m), and the lowest is in the center of Kolkheti Lowland (−1 m). The interaction of endogenic and exogenic processes has led to the distinctive major morphostructural zones (units): 1. Medium and high mountain landscape zone of the Greater Caucasus; 2. The intermountain lowland zone; and 3. The highland zone of southern Georgia (Fig. 7.1).

L. Tielidze (ed.), *Geomorphology of Georgia*, Geography of the Physical Environment,
https://doi.org/10.1007/978-3-319-77764-1_7

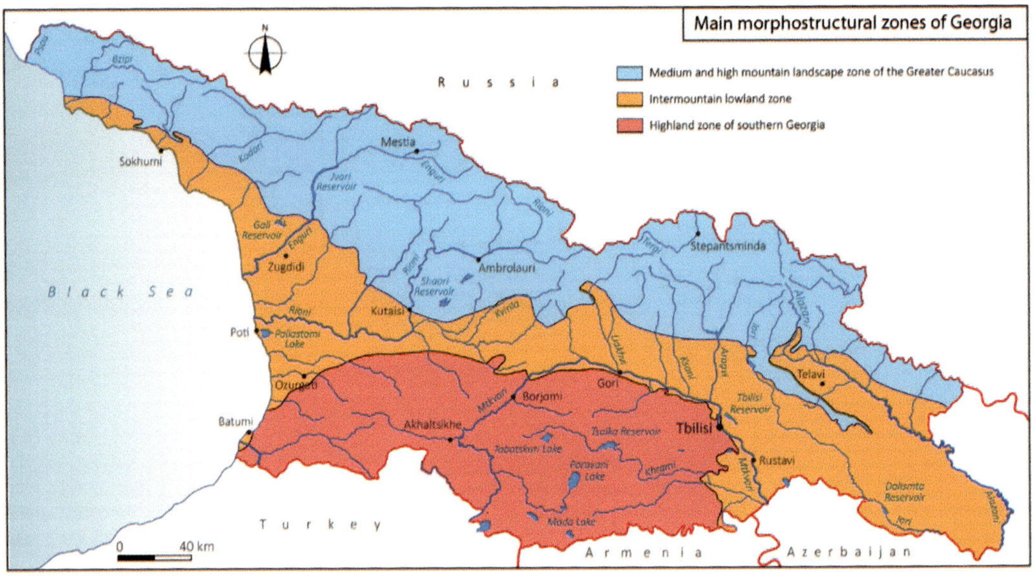

Fig. 7.1 The three major morphostructural zones (units) of Georgia

7.2 Medium and High Mountain Landscape Zones of the Greater Caucasus

Within Georgia, the Greater Caucasus includes these subzones: the southern slope of the Greater Caucasus—from Mt. Avadhara (2960 m) up to Mt. Tinovroso (3374 m) and the northern slope—from Mt. Vatsisparsi (2562 m) up to Mt. Shavikldе (3578 m). Hypsometrically, the highest is the central section of the Greater Caucasus within the Racha-Svaneti region (Fig. 7.2). The peaks with height over 5000 m are located on the "Pitsrula and Shkhara massifs. The Axis belt of the Greater Caucasus, from Mt. Avadhara up to Mamisoni Pass, is built on Paleozoic and Proterozoic plagiogranites, plagiogneisses, amphibolites, and crystalline slates. The Dariali gorge (the Tergi River gorge) is built on the same rocks. Jurassic, Cretaceous, and Tertiary suites participate in the construction of the southern slope of the Greater Caucasus folded system. The northern slope is built on Jurassic suites. The Racha-Lechkhumi syncline and its continuation in the form of a narrow strip up to Java are built on Tertiary sedimentary rocks. The Carbonate-Triassic suites form the center of the Svaneti Range and its continuation to the west and east.

The fault lines have a special role in the formation of the morphostructure of the Greater Caucasus, reflected in the relief in the form of stages. On the slopes of the Greater Caucasus, the lineaments occur on the edge of the main suites—between the crystalline suites and the Jurassic sediments, between the Bajocian and carbonate suites, and between the Tertiary and the Cretaceous sediments. The faults of the submeridian direction are also seen in the relief. Earthquakes are frequent in the knot of the faults crossing, and are particularly notable in the northeastern section of Racha.

In the Greater Caucasus, different belts experience different rates of movement. The Axial belt rises by 10–15 mm/year, inverse and semiinverse ranges constructed with carbonate flysch go up by 4–8 mm/year, while the areas located between them represent relative downward movement, sinking by 1–3 mm/year (Gobejishvili 2011). In the intensive uprising areas the fall of the riverbeds is high when erosion processes are in progress, and in the sinking zones, the fall is about 10–12 m/km, where mainly the accumulation and side erosion are underway in the

Fig. 7.2 The central Greater Caucasus in Svaneti (*photo by* L. Tielidze)

riverbed. Due to deep river erosion, while they straight cross the ascended morphostructures, narrow gorges and cliffs occur, especially in the Cretaceous suites (e.g., Bzipi, Gumista, Kelasuri, Kodori, Okumi, Ghalidzga, Enguri, Khobistskali, Tekhuri, Abashistskali, Tskhenistskali, Rioni, and Tergi rivers). The gorges, developed in the direction of morphostructures, are wide and they are represented in the form of hollows (e.g., Bzipi (Pskhu), Kodori, Enguri, Tskhenistskali, mountain Racha, Racha-Lechkhumi, Java, Truso, and Tusheti). In other cases, the form of gorges is V-shaped and the inclination of the slopes is different.

Besides the tectonic and erosive processes in relief genesis, the role of geodynamic processes is considerable, such as nival-glacial, karst, and gravitational processes. The formation of relief in high mountainous zone is associated with the Pleistocene and modern glaciation. Modern glaciers are widespread in the central Greater Caucasus (Svaneti-Racha section), in the western Greater Caucasus, from the Marukhi Pass up to the Dalari Pass, also on the Kazbegi massif. Large size (>5 km^2) glaciers of valley type (Lekhziri, Northern and Southern Tsaneri, Chalaati, Adishi, Khalde, Kirtisho, Boko, Eastern Suatisi, Gergeti, and Devdoraki) are common among the morphological types of glaciers. There are also many smaller cirque glaciers. Since the LIA maximum, till today the glaciers of the Greater Caucasus have retreated: the valley glaciers—by 1.0–2.5 km, and cirque glaciers—by 300–500 m (Tielidze 2017a). In the same period, the snow line in the Greater Caucasus has shifted up by several hundred of meters, accompanied as well by the shifting of natural zones.

Snow and snow avalanches have great impact on the formation of micro and mezzo forms of the landscape in the Greater Caucasus. During the Pleistocene glaciation, due to active nival and glacial processes, the terrain experienced the total denudation at the height of 2000 m (Gobejishvili et al. 2011; Tielidze 2017b) and currently, above 3000 m. In Würm (Late Pleistocene), the tongues of large glaciers descended to 600–1000 m, the indication of which are the glacier forms, moraines and trough valleys. In the high mountainous zones, there are cirques, glacial horns, ledges, debris, rock streams, and other nival-glacial forms.

In the periphery zone of the southern slopes of the Greater Caucasus, there are Cretaceous limestones, which facilitate the formation of surface and underground karst forms. Altitudinal karsts are well developed in the Gagra, Bzipi and Racha ranges, as well as in separate massifs (e.g., Rikhva, Okhachkue, Kvira, Migaria, Askhi, and Khvamli). Surface karst forms are represented by sinkholes, poljes, wells, dolinas, and shafts (Kipiani 1974). The underground caves are made up of stalagmites and stalactites, with various underground karst forms. The world's deepest abysses are located in the Arabica massif: the Krubera (2197 m), Sarma (1830 m), Ilyukhin (1275 m), Arabica-kuibyshev (1110 m) and Dzou (1090 m) (Tsikarishvili et al. 2010). The limestones are distinguished by precipices in the relief, on the slopes of which debris cones are formed.

The gravitational forms are intermittently represented in the Greater Caucasus, especially rock avalanches. In the area of their spread, modern lakes or the traces of lakes are found (Didi Ritsa, Patara Ritsa, Amtkeli, and Kvedi). The Khakhieti and Patsa Lakes resulting from the 1991 earthquake were filled up by solid river deposits.

The volcanic relief is represented in the areas of Mt. Kazbegi (Mkinvartsveri) and Keli Upland. The Pleistocene relief of volcanic cones and lavas are characterized by interesting forms. The Khorisari, Tkarsheti, Tsitelikhati and Khodzi volcanic flows, and volcanic cones are of Holocene age (Maruashvili 1981).

According to morphological and morphometric features, the Greater Caucasus Range is divided into three sections (subzones): The Western Greater Caucasus (from the Mt. Avadhara to Dalari Pass), Central Greater Caucasus (from Dalari Pass to the Mt. Vatsisparsi), and the Eastern Greater Caucasus (from the Mt. Vatsisparsi to the Mt. Tinovroso).

The main morphological regions of the Greater Caucasus Range are the sublatitudinal and latitudinal mountain ranges and river gorges located on the slopes of the Western and Central Greater Caucasus. Among the ranges, the following are noteworthy: Gagra, Bzipi, Chkhalta (Apkhazeti), Kodori, Svaneti, Egrisi (Samegrelo), Lechkhumi, Shoda-Kedela, Racha, and Germukhi. On the slopes of the Greater Caucasus within eastern Georgia, there are mostly the longitudinal ranges (except the Khokhi and Pirikita ranges) as follows: Kharuli, Lomisi, Kartli, Kakheti, Kuro, Shani, Kidegani, Khevsureti, and Atsunta. At the same time, these mountain ranges are the watersheds for the Greater Caucasus main river gorges as follows: Psou, Bzipi, Kodori, Enguri, Rioni, Liakhvi, Ksani, Aragvi, Iori, Alazani, Tergi, Asa, Arghuni, and Pirikita Alazani.

7.3 Intermountain Lowland Zone

The Intermountain lowland of Georgia is located between the Greater Caucasus and southern Georgia's highlands. It stretches from the Black Sea to the east—the downstream of the Alazani River (up to the Azerbaijan border). Its landscape is quite specific, although its highest places do not exceed 2000 m asl (Mt. Tsivi - 1991 m), on the background of low valleys, they are orographically distinguished in the relief (Gombori Range, Saguramo-Ialno Range, Imereti Highlands and Likhi Range). The flat valleys are surrounded by the foothill relief and the Iori Upland. The hilly relief is fragmented due to active impact of erosive and denudation processes. Accumulative and denudation processes and related forms prevail over the valleys (Fig. 7.3).

The intermountain lowland is a tectonic depression. The relief formation took a long period of time. It obtained its final look in the Pliocene and Pleistocene. The surface of modern relief experiences differential movements with maximum sinking in the neighborhoods of Poti (−6.0 mm/year); in conditions of general uprising, the Kolkheti Lowland, Shida and Kvemo Kartli valleys and the Alazani valley of graben character experience relative sinking

Fig. 7.3 Intermountain lowland of Georgia (between the Greater Caucasus and southern Georgia's highlands) (*photo by* G. Lominadze)

(1–2 mm/year); while the Gombori Range is uprising by 4–6 mm annually, and the Iori Upland by 2–3 mm/year (Gobejishvili 2011).

The landscape of the intermountain lowland, except the Dzirula massif, is constructed with late sediments. Kolkheti Lowland, Shida and Kvemo Kartli valleys, Alazani Valley, and the flatlands of Iori Upland are created from Pleistocene alluvial and deluvial sediments. The hilly line, the Iori Upland and the lower tier of the Gombori Range are represented by tertiary marine-continental molasses (sandstones, clays, conglomerates, clay sandstones, and marls in some areas). The axial line of the Gombori Range is constructed from Cretaceous carbonate suites. In western Georgia, the hilly belt (in the northern part) consists of Cretaceous limestones.

The Dzirula massif is different, having undergone peneplanation several times. The central stripe of the Dzirula massif and Likhi Range are constructed from gneisses, amphobolites, granitoids, quartz diorites, and Middle Jurassic granitoids. The old rocks around Dzirula massif are covered by Middle Jurassic (Bajos), Cretaceous and Tertiary sediments suites, lain mostly horizontally on the old substrate.

Along with the geologic and tectonic processes relief forming geodynamic processes occur, such as erosion, accumulation, mudflow and landslide processes, horizontal washing away, etc. Erosion processes occur mainly in the hilly relief; widely opened V-shaped gorges are formed in the Tertiary sediments. In the Chiatura neighborhoods, in the Cretaceous sediments, the Kvirila River gorge is a canyon. Accumulation processes occur in Kolkheti Lowland, Alazani Valley, Mukhrani-Tiriponi and Marneuli-Gardabani valleys and in the flatlands of Iori Upland and powerful debris cones are created by mudflow processes (Kakheti, Kartli). Landslide processes are active in the Kolkheti hilly belt and Gombori Range; landslide areas can be found in almost all hilly terrains. Karst processes are actively developing in the intermountain lowland, represented by caves, sinkholes, poljes and shafts, notably in the Tsebelda, Odishi, and Okriba subregions. The well-arranged caves are Tskaltubo, Sataplia, Navenakhevi and Akhali Athoni. The Apkhazeti and Imereti caves were human dwellings in the Paleolithic era.

The intermountain lowland is divided into two subzones—the western Georgia (Kolkheti) Plain and eastern Georgia (Iveria) Plain. These two subzones are separated from each other by the Likhi Range. The western Georgia (Kolkheti) Plain is inclined toward the west and entirely belongs to the Black Sea basin. Eastern Georgia (Iveria) Plain is inclined toward the east and its river network belongs to the Caspian Sea basin.

These two subzones differ in climate and natural conditions. The western Georgia (Kolkheti) Plain is mostly humid subtropical, while the eastern Georgia (Iveria) Plain belongs to the dry subtropics. This contrast is seen in the differences of landforms and geodynamic processes taking place within their borders.

In the western Georgia (Kolkheti) Plain, the following areas are distinguished: Kolkheti Lowland, the northern Kolkheti foothill hillocks, the southern Kolkheti foothill hillocks, Imereti Upland and the Likhi Range. The eastern Georgia (Iveria) Plain subzone includes the following areas: Shida Kartli Valley, Kvemo Kartli Valley, Iori Upland, Gombori Range and Alazani Valley.

7.4 Highland Zone of Southern Georgia

The highland of southern Georgia occupies the southern and south-western part of the country's territory. It borders the intermountain lowland zone from the south along its entire length.

The highland of southern Georgia is below the Greater Caucasus in elevation (Fig. 7.4). Here, the highest peaks are Mt. Didi Abuli (3301 m), Samsari (3285 m), Godorebi (3188 m), and Legli (3156 m). The watershed of the Black Sea and Caspian Sea basins stretches along the crests of the Achara-Imereti and Arsiani ranges, which leads to differentiation of natural conditions. The morphological units, located within the Mtkvari River basin, are characterized by relatively low precipitation, while the Achara-Imereti and Shavsheti ranges experience an abundance of precipitation. The most humid area in Georgia—the Mt. Mtirala is situated in the Achara-Imereti Range.

The landscape of the highland of southern Georgia is characterized by specific morphostructural and morphosculptural features, reflecting tectonic and lithological conditions of the formation suites. It consists of two subzones: 1. Achara-Trialeti-Loki mountain landscape and 2. Volcanic landscape of Georgia.

The folded landscape of Achara-Trialeti-Loki is mainly built from Paleocene and Eocene volcanogenic suites, located on the crystalline rocks

Fig. 7.4 Highland of southern Georgia (*photo by* N. Lomidze)

of the Cretaceous, Jurassic, and Loki massif. In the eastern part of the crests of Achara-Imereti and Trialeti ranges, the Cretaceous sediments are outcropped, mostly sandstones, marl limestone, volcanic breccia, and pink limestones. The Akhaltsikhe depression is constructed from coastal-sea sediments: sandstones, clays, conglomerates, aleuvolites, marls, and layers of lignites. The Eocene volcanic rocks—diorites and garbo are found in the form of islands.

The southern highland relief is constructed from Pliocene and Pleistocene (Goderdzi suite) lime-alkaline andesites, dacites, andesite-dacites, rhiolites, basalts, lacustrine conglomerates, sand, and clays. The Goderdzi suite's peripheris are constructed from Miocene and Pliocene continental sediments: tuffs, volcanic breccia, conglomerates, diatomites, lime-alkaline andesites and basalts (lower breccia part of Goderdzi suite) (Akhalkatsishvili 2006).

The Khrami-Loki area is constructed from Proterozoic crystalline rocks of crystalline schists, amphibolites, migmatites, Paleozoic quartz diorites, plagiogranites, and gneisses.

The formation of the southern Georgia's highland is dominated by erosive, tectonic, volcanic, gravitational processes and surface runoff. Differential tectonic movements are active in the formation of the morphostructures. The main morphostructures are the mountain ranges that are still being uplifted (Achara-Imereti, Trialeti, Loki, Shavsheti, and Arsiani ranges). Achara, Akhaltsikhe, Tori-Tadzrisi, and Tsalka synclinal depressions are relatively sinking. Volcanic relief covers the southern part of the zone. The main landforms are the volcanic cones, plateaus and flows of the Pliocene-Pleistocene and Holocene. Specific features include the Javakheti volcanic upland, the Erusheti volcanic massif and the volcanic plateaus of Persati, Bedena, Chochiani, Dmanisi, Gomareti, and Kvemo Kartli, as well as the Gujareti, Borjomi-Bakuriani, Dabadzveli and Pampula lava flows. A row of volcanic cones of longitudinal direction is represented in the Samsari and Javakheti ranges.

Landforms created by erosive processes are widespread. Deeply cut canyons are formed by the river erosion in volcanic plateaus. The longest canyons are the Mtkvari gorge from Aspindza (to the borders of Georgia), Paravnistskali gorge below Akhalkalaki, Khrami gorge in the Kvemo Kartli volcanic plateau, and the Mashavera gorge at Patara Dmanisi. The asymmetric gorges of Gujaretistskali, Borjomula, and Mashavera are formed at a contact of bedrocks and lava flow. The Achara-Imereti and Trialeti ranges are split by the erosive influence of the Mtkvari River when the crossing (antecendent) gorge was formed, stretchings from the villages of Atskuri to Tashiskari. The morphology of erosive gorges formed in the Achara-Trialeti Range is related to the lithology of the constructing rocks. In the firm rocks, the slopes of the gorges are strongly sloped and the bed is cascade. The slopes of the gorges formed in soft rocks and in the direction of structures are less steep (from Mtkvari—Aspindza to Atskuri, Gujaretistskali, Vere, Tana, Dzama, Tedzami, Kvabliani, Acharistskali, and Ktsia).

The impact of Würm glaciers is dominant in the formation of the high ranges of Samsari, Javakheti, Achara-Imereti, and Arsiani. Cirques and moraines remain and rock flows are created as a result of nival processes. Rock avalanches are found locally. However, landslides are widespread in Akhaltsikhe, Achara and Tori-Tadzrisi hollows (the active landslides of Kodiani and Dgvari). Landslides are also found in the Trialeti and Achara-Imereti sections.

Smoothed surfaces on folded structures are distributed step-wise on the slopes of the Achara-Imereti, Trialeti and Shavsheti ranges. Here can be distinguished six steps, the age of which is dated as the Miocene–Pliocene and Pleistocene (Tskhovrebashvili 1978).

As mentioned above, two subzones are distinguished in the highland of southern Georgia. Each of them includes several morphological units (areas).

1. *Achara-Trialeti-Loki mountain landscape* includes the following areas: Achara-Imereti Range, Borjomi gorge, Trialeti Range, Shavsheti Range, Achara cavernous (Acharistskali gorge), Arsiani Range, Akhaltsikhe hollow (cavernous), Khrami-Loki medium mountainous area.

2. *The volcanic highland of Georgia* includes the Erusheti massive (highland), upper gorge of the Mtkvari River, Javakheti Upland, Nialiskuri Range, Samsari Range, and Javakheti Range, Tsalka hollow and volcanic plateaus (Bedeni, Kvemo Kartli, Chochiani, Gomareti, and Dmanisi).

References

Akhalkatsishvili M (2006) Volcanic formations and glacial epochs age correlation in the Javakheti Upland. Ph.D. thesis, Tbilisi, p 90 (in Georgian)

Gobejishvili R (2011) Relief of Georgia. Monograph, Publishing House "Universali", Tbilisi (in Georgian)

Gobejishvili R, Lomidze N, Tielidze L (2011) Late Pleistocene (Wurm) glaciations of the Caucasus. In: Ehlers J, Gibbard PL, Hughes PD (eds) Quaternary glaciations: extent and chronology. Elsevier, Amsterdam, pp 141–147. https://doi.org/10.1016/b978-0-444-53447-7.00012-x

Kipiani Sh (1974) Karst of Georgia (attempting geomorphological characterization). Monograph, Tbilisi (in Georgian)

Maruashvili L (1981) Kavkasiis fizikuri geografia (Physical Geography of the Caucasus). Monograph, Publ. "Metsniereba", Tbilisi (in Georgian)

Tielidze L (2017a) Dynamics of the Glaciers of Georgia. In: Glaciers of Georgia. Geography of the physical environment. Springer. https://doi.org/10.1007/978-3-319-50571-8-5

Tielidze LG (2017b) Late Pleistocene and Holocene Glacier Extent in the Georgian Caucasus. Open J Geol 7:517–532. https://doi.org/10.4236/ojg.2017.74036

Tsikarishvili K, Barjadze Sh, Kvavadze E, Bolashvili N, Djanashvili R, Martkoplishvili I (2010) Speleology of Georgia: aspects of its current situation and perspectives. Cave Karst Sci 37(3):73–78

Tskhovrebashvili Sh (1978) Geometry of Achara-Trialeti Mountain Range. Monograph, Tbilisi (in Georgian)

Alexander Javakhishvili (1875–1973) was one of the oldest geographer and anthropologist in the Soviet Union. He was the founder of Georgian geographical school. His qualifications and awards include Doctor of Geographical Sciences (1937), Academician of the Georgian National Academy of Sciences (1944), an honorary member of the Soviet Union Geographical Society, and winner of a State Prize of the Georgian Soviet Socialist Republic.

Levan Maruashvili (1912–1992) was a prominent Georgian geographer. Doctor of Geographical Sciences (1954), Professor (1965). Honored Scientist of the Georgian Soviet Socialist Republic (1966). Honorary member of the Russian Geographical Society (1985). He produced over the 500 scientific and popular-scientific works in Georgian, Russian and English languages.

Ramin Gobejishvili (1941–2014) was a Doctor of Geographical Sciences, a geomorphologist–glaciologist. Until his last days, he was Professor of the Faculty of Exact and Natural Sciences, Ivane Javakhishvili Tbilisi State University and the chair of Geomorphology and Cartography. In 1996–2008 he was the president of the Geographical Society of Georgia, and in 2005–2007—the director of the Institute of Geography. He was the author and co-author of about 150 scientific works and ten monographs.

Levan Tielidze is a Ph.D. of Ivane Javakhishvili Tbilisi State University in the field of geomorphology–glaciology–cartography and a senior research scientist at Vakhushti Bagrationi Institute of Geography. He is also a researcher-scholarship holder of the Department of Earth Sciences of the Georgian National Academy of Sciences. The field of his research is mountain cartography and glacial-geomorphological study of the mountainous areas in the Quaternary (Late Pleistocene and Holocene). He is a member of several international organizations. In 2014/2015, he conducted a research stay at the Climate Change Institute of the University of Maine, USA, and in 2015/2016–2017, at the University of Northern British Columbia, Canada. He is the author of about 40 scientific papers, maps, and eight monographs.

8 Western Greater Caucasus

Ramin Gobejishvili, Levan Tielidze and Nikoloz Astakhov

Abstract

This chapter provides a geomorphological characterization of the subregion of the western Greater Caucasus. The orographic units include the main watershed range of the western Greater Caucasus, Gagra and Atsetuka ranges, Achibakhi (Rikhva) karst massif, the gorges of the Bzipi River and its tributaries, the Bzipi and Chkhalta (Apkhazeti) ranges, the gorges of the Kodori River and its tributaries, Kodori Range, and Okhachkue limestone massif. The text also describes the geological structure, tectonic conditions and current exogenic processes of the study region. The extensions of the modern and Late Pleistocene glaciation of the western Greater Caucasus are also considered. Each subchapter is followed by the corresponding images and maps.

Keywords

Western Greater Caucasus • Apkhazeti Gagra Range • Atsetuka Range • Achibakhi (Rikhva) karst massif • The gorges of the Bzipi River and its tributaries • Bzipi Range Chkhalta (Abkhazeti) Range • The gorges of the Kodori River and its tributaries Kodori Range • Okhachkue limestone massif Krubera karst abyss

8.1 Research Area

The southern slope of the western Greater Caucasus enters Georgia from Mount Avadhara (2960 m) to the Dalari Pass (3291 m). The western Greater Caucasus is far below the central and eastern Greater Caucasus in terms of morphometric indicators. Only one peak of the western Greater Caucasus—the Mount Dombai-Ulgen (4047 m) is over 4000 m. The western Greater Caucasus of Georgia is characterized by complex and interesting orography, which is due to the geological structure of the relief, tectonic movements, and the diversity of

R. Gobejishvili · L. Tielidze (✉) · N. Astakhov
Department of Geomorphology and Geoecology, Vakhushti Bagrationi Institute of Geography, Ivane Javakhishvili Tbilisi State University, Tbilisi, Georgia
e-mail: levan.tielidze@tsu.ge

R. Gobejishvili
Department of Geography, Faculty of Exact and Natural Sciences, Ivane Javakhishvili Tbilisi State University, Tbilisi, Georgia

L. Tielidze
Department of Earth Sciences, Georgian National Academy of Sciences, Tbilisi, Georgia

L. Tielidze (ed.), *Geomorphology of Georgia*, Geography of the Physical Environment,
https://doi.org/10.1007/978-3-319-77764-1_8

exogenous processes (erosive, nival-glacial, karst, denudative).

Basic orographic units, ranges, and gorges are of sublatitudinal direction. When cutting the structures and ranges, the gorges are of general meridional direction and narrow with steep slopes. The ranges of Gagra, Atsetuka (eastern part), Bzipi, Chkhalta (Apkhazeti), and Kodori are notable. At the same time, these ranges are the watersheds of the river basins. From the river gorges, the following are notable: Kodori, Bzipi, Psou, Ghalidzga, Gumista, Kelasuri, etc. (Fig. 8.1).

Paleozoic and Mesozoic rocks are involved in the construction of the relief of the western Greater Caucasus, extended in the form of strips. The dome of the anticlinorium of the Caucasus mountain range is built of Proterozoic and Paleozoic plagiogneisses, granites, amphibolites, and crystalline slates. In the south, it is followed by a narrow strip of Lower Jurassic soft rocks, schist, clay slates, aleurolites, gravelites, and sandstones, and in some places with middle layers of conglomerates and thick unit of thick-layered limestones.

The southern slope (the Gumista River basin) of the Bzipi Range and the northern slope of the Kodori Range are built from rocks of the same age. Middle Jurassic granitoids are involved in the construction of the southern slopes (the Jampali, Amtkeli and Kelasuri river basins) of the Bzipi and Chkhalta ranges. The main sections (especially the crests) of the ranges located on the southern slopes of the western Greater Caucasus are built of the suites of Bajocian porphyritic series and the Upper Jurassic clustic-limestone turbidites, argillites, sandstones, and clay slates (flysch suite). The southern part of the folding strip, namely the Gagra and Bzipi ranges (Astakhov 1973; Gobejishvili 2011), is built from Cretaceous sediments (limestones, dolomites, marls, carbonate clays, and sandstones).

The morphological look of the southern slope of the western Greater Caucasus is identified by the lithology of its building rocks together with tectonics. The relief of the crest of the Greater Caucasus is rocky and pointed and hypsometrically the highest. The role of nival-glacial processes is high in moulding landforms, resulting in widely distributed glacial, corries, moraines, trough gorges, karlings, debris, and rockfills. An active deep fault line passes between the old crystalline rocks and the Lower Jurassic, expressed in the form of a step in the relief.

The spread of Lower Jurassic sediments is associated with a soft, relatively low relief, including the passes of Khida, Adange, Amtkeli, Magana, and Akhuk-Dara. The gorges in a general Caucasian direction, such as the upper parts of the Bzipi, Chkhalta and Kodori river gorges, are also related to the same rocks. The serrated and cliff forms are characteristic of the relief built of Middle Jurassic porphyrites. The limestone relief is distinguished by the abundance of karst forms. Interaction of endogenous and exogenous processes and altitudinal distribution results in the formation of complex and strongly dissected relief.

In the high mountainous zone, over 3000 m, the surface of the relief suffers total denudation. Glacial processes identify the formation of meso shapes. The glacial relief is represented by modern glaciers in the high mountains. There are 164 glaciers in the western Greater Caucasus with an area of 44.2 ± 1.4 km^2 (Tielidze et al. 2015; Tielidze 2016). Most of the glaciers are located in the Kodori River basin, including smaller cirque glaciers. Several valley type glaciers are notable, such as Sakeni, Klichi, Marukhi, and Sopruju. The Würm glaciers are important in the modeling of mountainous relief, when the trough shape of the gorges of the Kodori and Bzipi tributaries were formed—the old glaciers descended to 600–700 m. The Würm Moraines are visible in the relief (Gobejishvili et al. 2011; Tielidze 2017a).

Erosion processes and forms are well developed in the western Greater Caucasus. Basic river gorges (the Kodori, Chkhalta and Bzipi gorges) are deep set in the relief. The bottoms of the gorges are often vast (in the relief built from soft rocks) and accumulation processes are underway. The depth of the gorges (Chkhalta-Kodori-Bzipi) is 2000–2500 m, in some areas—more than 3000 m (Chkhalta gorge

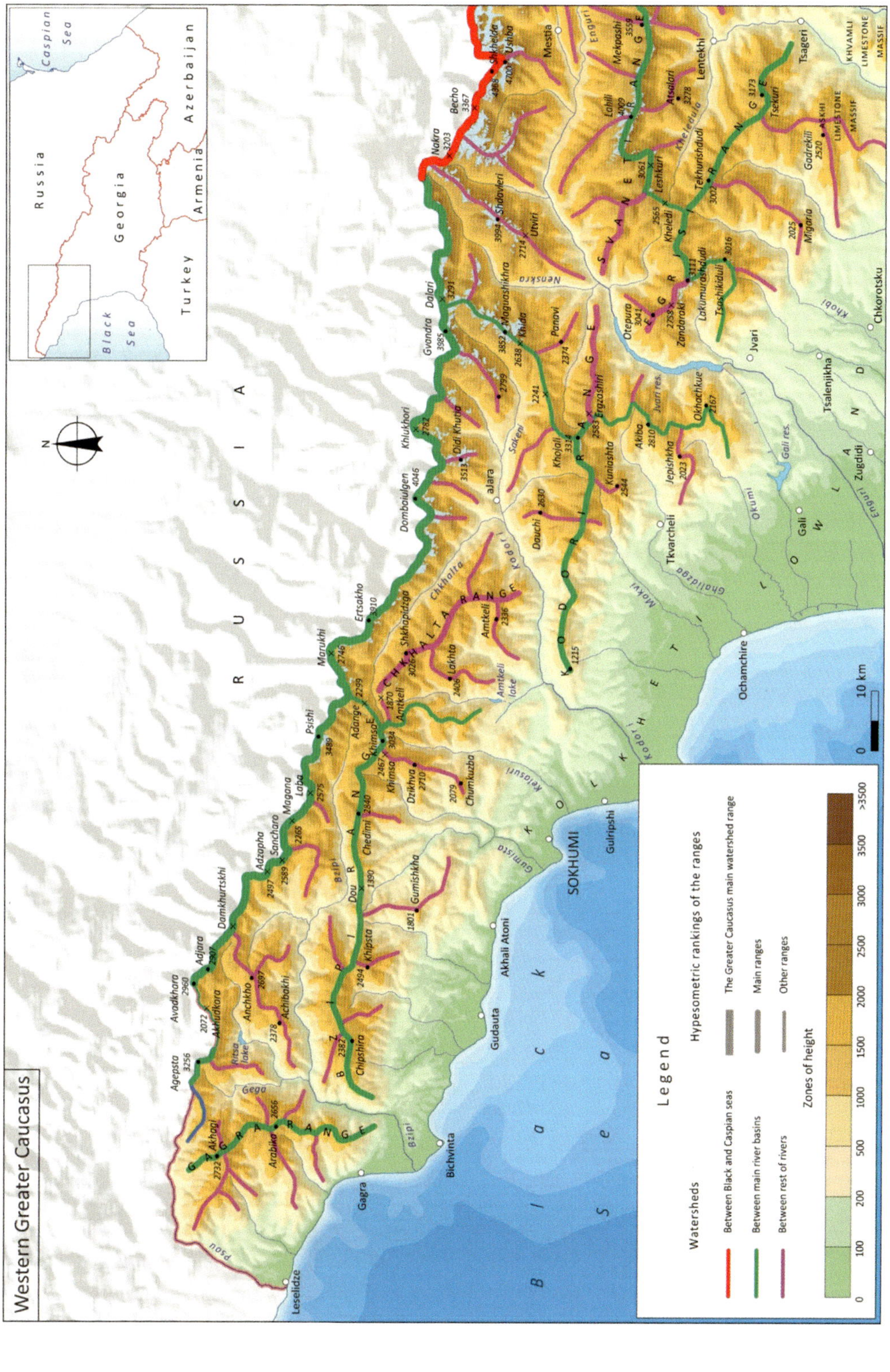

Fig. 8.1 The orography of the western Greater Caucasus

at Ptishi) (Gobejishvili 2011). In the formation of the landforms of the western Greater Caucasus, the landslides, rock avalanches, and debris play a role. The Amtkeli, Didi (Greater) Ritsa, and Patara (Smaller) Ritsa lakes formed by powerful rock avalanches.

Geomorphologically, the western Greater Caucasus is divided into the following orographic units: 1. The main watershed range; 2. Gagra Range; 3. Psou River gorge; 4. Atsetuka Range; 5. Gorges of the Bzipi River its tributaries; 6. Bazipi Range; 7. Chkhalta (Apkhazeti) Range; 8. Gorges of the Kodori River and its tributaries; 9. Kodori Range; and 10. Karst massifs of the western Greater Caucasus (Rikhva, Kodori, Tsebelda, Panavi, Okhachkue, and Gumishkha-Psirtskha).

8.2 Main Watershed Range of the Western Greater Caucasus

Within Georgia, the main watershed range extends from Mount Avadahara (2960 m) to Dalari Pass, in a sublatitudinal direction. Its length is approximately 125 km by direct line and 177 km along the crest. Within the western Greater Caucasus, the watershed range boundary is less winding than within the eastern Greater Caucasus. The watershed range descends toward the west, with the highest sector located between the mountains of Marukhi and Dombai-Ulgen. The heights of the mountains from the bottom of the Chkhalta River are more than 3000 m.

The following peaks are notable: Avadhara (2960 m), Marukhi (3800 m), Ertsakho (3910 m), Ertsogi (3863 m), Sopruju (3780 m), Juguturli-uchat (3896 m), Dombai-Ulgen (4046 m), and Ghvandra (3419 m). The main watershed consists of 15 passes, including Marukhi (2746 m), Klukhori (2782 m), Nakhari (2885 m), Sancharo (2589 m), and Chmakhra (2486 m). The lowest is the Magana Pass (2265 m).

The main watershed has many subranges. They are part of the structure of the main range and they are of erosive-tectonic origin. The subranges are as follows: Anchkhio, Chibisa, Adange, Khita, Khutia, Klichi, Ghvaghva, and Kharikhra, with the last four exhibiting higher elevation (>3500 m) and morphometry. The subranges are of submeridional or sometimes meridional direction.

The main watershed range is interesting and diverse by its geological structure. The complex morphotectonic structure and rock lithology are reflected in the morphological appearance of the relief. The watershed crest and southern slope are constructed from Paleozoic and Pre-Cambrian intrusive rocks: granites, gneisses, diorites, crystalline slates, and amphibolites. Crystalline rocks are represented as a narrow strip in the Bzipi River sector, while none are found in the area of the Sancharo-Magana passes. In the sector of the Kodori River gorge, crystalline rocks helped build the watershed crest and its southern subranges: Khita, Khutia, Klichi, Ghvaghva, and Kharikhra. Geological structure is reflected in the morphographic and morphometric nature of the relief of these orographic units (Geology of the USSR 1964). They are characterized by relief over 3000 m, intensive development of nival-glacial processes, and high mountain landforms, such as corries, karlings, troughs, moraines, debris cones, and rockslides. Erosion and snow avalanches are important in the transformation of the relief.

The southern slopes of the western Greater Caucasus are built in a narrow strip (wider in the Bzipi basin) of Lower Jurassic jasper slates and clay slates. These rocks are easily subordinated to denudative processes, and soft landforms are common. Low passes are located in the watershed, such as Adange, Magana, Amtkeli, Khida, and Kalmari.

The southern slopes of the watershed range are built of the Middle Jurassic granites and diorites. The deep fault line, passing the contact area between the crystalline and Jurassic rocks, is important in the relief formation (Gamkrelidze 1966).

Various genetic processes identified the formation of the landforms of the watershed range, namely, physical weathering, modern snow-glaciers impact, old glaciation, and erosion. The relief of the watershed range located above

3000 m lacks the vegetation cover and almost every sector of the topographical surface of the relief is exposed to exogenous processes.

This region is particularly characterized by abundant snow, exceeding 4–5 m; often the diurnal increase of snow thickness is more than 1 m. These conditions, along with the morphological nature of the terrain, cause strong snow avalanches. Due to the impact of snow avalanches, the slopes are fragmented by dingles and the forest cover is destroyed. The mass of the snow avalanche remains at the bottom of the gorges until August (Gobejishvili 2011).

Debris cones are one of the leading forms of the geomorphological landscape. Almost all of the heads of the gorges or the old corrie landforms and slopes of cornices are covered by young and old debris cones, which indicate the intense physical and frost weathering of the relief.

Despite the lower elevation of the main watershed, the glaciers are well developed. The crest of the main watershed range is located in the nival zone—above the firn line, and its morphological characteristics create good conditions for modern glaciers. Along with morphology, the abundance of solid precipitation is important in the formation of glaciers that create the infiltrative and congelation types of "Warm" glaciers in the nival zone (Gobejishvili 2011). The consequence excaration of modern and old glaciers is the presence of dense comb-shaped forms with pointed peaks (karlings), narrow and deep passes, with vast glacial corries.

Modern glaciers are located on the southern slopes of the main watershed range. Glaciers that extend from Magana Pass to the Adange Pass in the east belong to the Bzipi River basin, and the glaciers between the Adange and Dalari pass to the Kodori River basin. Glaciers are located between the Marukhi and Dalari passes. In this sector are the high peaks, such as Marukhi, Ertsakho, Sopruji, Dombai-Ulgen, Khakeli, Ghvandra, and the peaks of the Kharikhra Range: Kharikhra, Maguashirhka, and Okrilastavi; the height of these peaks is over 3700–3800 m.

There are 164 glaciers with an area of 44.2 ± 1.4 km^2 in the western Greater Caucasus (Tielidze et al. 2015; Tielidze 2016). The morphometric and morphographic peculiarities and climatic conditions cause the distribution of numerous, small size, cirque-type glaciers, which are the first by quantity and area. They make up 75% of the number and 40% of the area of the glaciers of the western Greater Caucasus. As the glaciers are mostly located on the southern slope of the western Greater Caucasus, their exposition is commonly southern. There are more numerous glaciers of the northern exposition in the subranges (Klichi, Ghvandra, Ghvaghva, and Kharikhra).

In the western Greater Caucasus, the valley and cirque-valley glaciers are numerous by area and length, namely the Sopruju, Southern Marukhi, Klichi, Sakeni, Ptishi, Memuli, and Chepara. By area, the cirque-valley glacier of Sopruju is the largest (3.66 km^2), and the longest is the valley glacier of Marukhi. The Marukhi Glacier has a well-visible firn, icefall, and tongue. Valley glaciers have retreated by 1.0–1.5 km after the Little Ice Age (LIA) maximum and the cirque glaciers by 300–400 m. The degradation of the glaciers of the western Greater Caucasus is actively still underway, related to natural global warming. Firn line stretches at 3000 m in the sector of the Bzipi River basin and at 3050–3150 m in the Kodori River basin (Tielidze 2017b). The typical glacial zone is almost unexpressed in the western Greater Caucasus, due to relatively low relief hypsometry.

The old glaciation contributed greatly to modeling the relief of the western Greater Caucasus. Numerous trough gorges, glacial layers, and excaration forms are well represented in the river gorges located on the southern slope. Particularly clearly preserved is the trace of the glaciations of Late Pleistocene (Würm) and Holocene.

The Late Pleistocene glaciation was of a mountain-valley type in the western Greater Caucasus. The main center of the glaciers was the watershed range of the Greater Caucasus. Despite the fact that the western Greater Caucasus is not hypsometrically high, the traces of old glaciation are preserved in the forms of cirques and moraines. The lower threshold of corrie forms is located at 1900–2100 m, indicating that in Late Pleistocene, the depression of the firn and snow lines in the river basins of the western Greater Caucasus were about 1000–1200 m. The largest glaciers descended to 600–700 m, when their lengths reached about 16–17 km.

During the Würm, the glaciers descended to the trough gorges of the rivers of Sakeni (~25 km), Klichi (~20 km), Ptishi (~14 km), Marukhi (~17.3 km) and Khetskvara. The Ptishi Glacier ended at the lowest elevation (~600 m) (Gobejishvili et al. 2011; Tielidze 2017a). The Klichi River gorge has a trough shape at about 700 m, along with the glacier tongue. There are several ledges developed on the bottom of the gorge. Lateral moraines are well preserved in the territory of the Ghvandra Village. Würm and Holocene moraines are well preserved in the Sakeni River gorge, as well as in the territories of the rivers of Martskhena (left) Ptishi, Khetskvara, and Omarishara. There was no united glacier in the Chkhalta River gorge. The tongue of the Marukhi glacier ended at 1380 m, where the layers of the Würm glacier are visible on the right slope below the rivers of Adange and Marukhi. The tongues of the individual glaciers (Sopruju, Adeba, and Atsiashi) descended in the Chkhalta River gorge and obstructed the river in some places. The nival processes created the following forms—rock fills, rockslides, debris, and debris cones. They are widely distributed in the heads of the rivers and in the old glacial forms, in the corries and troughs.

8.3 Gagra Range

The Gagra Range begins at the Kaldakhvara Village near the Black Sea coast. It first goes to the north and includes the Arabika massif, and then goes to the northwest and includes the Akhagi massif. In fact, the Gagra Range includes the Arabika and Akhagi massifs and their sub-ranges in the west—Tepebashi (Katsirkhva) and in the Zhvavakvara. The Arabika and Akhagi massifs are separated from each other by the Sandripshi Pass (2177 m). The highest peak of the Gagra Range is located in the Arabika massif (2757 m). The high peaks are Akhagi (2732 m) and Arabika (2656 m). The Gagra Range is separated from the Atsetuka Range by the Gega Pass (1954 m), through which the gorges of the Psou and Gega are connected to each other.

The Upper Jurassic and Cretaceous age suites participate in the construction of the Gagra Range. The southern sector of the Akhagi massif is built of Bajocian porphyrites, which is separated from the limestone zone by the fault line visible in the relief. The morphological appearance of the Gagra Range was formed by the impact of glacial, karst and erosive processes.

Erosion forms are widely spread in the Gagra Range (especially on the slopes). The eastern and northern slopes of the range are dissected by the right tributaries of the Bzipi River, and the western slope is dissected by the left tributaries of the Psou River. As for the southern slope, it is dissected by the tributaries of the rivers of Tsivtskala, Khashupse, Zhvavakvara, Gagripshi, and Tsikherva. The gorges developed in the limestone relief are narrow and deep; the gorges generated in the Jurassic sediments are V-shaped, sometimes with strongly inclined slopes and relatively broader bottoms. The Arabika karst massif completely lacks the surface waters and erosion landforms; only in some places are there weakly expressed dry ravines.

The Late Pleistocene (Würm) glaciations contributed greatly to the formation of a high crest of the Gagra Range (above 2000 m). The glaciers were developed on the slopes of the peaks of Arabika, Speleologists Peak, and Akhagi. The tongues of the valley glaciers descended to the height of 1700–1900 m. Glaciers were better developed on the slopes of the massifs with northern exposition. The glacial cirques and modified troughs are still well preserved in the northern part of the Akhagi massif. Numerous small size cirque glaciers were spread next to the valley glaciers.

Karst processes and related forms are widespread in the Gagra Range. Especially noteworthy is the Arabika massif built of powerful Upper Jurassic and Cretaceous limestones. Arabika has no analog in the karst regions of the world's mountainous areas by the depth of water movement, complexity of hydrodynamic zones, and outlets of the watery underground rivers. The thickness of the limestones that built the Arabika massif is over 2200 m; highly fractured limestones descend below sea level, even by several hundred meters. In such a case, the maximum depth of the karst process in the Arabika massif varies between 1800 and 2500 m. 159 vertical caves have been investigated in Arabika. Of these, five caves with the depth of over 1000 m (3.1%) are abysses. These include the world's №. 1 cave-abyss—Krubera (2197 m deep) and the abysses of Sarma (1830 m), Iljukhina (1275 m), Arabika-Kuibyshev (1110 m) and Dzou (1090 m) (Tsikarishvili et al. 2010).

High, medium and low karsts are distinguished in the Gagra Range. High karst is developed over 2000 m and is presented in various forms. It belongs to the bare or the surface karst type. There are almost all the conditions for karst phenomena to be developed intensively in the mountainous part of the Gagra Range, the presence of chemically pure limestones or calcium carbonate rocks, southern inclination of slopes, large thickness and intense cracks, deep location of karst erosion basis and sharp separation of the range from neighboring areas. All these conditions created an opportunity for the intensive development of karst processes.

The following karst forms are widespread in the high mountainous zone of Gagra: corries, dolinas, and uvalas, as well as the karst wells transition to underground forms. Circulation of surface waters, erosion and corrosion are important in the forming of dolinas. The sinking dolinas with a diameter of 12–20 m are common. Often, the uvalas are formed by the combination of dolinas, especially in the trough gorges of the Arabika massif (Kipiani 1974).

Karst wells and abysses are widespread in the Gagra Range. The Vakhushti Bagrationi, Krubera, Martel, and many other abysses were investigated and studied here.

The wide distribution of dolinas and abysses in the high mountainous part of the range is stipulated by the existence of deep karst along with other factors; here the vertical circulation zone of the underground waters is very deep. This must be related to the outlets of deep circulation waters observed in the area of the Black Sea.

In the Gagra Range, below 2000 m, the medium and low mountain karst is widespread. It occupies the western part of the Gagra Range and the small ranges and troughs (Liukovokhu, Zirkhu, Zhovekhu, Mamdzishkha, etc.) located in the south, along with the subranges directed toward the Gega and Bzipi river gorges. These territories are covered with soil-vegetation over the hidden type of karst. The karst surface forms are widely represented in these areas, distinguished by less depth and soft shapes (dolinas in the areas of Mamdzishkha and Akhachi). It is notable that the corries are less developed here than in the high mountain karst. Wells and shafts are less represented. The caves are not yet opened at the Black Sea level. During the Würm glaciation, when the sea level was 30–50 m lower, the caves' exits were opened. But today, they are covered by sea and it is natural that the freshwater exits in the sea are not observed.

8.4 Atsetuka Range

The Atsetuka Range is of sublatitudinal direction and its southern slope lies within Georgia. It extends 14–16 km from the head of the Psou River to the Avadhara River gorge. The highest peak is Mt. Agepsta (3256 m). The crestal part of the range is built of the Bajocian tuffo-porphyrites suites. The Würm glaciations contributed greatly to sculpting relief. The length of the valley and cirque glaciers was 2–4 km. The glacier forms (cirques and troughs) are well preserved on the northern slope and especially, around the Agepsta Peak, but are only slightly visible on the southern slope. On the northern slope of Mt. Atsetuka (2538 m), small lakes are developed in the cirque forms. The slopes of the Atsetuka Range are dissected by the right tributaries of the Iupshara River. The Atsetuka Range is separated from the main watershed range by the Akhukdara Pass (2072 m), through which the Avadhara River gorge is connected to the Mzimta River gorge.

8.5 Achibakhi (Rikhva) Limestone Massif

The Achibakhi massif is located between the canyon gorges of the Bzipi and Iupshara rivers. The northern border is conditional; it passes at the contact of limestones and porphyrites. There is a deep erosive gorge of the Pshita River in the east. In the west, it is separated from the Arabika karst massif by the Iupshara-Gega deep canyon, running in a southwestern direction. The length of the plateau is 7 km, and the width reaches 4 km. The eastern part of the karst massive is occupied by the latitudinal anticlinal Achibakhi Range with height over 2000 m. The southern slope of the range is steep and rocky, while the northern slope is less inclined (100 m/km). The flat surfaces of the plateaus are complicated by the surface karst forms. The slopes located on the plateau edges are dissected by the tributaries of the rivers of Bzipi and Iupshara. The highest peak of the massif is Achibakhi (2378 m). The length of the massif is ~17 km, about 8 km width and the area is about 120 km^2.

The plateau surfaces of Achibakhi massif are constructed from dolomitized and pure limestones of Cretaceous age. The lower parts of the steep slopes are represented by the reef, breccias and layered limestones of the Lusitanian period. In the north of the massif, the Tertiary clays, marls, and limestones stretch as a small strip (Tatashidze 1997).

Karst processes are intensively carried out in the plateaus built of the Cretaceous limestones. Among the surface karst forms, there are different size karst sinkholes, flatlands (Iashta, Jimasku, Rikhva, and Satibi) and polje (Kuuba). The length of the flatlands is 1.5–2.0 km, and a relative height of 150–250 m. Several karst sinkholes are located at the bottom of the flatlands. The waters absorbed in the sinkholes of the surface of the massif flow out from the bottom of the steep slopes at the contact line as vaucluses. Corrie surfaces are represented in the karst plateau of Achibakhi. The Kuuba polje is different in its form and dimension from other karst poljes in western Georgia.

8.6 Gorges of the Bzipi River and its Tributaries

The erosive gorge of the Bzipi River and its tributaries is located on the southern slope of the western Greater Caucasus and is the main orographic unit. The gorge is mainly of latitudinal and general Caucasian direction and is formed between the Bzipi Range and the watershed range of the Greater Caucasus. The relief of the Bzipi gorge is complex and diverse. The leading role in its formation, along with erosive processes, belongs to structural, lithological, and tectonic conditions of latitudinal direction. The length of the gorge is about 105 km. The width of the left slope is about 4–6 km and it is steep and in some areas vertical. The right slope is broad and is dissected by its tributaries (Gega, Iupshara-Avadhara, Pshitsa, Bagiu, Gribza, and Psishi). At the same time, these constitute the internal watersheds of the tributaries: Atsetuka, Anchkho, Chibiskha, Ashimkhuri and karst massifs of Shegishkha, and Rikhva.

Bzipi gorge is divided into two main parts according to the morphological signs: one is of latitudinal direction from the head to the Blue Lake and another is of meridional direction from the Blue Lake to the sea. In turn, they are also divided into separate sections. The bed falls 19 m/km from the head to the Adange (2300 m)—Reshevie depression (650 m), 16 m/km from Reshevie to Rikza (450 m), 11 m/km from Rikza to the Blue Lake (110 m), and 4 m/km from the Blue Lake to Akvara (50 m) (Gobejishvili 2011). Separate sections from Akvara to the sea differ from each other by orographic, morphometric, geological structures, and geodynamic processes.

The Bzipi River gorge is the V-shaped erosive form from its heads to the Reshevie. The river gorges flowed down from the Greater Caucasus morphologically represent the troughs transformed due to erosion; the troughs are created by the action of the Late Pleistocene (Würm) glaciers. But the form of the gorges of left tributaries is a trough. The troughs are well preserved by the Bajocian porphyritic suites and the excaration of the powerful glaciers developed on the northern slope of the Chadimi Range. The bottom of the Bzipi River gorge is constructed from the Lias slates, turbidites, and clay slates that are easily subject to erosion-denudation (Maruashvili 1971). As a result, quite deep and in some areas wide bottomed gorges are developed here. The V-shaped gorge, symmetrical and built of slates, has lowered areas between the Greater Caucasus and Bzipi Range and it is a relative sinking zone. There is a 50-m-wide floodplain in some areas, especially, on the right side of the gorge, where the floodplain (1–2 m) and upper floodplain (I—8–10 m, II—20–25 m) terraces are developed (Gobejishvili 2011). The gorge is constructed from porphyrite suites up to Reshevie after joining Khimsa, and in these sections, the gorges of the Bzipi River and its tributaries are morphologically narrow and deep; they are almost inaccessible.

The tectonic-erosive depression of Gribza of the right tributary of Bzipi covers the middle part of the Bzipi River gorge. The depression is built of schistous slates and is bordered by the relief built of Bajocian porphyrites in the south. The Grizba River develops the canyon gorge from the depression to Bzipi. The depression itself is about 2.5 km long and 400–600 m wide. The falling of the Gribza River bed in the depression is 12 m/km. The river meanderings in the bottom of the gorge, and the debris cones are developed on the banks.

The gorge is relatively wide from Reshevie to Rikza and the rock lithology is well reflected in the morphology of the gorge. The gorge of the Pskhu River, the right tributary of the Bzipi River, is a depression morphologically. The depression is characterized by a symmetrical development that conventionally can be divided into two parts: Pskhu and Aguripsta. There is a relatively broad plain represented in the Pskhu depression with the width of 1.5–2.0 km in the area of Bzipi. The plains are built of alluvion. The Pskhu Plain is located above 50–100 m from the Bzipi-Pskhu riverbed. The largest village of Pskhu is built in this plane. The genesis of the depression is quite complex; the main role in its formation belongs to differential tectonics, and below Rikza the rock avalanche descended from the right side of the gorge would have created a water reservoir that has contributed to the accumulation of powerful (20–30 m) alluvial material.

From Rikza and the Blue Lake the Bzipi River gorge is quite narrow, often with steep slopes; the depth of the gorge is 1500–2000 m. In this section of the gorge (at a height of 150 m), the longest river of the basin Gega conflows Bzipi from the right side. The Gega River is composed of two rivers, Gega itself and its tributaries—Iupshara and Lashipse. The head of the Gega River morphologically looks like a cirque, surrounded by the heavily dissected southern slope of the Atsetuka Range and northeastern slope of the Gagra Range. The lower parts of the Gega and Iupshara river gorges are formed perpendicularly to the building suites of Cretaceous age. The gorges are deep and narrow and the bed is strongly sloped. The 13-km-long section is the canyon, the depth of which is 1500–2000 m. Near Bzipi, the bed of the Gega gorge falls down by 150 m at a height of 400 m. Due to this, the Gega gorge is hanging in shape.

Two lakes—the Didi (Greater) Ritsa and Patara (Smaller) Ritsa are created in the Iupshara gorges above the Pshegishkha limestone massif, as a result of the powerful rock avalanche.

8.7 Bzipi Range

The Bzipi Range is extended from the Amtkeli Pass to the downstream of the Bzipi River in a latitudinal direction. The length of the range is ~75 km, and width varies, in the east it is ~26 km, in the central part—15 km and in the west it is extended to ~21 km. The average height of the Bzipi Range is 2300 m. From the north, it is sharply bordered by the left steep slope of the Bzipi River, and the height of the southern slope gradually decreases to 800–1000 m. The range is divided into three parts by morphological, morphometric, and morphostructural characteristics as follows: 1. The eastern section, known as the Chedimi Range; it is the highest and wider (width ~20 km); the highest peak of the range, the Mt. Khimsa

(3034 m), is located here; 2. The central section is narrow (width ~15 km) and low; and 3. The western section is quite high (up to Mt. Mukhursha—2207 m) and wide. The highest peak is Mt. Kvansha (2652 m).

The Bzipi Range is built of Mesozoic sediments, effusive and intrusive rocks. The northern slope of the range and the highest eastern part are built of the Bajocian porphyritic series, andesite, basaltic and dacite lavas, lava breccias, tuffites, and pyroclastolites. They are topped by tuffo-conglomerates, tuffo-sandstones, sandstones, and clays. The central lower section of the range is built of Upper Jurassic shallow sea deposits—limestones, gravelites, clay slates, sandy clays, marls, and carbonate sandstones. In the western section, the relief of the Bzipi Range is built by the alternation of Cretaceous age Urgonian facies limestones, ammonite limestones, dolomitized limestone, dolomites, and marls. In the easternmost Bzipi Range, in the Kelasuri River gorge, the relief is built of Middle Jurassic intrusive rocks granitoides (Gobejishvili 2011). Geological structure is reflected in the relief. The eastern part of the Bzipi Range, the Chedimi Range morphostructurally is horst-synclinal and is a structural continuation of the Chkhalta Range with average width of 7–10 km. The central part is a synclinal depression, and the western section is a synclinal-block massif.

The western section of the Bzipi Range belongs to one of the highest and unique karst massifs in the world and is known as the Bzipi massif. It is separated from the Arabika massif by the Bzipi River gorge. The eastern boundary is a deeply cut meridional gorge of the Aapsta River and then the gorge of the Reshevie—the tributary of the Bzipi River. The southern border is also clear; it stretches along the 300–400 m horizontal (the areas of the villages of Kaldakhvara, Otkhari, Khopi, Duripshi and Khabio) on the northern slope of Kolkheti hilly relief. The length of the western section of the Bzipi Range is about 35 km, the maximum width is 15–20 km, and the area is ~560 km^2. By area it exceeds the Arabika massif, but is far behind the Racha limestone massif. The array has an asymmetric structure. The western and northern slopes are cliffy and of gravitational nature (width 2–5 km). The southern deluvial slope is long (with an average width of 12 km) and gently inclined.

The Bzipi massif is built of 1100–1300-m-wide Upper Jurassic and 2000-m-wide Cretaceous carbonate sediments. The morphology and morphography of the western part of the Bzipi Range are identified by a spread of karst rocks (Tatashidze 1988). Here is a complete representation of both the surface karst forms and underground huge karst forms—caves, wells, sinkholes, dolinas, and poljes. Two types of karsts are sharply distinguished here—the bare and hidden karsts.

The bare karst covers the crestal areas of the Bzipi massif above 1800 m. Plain surfaces of karst plateau are complicated by anticlinal lower hills and depressions of latitudinal direction, the slanting slopes of which are sieved with karst sinkholes.

In this respect, the Ahva cirque and the bottom of the gorge, are distinguished from the Akugra and Chipshira corrie fields. Their area ranges from hundreds of meters to several km^2. There are up to 150–160 sinkholes per km^2. Due to uneven conditions of karst formation, the morphology of sinkholes is different. The absorbing, sinking, asymmetric and cone shapes are common here. Interesting shapes have inserted sinkholes, the number of which sometimes reaches 3–4, which is the indicator of their gradual deepening. The sinkholes are ended with open ponors.

Karst hollows and dead gorges are notable among the large surface karst forms of the Bzipi massif. The hollows (relative fragmentation depth of 150–500 m) are spread on the northern and southern slopes of the Mt. Akugra in the areas of Ahva, Halki, Batakhi, and Chipshira. The snow is collected in the hollows thus increasing the intensity of karst formation.

The total number of karst cavities in the Bzipi massif is over 450 so far. Vertical karst forms with depth less than 50 m are quite common in the Bzipi massif, along with snowy wells with depth less than 20 m. Their total number is more than 180 today (Gobejishvili 2011). Here are formed the following deepest cave systems:

Snowy-Mezhen, Pantyukhin, Napra, and Forelnaia (Trout) karst abysses. The Snowy-Mezhen system (1753 m depth) is commenced in the heavily cracked limestones and dolomite suites of Lower Cretaceous age. The morphology of the snowy corridors and halls is controlled by the prevailing cracks network, namely stepwise distributed faults.

In the higher karst caves of the Bzipi massif, the formation of stalactite-stalagmite deposits is not as intense as it is characteristic to plains and foothill zones. Seismo-gravitational formations are widespread in the caves. The capacity of the rock-avalanche mass sometimes reaches 100 m, and the volume of individual boulders reaches 1000 m^3 (Tatashidze 1988). Snow-ice layers are worth mentioning in the caves of the Snowy abyss. Snow, firn, and underground glacier are represented here, which are about 5000 years old with volume of 60 thousand m^3. The underground glacier is stable. It is modified slightly in summer during the liquid precipitation inflow (Gobejishvili 2011). Due to the falling of a water stream on the cone, ice wells and ice stalactites and stalagmites occur.

The hidden or the "Forest karst" occurs between 400 and 1800 m. The surface is covered with soil-vegetation cover, slowing down the intensity of the karst processes. In the section of the massif, on the slightly inclined slopes, there are well-developed karst sinkholes and cavities. The Alpine zone ranged from 800 to 1000 m at the time of the Würm glaciation and naturally, at this height, the forming of large karst forms would have occurred in this period.

Hypsometrically the Bzipi massive is of medium height, with only some peaks above 2500 m—Dzishra (2623 m), Akugra (2511 m) and Kvansha (2652 m). Such height of relief allowed the existence of small cirque and cirque-valley glaciers, and numerous snow spots deeply set in the karst forms facilitated the process and extension of karst forms. Physical weathering is active in the mountainous areas, and chemical and organic weathering in the forest zone. The northern and western slopes of the karst sector of the Bzipi Range are steep and slightly dissected. The ravines are mostly dry; the springs flow out from their lower parts, from deluvium. The southern slope is more dissected by the tributaries of the rivers of Mchishta and Khipsta. It is worth mentioning the heads of the Khipsta River, which are intruded up to the crest of the massif and divides it into two parts—Khipsta in the east and Batakhi in the west. The upstream of the Khipsta River is deeply cut into limestones: its depth is 1300–2000 m and width 4–5 km. In the southern cliffy slopes of the karst massive, the ravines are almost dry; in the bottom of the slope, the sources with abundant debit flow out, which feed the rivers of Mchishta and Khipsta. Among the sources, the Mchishta vaucluse is characterized by high debit, and there is an opinion that it feeds from the Khipsta massif. The drainage of karst waters occurs under the Khipsta River bed.

The central section, based on the lithology of the relief building rocks, is hypsometrically low and characterized by a soft relief. Its weakly expressed peaks cannot reach 2000 m. This is why the lowest passes on the Bzipi Range are located here: Dou (1390 m), Gudauta (1566 m) and Achavchara (1795 m). Due to the fact that the southern slope of the range and partially its northern slope are built of Jurassic soft rocks, they are heavily dissected by the rivers of western Gumista and Aapsta in the south and by the tributaries of the Reshevie River in the north. The cross-section gorges are deeply cut in the relief. The crest of the central section and the northern slope is constructed from the Bajocian porphyritic suites and is distinguished by a rocky relief and weak dissection. On the southern slope, the watershed ranges create the skeleton of the relief including the rocky range (Mt. Gumishkha, 1800 m).

The eastern sector of the Bzipi Range has a latitudinal direction. It is known as the Chedimi Range. The relief built of Bajocian porphyrites in the crestal part is characterized by sharp and rocky forms. The high peaks are Chedimi (2840 m), Khaka (2694 m), Khimsa (3033 m), and Dzikhva (2710 m). The northern slope of the mountain range is dissected by the left tributaries (Psikva, Khimsa, and Mura) of the Bzipi River; the southern slope is relatively long and deeply dissected by the tributaries of the eastern Gumista and Kelasuri river basins.

The leading role in the morphology of the crestal part of the Chedimi Range belongs to the nival-glacial processes; the most important is the trace of the action of Würm glaciers. The Khimsa massif is located here with six modern small sized cirque type glaciers. Glaciers are located on the northern slope; only one of them is located in the heads of the Kelasuri River. The firn line passes at the Bzipi Range at about 3000 m.

The trace of Late Pleistocene glaciation is well preserved on the northern slope in the forms of cirques, moraines, and modified troughs. The lower threshold of the cirques is at a height of ~2000 m. The length of the glaciers reached ~6–8 km and descended almost to the bottom of the Bzipi River. Smaller lakes have been developed in the corrie shapes on the crest of the range as a result of the excaration action of glaciers, such as the Anna, Mitsara, and Psikva Lakes.

The traces of the Late Pleistocene glaciation have been preserved in the relief of the Dzikhva massif built of porphyrites. It is the southern subrange of the Bzipi Range and the watershed of the eastern Gumista and Kelasuri river basins.

8.8 Chkhalta (Apkhazeti) Range

The Chkhalta Range follows the Caucasian direction (from the northwest to the southeast). In the west, it is separated from the Bzipi Range by the Amtkeli Pass, and in the east, it descends to the Kodori River gorge in the form of two subranges (Guarapi-Kvabchara and Zhurgia-Zhvirgva) in the areas of the villages of Chkhalta and Kvabchara. The range is located in the Kodori River basin and it is a watershed of the Kodori River right tributaries—the rivers of Chkhalta and Amtkeli-Jampali. Its length is ~40 km by direct line and ~45 km along the watershed. The highest peak is Mt. Shkhapidzga (3026 m), located in the north of the watershed. The highest peaks are Amtkeli (2893 m), Shoudidi (2755 m), Atsgara (2735 m), Guarapi (2742 m), and Zhvirgva (2294 m). Morphologically the Chkhalta Range is asymmetric. The northern slope is steep and short and it is dissected by the right tributaries of the Chkhalta River. The width of the slope from the crest of the range to the bottom of the Chkhalta River is ~5–7 km. There are many small size subranges (~2–3 km long) on the northern slopes. Among them, the Sibista and Shoudidi river watersheds have a length of ~4–5 km. The southern slope is wider, ~12–14 km, and is divided by the right tributaries of the Kodori River, including the Amtkeli and Jampali. Their basins are located in the central and western sections of the Chkhalta Range. The Lakhta and Amtkeli subranges are located on the southern slope. The Lakhta subrange is about 15 km long and is the watershed of the Amtkeli and Jampali rivers. The Amtkeli subrange separates the Jampali River gorge from the basin of the right tributaries of the Kodori River.

At the peak of the Belai (2560 m), the southeastern subrange of Guarapi of the Chkhalta Range is divided into three ranges as follows: in the west—the Amtkeli Range, in the south—the Achamkhara Range and in the southeast—Kvabchara. The crest of the Chkhalta Range is uplifted by 1400–1600 m from the thalweg (from the bottom of the gorge) of the Chkhalta River. The relative height of the southern slope is quite variable between 1000 and 2000 m.

Morphostructurally, the Chkhalta Range is of horst-synclinal nature (along with the eastern part of the Bzipi Range); it is a tectogenic structure in the common Caucasian direction with width of 7–8 km. Deep fault lines are developed on both slopes of the horst-synclinal range, which is reflected in the relief (especially vividly on the northern slope).

Due to its tectonic nature, the higher part of the crest of the Chkhalta Range is built of Bajocian porphyritic series from the Amtkeli Pass to the Chkhalta Village. In the south and north, the relief is built of the flysch alternation of Lower Jurassic sandstone turbidites, clay slates, and argillites. A short profile of the northern slope of the Chkhalta Range, proximity of erosion basis, and tectonics led to the creation of 4–7 km long simple river gorges. The lowered sections of the range in the center, the heads of the rivers of Shoudidi, Sibista, and Lakhta and in the east, the heads of the rivers of Jampali and Kvarashi are built of Lower Jurassic loose sediments. The areas above the Amtkeli Lake and the heads of the Amtkeli River, as well as the heads of the Jampali River, are built of Middle Jurassic intrusions, granites, diorites, and granodiorites. The relief built of these rocks is known as the Guarapi massif. The northern sector of the Amtkeli Range is built of the Upper Jurassic layered limestones, sandstones, marls, and flint argillites in the form of a narrow strip.

Nival-glacial processes along with neotectonic movement are important in modeling the relief of the Chkhalta Range above 2000 m. The morphology of the Chkhalta Range shows that even today and especially in Late Pleistocene and Holocene, the glaciation was of asymmetric character.

Today, there are five small glaciers in the Chkhalta Range; two are in the head of the Amtkeli River, and another three are located on the northern slope of the highest peak of the range Mt. Shkhapidzga (3026 m). Trace of impact of the Late Pleistocene glaciers is well preserved in the relief built of Bajocian porphyrites in the forms of cirques, moraines, troughs, karlings, and ledges. By morphometric and morphological signs, the Late Pleistocene glaciation was nearly the same as it is today in the watershed range of the western Greater Caucasus from the Marukhi Pass to the Klukhori Pass. The powerful glaciers were located around the massifs of the rivers of Shoudidi, Sibista, Amtkeli, Gorapi, Kvarashi, and Shapidzgha. The corrie forms are visible in the southwestern subrange of the Chkhalta Range in the Gorabi massif. The relief is quite high (2500–2700 m), which provides the feeding of the valley and cirque glaciers.

The consequence of the excaration impact of old glaciers is the presence of numerous small lakes in the cirques (Gorapi, Jugi, Adzmakhi, Atsgara, etc.). The tongues of the valley glaciers almost reached the bottom of the Chkhalta gorge. Particularly interesting was the glacier formed in the head of Amtkeli. In the Late Pleistocene, the heads of the Amtkeli River belonged to the Adange River basin; the mainstream of the tongue of the powerful valley glacier hanged off toward the Adange gorge, and part of it in the Amtkeli gorge. The Amtkeli River caught the heads of the Adange River by backward erosion (relief built of flysch suites) (Gobejishvili 2011).

In Würm, the Amtkeli-Adange Glacier ended at a height of ~1700 m, as indicated by the presence of lateral moraines. The length of the old glacier was 7–8 km. The cirque and cirque-valley glaciers created the landscape in the Chkhalta Range. First, the granitoid boulders were brought by glaciers to the gorge from the Gorabi (Guarapi) massif at 1700–1800 m, and then they were brought and deposited by the action of Jampali mudflows to the area of the villages of Jampali and Chini.

Erosion processes stipulate the formation of morphosculptural landforms on the slopes of the Chkhalta Range below 2000 m. The northern slope of the range is dissected by the tributaries of the Chkhalta River. Dissection depth in the middle stream of the rivers is 300–500 m, the V-shaped gorges are developed in the Jurassic loose suites; the slopes are characterized with soft profile and less inclination. In the upper stream of the gorge, the troughs present are modified by erosion.

The ravines and gorges on the northern slopes of the Chkhalta Range are of submeridional direction (SW–NE) and are developed perpendicular to morphostructures. The fall of the individual riverbed from the crest to the Chkhalta River varies from 150 m/km up to 350 km/km.

The southern slope of Chkhalta is dissected by the tributaries of the rivers of Amtkeli and Jampali in a southern direction. Cutting depth and morphology of the ravines and gorges in relief depends on geological structure. The erosion cutting depth in the middle stream is 500–1000–1500 m. The slopes of the ravines are V-shaped and quite inclined. Narrow, deep and sometimes antecedent gorges are formed in the relief built of the limestones and granitoid rocks. The main rivers are Amtkeli and Jampali; the length of the Amtkeli River gorge up to the Amtkeli Lake is about 25 km before the lake; the gorge is erosive and V-shaped below 2000 m. The depth of the cutting in the relief is 1000–1500 m. The Amtkeli Pass (1870 m) is located in its heads. Genetically, it is a glacial formation with the participation of denudative processes. The Amtkeli Pass separates the Bzipi and Chkhalta ranges from each other, which are structurally the same. In the Amtkeli gorge, at a height of 507 m, above the limestone zone is located the Amtkeli Lake (length ~2 km, width ~800 m). Amtkeli Lake appeared in 1891 as a result of rock avalanches descended in the gorge after a strong earthquake.

The Jampali River gorge is located on the southwestern slope of the Chkhalta Range, from the Shoudidi Pass to the peak of Luncha (2425 m). Its watershed ranges are Lakhta and Amtkeli. The gorge is opened like a fan toward the crest of the Chkhalta Range. Narrow gorges are formed in the relief built of limestones and granitoids.

The karst landforms are spread in the subranges of Lakhta and Amtkeli built of limestones. The Skhapacha karst massif is represented in the Lakhti Range. Its height is 1800 m, and reduces to 300–400 m in the southeast. There are numerous karst sinkholes formed in the Skhapacha massif, as well as the Tsivi cave (660 m asl) with length of ~80 m and karst wells, Apushta (5 m deep) and Amtkeli (36 m deep).

8.9 Gorges of the Kodori River and its Tributaries

The Kodori River basin is located on the southern slope of the western Greater Caucasus. Due to the morphological and lithological-tectonic peculiarities of relief, the Kodori gorge is divided into several sections from its heads to the confluence, such as: 1. The heads of the Kodori River and its right tributaries, located in the main range (their description is given in the subchapter about the Kodori Range); 2. The intermountain (Chkhalta-Kodori) depression, which has a latitudinal direction; 3. The Kodori gorge of submeridional direction; 4. The terraced gorge; and 5. The delta of Kodori. The last two units will be discussed in the zone of intermountain plain.

The gorges (Ptishi, Khetskvara, Klichi, Ghvandra, and Sakeni) of the Kodori River tributaries located on the southern slope of the Greater Caucasus watershed range are of meridional or submeridional direction and are deeply set in the relief. Morphologically, they are trough gorges and are slightly modified by erosive processes. Moraines, glacial deposits are located in the bottoms and on the slopes of the gorges. Erosion forms are weakly represented (debris cones). Numerous snow avalanches participate in the modeling the gorges.

The Kodori-Chkhalta depression is located between the Greater Caucasus Range from the north and Chkhalta-Kodori ranges in the south. It is conventionally stretched between the Adange and Khida Passes and is located in the relative sinking zone. Its western section in the Chkhalta gorge is of a graben nature.

The depression is built of Lower Jurassic clay-slates, slates and sandstones. Tectonically, the depression building suites are separated by fault lines from the crystalline and Bajocial suites of the Greater Caucasus, which is well reflected in the morphology of the relief.

The morphological peculiarity of the Chkhalta gorge is its asymmetry in its middle and lower parts. The left slope of the gorge is part of the Greater Caucasus Range. Morphologically, it is very high, steep, and less fragmented and creates a sharp step in the relief. The right slope of the gorge is more inclined and low. Due to the shape of the gorge, there are normal erosion forms developed on the right side of the gorge. The bottom of the gorge is mainly covered by debris cones. The rivers on the left slope of the Chkhalta gorge are characterized by their high fall. Erosive-glacial forms are also weakly expressed (small length and width). The heads (above 1400 m) of the Chkhalta River gorge are notable. Above the confluence of the rivers of Adange and Marukhi, the gorge is expanded, the bottom is wide and covered with fluvio-glacial deposits. The floodplain and the terrace of the upper floodplain are well developed. The trace (there are glacial deposits on the right side of the gorge) of impact of Marukhi Würm glacier is clearly visible in the relief. In the Adange gorge, the impact of Würm glaciers is fixed by the lateral moraines (at a height of 1700 m).

The Sakeni gorge is geomorphologically complex and interesting. The middle and lower section of the gorge is a modified trough. Moraines are represented by the granite boulders in fragments in the area of the Sakeni Village. The Sakeni River gorge in its middle part at heights of 1400–1500 and 1800–1950 m is widened with a straight and wide bottom. The river with bayous flows slowly; these sections are the former lakes. Their formation is related to the Holocene stadial glaciation. There are six stadial moraines in the Sakeni gorge that are located in three pairs.

The Kodori River gorge has a latitudinal direction from Chkhalta to Sakeni at a height more than 500 m. The gorge is wide here and it has a specific form of hollow. Its length along the Kodori River is 18–20 km and the width is 4–5 km. The bottom of the gorge is built of loose sediments of the Late Pleistocene age. There are weakly developed debris cones on the left side of the gorge, and on the right slope, almost all of them are located in the gorge. Particularly distinguished are the debris cones of Tvibi, Tvibrasheni, and Khetskvara. The mudflow streams of Tvibrasheni often blocked the river, which is reflected in the morphology of the gorge. Fluvio-glacial material is also involved in the construction of the Tvibrasheni cone, as in the heads of this gorge the glaciers conducted active excaration action. Above the cone, the bottom of the Kodori gorge is filled with fluvio-glacial sediments, the thickness of which is reduced compared to the Klichi and Ghvandra River Gorges. The thickness of the alluvial and fluvio-glacial sediments is 150–160 m at Marjvena Gentsvishi (Gobejishvili 2011). These sediments are completely cut by the Kodori River. The terrace steps are formed in the gorge from Tvibrasheni to Chkhalta. The terraces are found in basic rocks and in debris cones. There are three fluvio-glacial terraces formed in the surroundings of the Chkhalta Village: I—0–15 m, II—70–75 m and III—150 m. The main villages of Zemo Kodori are located in the river terraces and debris cones.

The Kodori River gorge is sublatitudinal at a distance of 25 km from the Chkhalta Village (500 m) to the Bagadka Village (200 m) and is in a western direction. The gorge is antecedent and crosses the relief building suites perpendicularly, sometimes indirectly. The gorge is deep and V-shaped between the villages of Chkhalta and Zemo Lata and its slopes are dissected by the Kobshara and Zima rivers (the right tributaries of the Kodori River) and Shabatkvara and Vaiamshi (the left tributaries of the Kodori River). In this part, the depth of the erosive cut of the tributaries is 900–1000 m, and the Kodori River itself is cut at a depth of 1500–2000 m. The width of the bottom of the Kodori River gorge is 100–150 m in average. The erosive terraces are visible on both the sides of the gorge as fragments. Here are the two terraces of 5–7 m and 25–30 m height. The width of the terraces is 100–150 m in some places.

The Kodori River gorge is of latitudinal direction from the Zemo Lata Village to the confluence of the Amtkeli River at the Bagadka Village and is 16–17 km long. The narrow gorge is asymmetric. The left slope is strongly inclined and is weakly fragmented. The right slope is relatively wide (6–8 km) and is dissected by the rivers located at the Amtkeli Range, such as Arghunia, Uchkuria, and Khinga. In this section, the bottom of the Kodori River gorge is variable and sometimes reaches 1 km. There are villages constructed in the lower terraces (the 1st terrace 5–7 m and the 2nd 30–35 m) formed in the gorge, and the high terraces are represented as fragments. The height of the 3rd terrace is 50–55 m, the 4th is 150–155 m and the 5th 350–370 m. The Amtkeli River, which flows through the Amtkeli karst relief, confluences the Kodori River in this section from the right side. The upper part of the gorge is opened and the lower part is deeply cut in the relief. The high erosion terraces are formed in the gorge in the area of Jampali, Chini, and Pali, the relative heights of which are 250 and 500 m. In the terrace structure, the granite material of Gorabi is also involved. The Amtkeli Lake is the geomorphological element of the Amtkeli River gorge, which was formed in 1891 as a result of earthquake; the powerful rock avalanche descending from the slope of the Mt. Skhapchi blocked the Amtkeli River gorge at a length of 1 km.

8.10 Kodori Range

The longitudinal Kodori Range is located between the Kodori and Enguri gorges (Khaishi-Amzara). It is a well-expressed orographic and morphostructural unit. Its length is ~65 km toward the watershed (Fig. 8.2); the highest peak is Khojali (3313 m), which is located in the central part of the range.

The western and central sectors of the Kodori Range are watersheds of the rivers of Kodori (Mramba, etc.) and Ghalidzga-Mokvi river basins, while the eastern sector is a watershed of Larikvakva and Darchi-Ormeleti—the rivers of the Enguri River basin. Hypsometrically, the central part of the Kodori Range is the highest, located in the alpine and nival zones. The high peaks are Bashkabsara (2730 m), Khojali (3313 m), Okrosmta (3102 m), Aduada (2775 m), and Vovtske (2414 m). The Kodori Range is low (1200–1000 m) in the west from the Mt. Vovtske and in the east, it gradually descends toward the Enguri gorge (the Khaishi Village).

The orographic appearance of the Kodori Range is created by its subranges: Ashdu, Mukajikvara, Dauchi, and Anarva (Garva) in the north and Bakhujara, Kuniashta, Uokhvtskia, Otirda, and Kada in the south. The Ergzashira Pass is notable among the passes. Subranges exist in the watersheds of those rivers originating in the Kodori Range (Mukvajikvara, Lomkatsa, Mchamba, Darchi-Ormeleti, Larikvakva, Ghalidzga, Mokvi, and Dgamishi). Dauchi, Akharva, Bakhunjara (watershed of the rivers of Larikvakva and Ghalidzga), and Kuniashta (Ghalidzga-Mokvi watershed) are distinguished by their orography and hypsometry among the subranges. These ranges are higher than 2000 m. The 10–12 km long southern subranges are well depicted in the relief. The southern subranges are relatively short (7–8 km). The nival-glacial and erosive processes together with the tectonics are important in the formation of these landforms.

The morphographic and morphometric peculiarities of the relief of the Kodori Range are conditioned by geological structures. The relief is mainly built of the Middle Jurassic suites, namely, the Bajocian porphyritic series (limestone-alkaline, basalts, andesite and rhyolithic lavas, lava breccias and pyroclastolites). The northern slope of the Kodori Range (in the Bramba River basin and to the west) is built of the Lower Jurassic loose sediments. The flysch alternation of sandstones, clay slates, and argillites is common here. The lower parts of the southern slopes are represented by a narrow strip of Cretaceous limestones.

The orography of the Kodori Range corresponds to the morphostructure of the relief; it is a semi-inverse range and is separated from the Greater Caucasus Range by the Kalmari-Khida lowered part (2200–2600 m).

The tectonic and erosive processes are important in the formation of the morphosculptural forms of the relief of the Kodori Range, and the nival-glacial processes in the formation of the crest of the range. With the excaration action

Fig. 8.2 Kodori Range (*photo by* L. Tielidze)

above 2200 m, the Modern and Late Pleistocene glaciers created the following glacial landforms: corries, troughs, ledges, and moraines. Modern glaciers are represented only on the northern slope of Khojali, the highest peak of the mountain range reaching into the nival zone. Despite the relatively low hypsometry, the formation of small glaciers is related to abundant atmospheric precipitation and old glacial forms well preserved in porphyrites. Here are nine glaciers, four in the Bramba basin, one in the Larikvakva River basin and four in the Darchi-Ormeleti River basin. The glaciers are distributed in the central sector of the Kodori Range and on the slopes of the peaks of Khojali and Okro. All glaciers are of cirque type and are distinguished by small size. Their exposition is northward, northwestward, northeastward, and eastward.

A special role in modeling the relief of the Kodori Range belongs to the action of Late Pleistocene and Holocene glaciers. Würm glaciers were well developed here. The glacial cirques, troughs, ledges, horns, and moraines are found in the relief. The glaciers descended from the Kodori Range fed the rivers of Bramba (The Kodori River tributary), Darchi-Ormeleti, Larikvakva (the Enguri River tributaries), Ghalidzga and Mokvi. Large glaciers were developed in the vicinities of the peak of Khojali. The peak itself was located in the glacial zone and glaciers of different exposition were fed from the glacial cap developed in its massive. The largest glaciers were in the gorges of Bramba, Darchi-Ormeleti, and Larikvakva. The length of the glaciers reached 6–8 km and ended at 1600–1800 m (Tielidze 2017c). In the old corrie forms created by the excaration action of glaciers, various size lakes are developed (Aduedaadzishi, Didi Okrostskali, and Patara Okrostskali).

Erosion is the main factor of morphogenesis in the Kodori Range. Both slopes of the mountain range are heavily dissected by the erosive effect of rivers. The gorges are symmetrical, mostly V-shaped. In the upper part of the ravines, troughs modified by erosion are widespread. The depth of the gorges is 500–1000 m; sometimes they reach 1000–1500 m in the gorges of Darchi-Ormeleti and Larikvakva. The gorges are characterized by narrow bottoms. In some gorges, young terraces are developed. In the southern part of the Kodori Range, which is built of limestones, the river gorges are narrow and canyon-shaped. The relief, built by the limestones, is fragmented in separate karst massifs by rivers.

8.11 Okhachkue Limestone Massif

Okhachkue is a medium-sized limestone massif, with average height 1600–1800 m; the highest point is Mt. Okhachkue at 2156 m. The surface of the massif is inclined from the north to the south.

Significant transformation of the massif's surface was caused by erosion processes. The fragments of the old river bed are preserved in the relief on the eastern slope faced toward the Enguri River gorge. Intensive karst processes were developed in the limestone rocks after washing away the Tertiary clay-sandy layers that were promoted by tectonic movements. The unforested crests and densely forested slopes are covered with sinkholes. The sinkholes, wells, deep abysses, and cracks are developed in the limestones of several hundred meters thick.

Precipitation flows out in the form of springs in the peripheries of the massive at a height of 250–300 m. The karst forms are particularly well represented in the relief built from Urgonian limestones. The depth of the sinkholes ranges from 5–10 to 50–60 m. In the Okhachkue massif, the polje type closed depressions stretched longitudinally are common. Out of them, the Didi Opuje polje is formed in the 2.0–2.5 km long synclinal depression by merging karst sinkholes. From the karst wells, the Shusheli, Jakonia and Okhachkue freezers attract attention.

References

Astakhov NE (1973) Structural geomorphology of Georgia. P/H "Metsniereba", Tbilisi (in Russian)

Gamkrelidze PD (1966) The main features of the tectonic structure of the Caucasus. Geotectonics, № 8

Geology of the USSR (1964) Vol. X. Georgian SSR. Part I.—M. Nedra, Moscow (in Russian)

Gobejishvili R (2011) Relief of Georgia. Monograph, Publishing House "Universali", Tbilisi (in Georgian)

Gobejishvili R, Lomidze N, Tielidze L (2011) Late Pleistocene (Wurmian) glaciations of the Caucasus. In: Ehlers J, Gibbard PL, Hughes PD (eds) Quaternary glaciations: extent and chronology. Elsevier, Amsterdam, pp 141–147. https://doi.org/10.1016/b978-0-444-53447-7.00012-x

Kipiani Sh (1974) Karst of Georgia (attempt of geomorphological characterization). Monograph, Tbilisi (in Georgian)

Maruashvili L (1971) Edit. Geomorphology of Georgia. Pub. House "Metsniereba", Tbilisi (in Russian)

Tatashidze Z (1988) Bzipi massif karst and caves. Publishing house "Metsniereba", Tbilisi (in Georgian)

Tatashidze Z (1997) The dynamics of the karst caves relief of Georgia and their slopes evolution. Tbilisi

Tielidze LG (2016) Glacier change over the last century, Caucasus Mountains, Georgia, observed from old topographical maps, Landsat and ASTER satellite imagery. Cryosphere 10:713–725. https://doi.org/10.5194/tc-10713-2016

Tielidze LG (2017a) Late Pleistocene and Holocene Glacier Extent in the Georgian Caucasus. Open J Geol. https://doi.org/10.4236/ojg.2017.74036,7,517-532

Tielidze L (2017b) The morphological types, exposition, snow, and firn line location of the glaciers of Georgia. In: Glaciers of Georgia. Geography of the physical environment. Springer. https://doi.org/10.1007/978-3-319-50571-8-4

Tielidze L (2017c) Late Pleistocene and Holocene glaciation. In: Glaciers of Georgia. Geography of the physical environment. Springer. https://doi.org/10.1007/978-3-319-50571-8-6

Tielidze LG, Gadrani L, Tsitsagi M, Chikhradze N (2015) Glaciers Dynamics Over the last One Century in the Kodori River Basin, Caucasus Mountains, Georgia, Abkhazeti. Am J Environ Prot. Special Issue: Applied Ecology: Problems, Innovations. 4(3-1):22–28. https://doi.org/10.11648/j.ajep.s.2015040301.14

Tsikarishvili K, Barjadze Sh, Kvavadze E, Bolashvili N, Djanashvili R, Martkoplishvili I (2010) Speleology of Georgia: aspects of its current situation and perspectives. Cave Karst Sci 37(3):73–78

Ramin Gobejishvili (1941–2014) was a Doctor of Geographical Sciences, a geomorphologist–glaciologist. Until his last days, he was Professor of the Faculty of Exact and Natural Sciences, Ivane Javakhishvili Tbilisi State University and the chair of Geomorphology and Cartography. During 1996–2008, he was the president of the Geographical Society of Georgia, and in 2005–2007—the director of the Institute of Geography. He was the author and coauthor of about 150 scientific works and ten monographs.

Levan Tielidze is a Ph.D. of Ivane Javakhishvili Tbilisi State University in the field of geomorphology–glaciology–cartography and a senior research scientist at Vakhushti Bagrationi Institute of Geography. He is also a researcher-scholarship holder of the Department of Earth Sciences of the Georgian National Academy of Sciences. The field of his research is mountain cartography and glacial-geomorphological study of the mountainous areas in the Quaternary (Late Pleistocene and Holocene). He is a member of several international organizations. In 2014/2015, he conducted a research stay at the Climate Change Institute of the University of Maine, USA, and in 2015/2016–2017 at the University of Northern British Columbia, Canada. He is the author of about 40 scientific papers, maps and eight monographs.

Nikoloz Astakhov (1911–1993) was a Doctor of Geographical Sciences, Professor (1965). He was researcher at the Department of Geomorphology of the Vakhushti Bagrationi Institute of Geography. His research area was structural geomorphology.

9 Central Greater Caucasus

Levan Tielidze, Ramin Gobejishvili, Levan Maruashvili and Nikoloz Astakhov

Abstract

This chapter provides the geomorphological characteristics of the central Greater Caucasus sub-zone. Orographic units are reviewed in detail, such as: the central Greater Caucasus main watershed range, the Enguri River and its tributaries, Egrisi (Samegrelo) and Svaneti ranges, the Tskhenistskali River and its tributaries, Lechkhumi Range, the Rioni River and its tributaries, Shoda-Kedela and Racha ranges, the Didi Liakhvi and Patara Liakhvi river gorges, Germukhi and Kharuli ranges. The description of the limestone massifs (Tsulishi, Kvira, Gauchi, Migaria, Askhi, and Khvamli) of the central Greater Caucasus is presented as well. The text also provides a geological structure, tectonic conditions and current exogenic processes of the research region. The extents of modern and Late Pleistocene glaciation of the central Greater Caucasus is also considered. Each subchapter is accompanied by appropriate images and maps.

Keywords

Central Greater Caucasus • The gorges of the Enguri River and its tributaries • Egrisi (Samegrelo) Range • Svaneti Range
The gorges of the Tskhenistskali River and its tributaries • Lechkhumi Range • The gorges of the Rioni River and its tributaries • Shoda-Kedela Range • Racha Range
Didi liakhvi and Patara Liakhvi river gorges
Germukhi Range • Kharuli Range • Tsulishi limestone massif • Kvira limestone massif
Gauchi limestone massif • Migaria limestone massif • Askhi limestone massif • Khvamli limestone massif

9.1 Study Area

The highest middle section of the Greater Caucasus Range is called the central Greater Caucasus. Conventionally, its boundary stretches

L. Tielidze (✉) · R. Gobejishvili · L. Maruashvili · N. Astakhov
Department of Geomorphology and Geoecology, Vakhushti Bagrationi Institute of Geography, Ivane Javakhishvili Tbilisi State University, Tbilisi, Georgia
e-mail: levan.tielidze@tsu.ge

L. Tielidze
Department of Earth Sciences, Georgian National Academy of Sciences, Tbilisi, Georgia

R. Gobejishvili
Department of Geography, Faculty of Exact and Natural Sciences, Ivane Javakhishvili Tbilisi State University, Tbilisi, Georgia

L. Tielidze (ed.), *Geomorphology of Georgia*, Geography of the Physical Environment,
https://doi.org/10.1007/978-3-319-77764-1_9

along the meridian of the peaks of Elbrus (5642 m) and Mkinvartsveri (Kazbegi) (5047 m). But Maruashvili (1981) highlights this boundary more precisely: the western border of the central Greater Caucasus passes through the watersheds of the Enguri and Kodori Rivers, and the eastern border passes through the bottom of the Tergi, Bidara and Aragvi river gorges. Thus, according to Maruashvili (1981), the central Greater Caucasus extends from the Dalari Pass (3291 m) to the Jvari Pass (2379 m).

From our point of view, such a division of the Greater Caucasus Range is partially acceptable within Georgia; we consider the conventional boundary between the western and central Greater Caucasus. As for the border of the central and eastern Greater Caucasus, we think that there are some shortcomings here, namely: if we draw the border according to Maruashvili (1981), it means that the gorges of the rivers of Tetri Aragvi, Bidara and Tergi will be divided and one of their slopes will get in the central Greater Caucasus zone and the other in the eastern Greater Caucasus zone. In addition, according to the same division the heads and upper stream (Truso cavernous) of the Tergi River will get in the central Greater Caucasus zone, while the rest in the eastern Greater Caucasus zone. The same problem occurs in the case of the glaciers in the Tergi basin, when one part of the glaciers get in the central Greater Caucasus zone, and the second part in the eastern Greater Caucasus zone. Consequently, from our point of view, drawing a border between the central and eastern Greater Caucasus along the bottoms of the rivers gorges is not justified.

We define Mount Vatsisparsi (3573 m) as a conventional boundary for the central and eastern Greater Caucasus, which is located in the watershed of the rivers of Tergi, Didi Liakhvi and Ardon (northern Caucasus). From Mount Vatsisparsi to Mount Shaviklde (3578 m) the northern slopes of the Greater Caucasus with its sub ranges and river gorges lie within the boundaries of Georgia. The conditional border from Vatsisparsi goes first along the main watershed range in the southeast, then to the south and passes through the Laghati Range, which at the same time crosses the Keli volcanic upland and is a watershed of the Didi Liakhvi and Ksani-Aragvi river basins. After that, the border passes through the crest of the Kharuli Range (Fig. 9.1).

The central Greater Caucasus is higher than the western and eastern Greater Caucasus and more complicated morphologically and geologically. The highest peak of Georgia, Mount Eastern Shkhara (5203 m) is located there. It should be noted that the heights of the four peaks in the Shakhara massif exceed 5000 m, and five more peaks in the "Pitsrula" (The Bezingi Wall). Tectonics, glaciers impact, erosion, weathering, gravitational processes have played a major role in the formation of the landforms of the central Greater Caucasus. In the easternmost part the role of volcanoes is great, and in the limestone regions karst processes are developed.

The orography of the relief of the central Greater Caucasus is very complex and interesting. Tectonic-stratigraphic feature of the southern slope of the Greater Caucasus is reflected in the morphostructural nature of the relief. The basic orographic units, ranges and gorges are sublatitudinal and parallel to the general direction of the Greater Caucasus Range. The gorges have a sub meridional direction while crossing the main structures and ranges. The gorges are narrow and often of canyon type. In the terrain constructed of carbonate and non-carbonate flysch, the gorges are of sublatitudinal direction and wide. They are linked to each other in the heads with lower passes (Kalmari-Khida, Zagari, Vatsistsveri, Mamisoni, Ertso, Truso and Jvari). From these ranges the following are notable: the eastern section of the Kodori Range, the Svaneti, Egrisi (Samegrelo), Lechkhumi, Racha, Shoda-Kedela, Germukhi, Kharuli and Lomisi ranges. At the same these ranges are high rate watersheds of river basins, namely, the Svaneti Range (Enguri-Tskhenistskali), Lechkhumi Range (Rioni-Tskhenistskali), Racha Range (Rioni-Kvirila) and Germukhi Range (Didi Liakhvi and Patara Liakhvi). From river gorges the Enguri, Khobistskali, Tekhuri, Tskhenistskali, Rioni, Liakhvi and Lekhura are notable.

The length of the central Greater Caucasus is about 220 km from the Dalari Pass to Mount

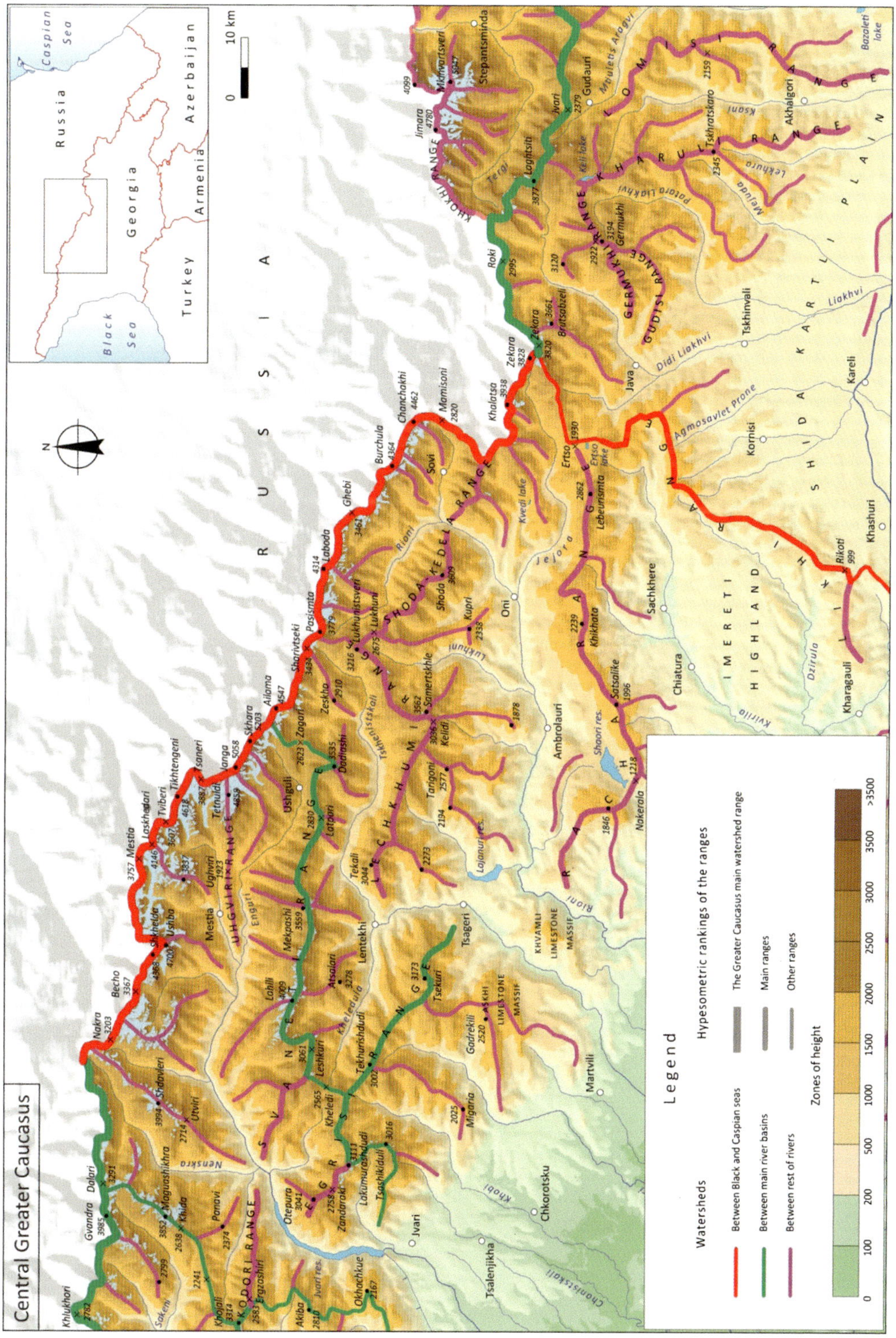

Fig. 9.1 The orography of the central Greater Caucasus

Vatsisparsi. Its southern slope is heavily dissected by erosion. River gorges are deeply inserted in the relief. The depth of the insertions is different and decreases towards the southern direction. Insertion depth in the upper streams of the main rivers is 2500–3000 m, in the middle streams from 1500 to 2500 m and in the limestone strip 500–1000 m.

The geological structure of the central Greater Caucasus is complex. Rocks of all ages are found from Proterozoic to Tertiary. The structural suites are located in parallel to each other and become younger from the axial strip to the peripheries. The Greater Caucasus is the anticlinorium, and therefore, such alignment of sediments is natural. The oldest rocks of Proterozoic and Paleozoic ages construct the main watershed; they are represented by plageogranites, gneisses, amphibolites, granites and granodiorites. The axial strip of the Svaneti Range and its extension in the Enguri and Tskhenistskali gorges (from Sasashi to Chuberi) are composed of Triassic and Devonian layers (quartzites, phyllites, sandstones, slates, gravelites, and conglomerates).

The central Greater Caucasus southern slopes are characterized by wide distribution of Jurassic deposits. The southern part of the main watershed is represented by Lower Jurassic layers, with a width of 5–10 km and constructed of soft rocks: clay slates, sandstones, gravelites, aleurolites and argillites. The deep fault line passes through the contact of these rocks and crystalline rocks. The fault line creates a sharp step in the relief (Gobejishvili 2011). The Svaneti Range, the Tskhenistskali, Rioni and Enguri gorges are built of the same suites. The soft and low relief is related to the rocks of this age.

The Middle Jurassic suites are represented by the Bajocian volcanogenic porphyritic series—andesites, basalts, tuffo-breccias, tuffo-conglomerates, tuffites and tuffo-sandstones. The Kodori, Egrisi (Samegrelo), Lechkhumi, Samertskle-Kupri ranges and the Enguri, Tskhenistskali and Rioni gorges (from Khojali to Liakhvi gorge) are built of the deposits of porphyritic series.

Upper Jurassic sediments (limestones, clay slates and sandstones) are spread in the form of a narrow strip in the contact of porphyritic series and Cretaceous sediments. Its sediments are close to the sediments of the Cretaceous carbonate flysch and are often considered together. A parallel zone (Mestia-Tianeti zone) close to the watershed range is built of the sediments of the Cretaceous age. It is represented by limestones, sandstones, clay slates, marls and clays (carbonate flysh). The periphery of the southern slope of the central Greater Caucasus is built of the same age sediments represented by massive limestones, layered limestone, marl limestones and marls. Karst forms are well developed in these sediments; the karst terrain is represented as separate massifs and ranges (Askhi, Migaria, Kvira, Khvamli and Racha Range). Tertiary sediments—the sandstones, clays, and clay sandstones (marine and continental molasses) are common in the Racha-Lechkhumi syncline and to the east as a narrow strip up to Java.

In the formation of the central Greater Caucasus landforms together with the exogenic processes a certain role is played by the lithological zonality of suites and neotectonics. The high mountainous terrain is constructed from crystalline and volcanic rocks and is formed by the active influence of nival-glacier processes throughout the Pleistocene. High mountainous terrain is characterized by the forms of different origins, such as corries, troughs, karlings, pointed peaks, gendarmes, and debris cones. Glacial relief is represented by modern glaciers and rock glaciers.

The glaciers are mostly located in the Racha-Svaneti Caucasus section. The glaciation centers are on the slopes of northern exposition of the Svaneti, Samegrelo, Shoda-Kedela, and Lechkhumi ranges, as well as in the Dvaleti Caucasus section. The large glaciers include Lekhziri, Northern and Southern Tsaneri, Chalaati, Adishi, Khalde, Kvishi, Kirtisho, Boko.

In the central Greater Caucasus relief the following regions can be distinguished according to their morphological and genetic characteristics: the main watershed range, the gorges of the Enguri River and its tributaries, Svaneti Range, Egrisi (Samegrelo) Range, the gorges of the Tskhenistskali River and its tributaries,

Lechkhumi Range, Shoda-Kedela Range, the gorges of the Rioni River and its tributaries, Racha Range, the gorges of the Liakhvi River and its tributaries, Germukhi and Kharuli ranges, the gorges of the rivers of Mejuda and Lekhura and karst massifs (Tsulishi, Kvira, Migaria, Askhi and Khvamli). The relief of the central Greater Caucasus will be described by these morphological units.

9.2 Central Greater Caucasus Watershed Range

The main watershed range is of sublatitudinal direction and reiterates the general direction of the Caucasus from northwest to southeast. From the geomorphological and geological point of view it is divided into two sections: (1) in the west, the axial Racha-Svaneti section of the main range constructed from old crystalline rocks with high hypsometric distribution and glaciers stretching from the Dalari Pass to the Mamisoni Pass; and (2) in the east, from the Mamisoni Pass to Mount Vatsisparsi. The eastern section is constructed from the sediments of Jurassic and Cretaceous ages. It is divided into the Racha (the eastern section of the Shoda-Kedela Range) and Dvaleti (from the Mount Zekari to the Mount Vatsisparsi) sections. Small glaciers are common here.

The Racha-Svaneti's Caucasus Range is the highest in morphometry. The Eastern Shkhara (5203 m), the highest peak of Georgia is located here (Fig. 9.2). It is noteworthy that several peaks in the Shkhara massif are over 5000 m, these are as follows: the Western Shkhara 5068 m, the Central Shkhara I 5045 m, II 5106 m and III 5158 m. In general, the crest of the middle and east sections of the Shkhara massif is over 5000 m. There are several other peaks in the "Pitsrula" (Bezingi Wall) higher than 5000 m (Jangha, Pushkin Peak, etc.). The passes in the Svaneti-Racha section are located over 3000 m asl. One can cross them mostly on foot. In the historical past only some of them were used for cattle transfer to northern Caucasus (e.g. Nakra, Mestia, Ghebi, Gurdzievtseki, Geze).

The eastern section of the central Greater Caucasus watershed range is much lower. No

Fig. 9.2 "Pitsrula" (Bezingi Wall) and Shkhara (5203 m), the highest peak of Georgia (*photo by* L. Tielidze)

peak reaches 4000 m here. In the Dvaleti section, the Roki tunnel is constructed (the only tunnel in the Greater Caucasus Range), that connects the northern and southern Caucasus.

The relief of the Svaneti-Racha section is mainly constructed from Proterozoic and Lower Paleozoic plagiogranites, plagiogneisses, quartz diorites and crystalline slates. Lower Jurassic clay slates, schists, sandstones, aleurolites, gravelites, basal conglomerates, and quartzites stretch along the intrusives as a narrow strip in the south.

The watershed range of the central Greater Caucasus within the borders of eastern Georgia is built of Upper Jurassic and Lower Cretaceous carbonate flysch suites. Here are distinguished the Lower Cretaceous sandstones, limestones, marls, clay slates, argillites, dolomites and Upper Jurassic pelagic marls, argillites, clustic limestones, aleurolites and clay slates (Geology of USSR 1964).

The sub-ranges descend from the watershed range of the Greater Caucasus in a southern direction, out of which notable are the Kharikhra, Shdavleri, Tsalgmili, Bali, Dalakora, Gvaldi, Lakchkhildi, Ukreni, Lamaria, Atkveri, Kareta, Vatsistsveri, Bodurashi, Mkhvreleti, Molisi, Gormaghali, Leta and Java subranges. Through the subranges the main range is connected to the Kodori, Svaneti, Lechkhumi and Racha lateral ranges, and is separated from them by lower passes (Khida-Kalmari, Zagari, Vatsistsveri and Ertso).

The morphological look of the watershed range of the central Greater Caucasus is formed as a result of modern snow-glaciers influence, intense physical weathering, excaration action of Late Pleistocene glaciers and river erosion.

In the glacial-nival zone, the coordination of processes leads to the formation of bare, rocky, sharp, strongly inclined and jugged karling terrain. Notable pointed peaks include Chanchakhi, Ushba, Shkhelda, and Dalari. The high, rocky, and pyramid peaks are Shdavleri, Tetnuldi, Shkhara, Jangha, Tikhtengen, Jimara, Khalatsa, Burchula, and Zeskho. Among the karlings notable are: Lakchkhildari, Dalakora, Lekhziri, Tskhvandiri, Dalari, Baki, Kareta, and Dombura. In case of different lithological conditions, the stepped terrain, which is the area of action of the nival processes, and the debris cones, rockfills and stone flows are developed at the foot of the step. The relief of the Greater Caucasus represents the total denudation zone above 3000 m.

The geological structure of the western and eastern sections of the central Greater Caucasus contributes to the formation of different meso and microforms. The morphologically well-formed landforms on the Racha-Svaneti section of the central Greater Caucasus such as the steep slopes, karlings, large-fractured debris cones, corrie forms, ledges and horns are related to the distribution of crystalline rocks. As for the relatively soft forms—the inclined slopes covered with fine deluvial trails, deformed glacial cirques, smoothed crests—are related to the distribution of sedimentary rocks of Mesozoic age.

Glaciers are widespread in the watershed range of the central Greater Caucasus. Modern glaciers are formed in the nival-glacial zone (over 3000 m) and their tongues end in the alpine zone. The glacial zone is located above 4200 m and is only expressed in the Racha-Svaneti section. The firn line rises from west to east and extends to a height of 3200–3500 m. Among the morphological type of glaciers, the valley, and corrie types of glaciers are notable. Smaller corrie-type glaciers prevail according to the number, while the valley glaciers have greater area.

The firn line ascended by 150–200 m in the nineteenth and twentieth centuries. The temperature increased by 0.1–1.5 °C, followed by the altitudinal rising of natural boundaries (Gobejishvili 2011; Tielidze 2017a).

The valley glaciers are only widespread in the Racha-Svaneti section of the central Greater Caucasus (Fig. 9.3). The glaciers of Georgia have retreated since the LIA maximum (1840s) up to present due to climate global warming. The glaciers' area has decreased by 30–40%, while the glaciers' tongues decreased in length by 1.0–2.5 km. The valley glaciers are characterized with greater parameters of reduction (Tielidze 2017b).

During the twentieth century, the complex valley glaciers disintegrated into simple glaciers, such as the Tviberi, Lekhziri, Kvishi, Ushba, and

Fig. 9.3 Tbilisa Glacier at the Racha Caucasus section (*photo by* L. Tielidze)

Tsaneri glaciers in the Enguri River basin, as well as the Zopkhito-Laboda glaciers in the Rioni basin. The morphology of a complex valley glacier is still maintained by the Chalaati and Buba glaciers.

The table below shows the parameters and variability of some larger glaciers of the central Greater Caucasus in the last half a century (Tielidze and Wheate 2018), (Table 9.1).

The main role in the formation of the landforms of the watershed of the central Greater Caucasus and its subranges belongs to the old glaciation. During half a million years, especially in the Late Pleistocene, the terrain morphology of the axial range was formed under the influence of glaciers (Gobejishvili 2011). The following glacial forms are well represented in the relief: corries, ledges, troughs, and moraines. Rock lithology identifies the formation and maintenance of glacial forms. Old glacial forms are particularly well preserved in the Racha-Svaneti section built of granite and crystal rocks. The old glacial forms are poorly preserved due to active impact of erosive and denudative processes on the central and eastern Greater Caucasus watershed range built in relatively soft rocks.

In the Late Pleistocene, in the Racha-Svaneti Caucasus the complex valley type of glaciers were developed. They descended at the lower heights and completely occupied the river gorges. Some glaciers entered the main river gorges, such as Enguri, Rioni, and Tskhenistskali and blocked them, as indicated by the distribution of lacustrine sediments (Mulakhi, Tskhumari, and Ipari). Such glaciers were developed in the gorges of the Enguri River tributaries—Dolra, Mulkhura, Khalde and Shkhara, and Rioni River tributaries—Zopkhito, Chveshura and Chanchakhi (Buba-Boko). In the Racha-Svaneti Caucasus the glaciers descended to the lowest heights in the Nenskira River gorge (650 m asl) (Gobejishvili et al. 2011; Tielidze 2017c).

Nearly all the gorges have been traced by the action of the Late Pleistocene glaciers. Trough gorges with wider bottoms are well preserved in the gorges of the rivers of Nenskra, Dolra,

Table 9.1 Central Greater Caucasus valley glaciers variability over the last half a century

Name	River basin	Area (km^2), 1960	Area (km^2), 1986	Area (km^2), 2014	Terminus retreat 1960–1986		Terminus retreat 1986–2014		Terminus elevation (m), 2014
					m	m $year^{-1}$	m	m $year^{-1}$	
Lekhziri	Enguri	35.80 ± 0.45	33.95 ± 0.47	23.76 ± 0.36	859	33.0	736	26.3	2345
Tsaneri southern[a]	Enguri	28.26 ± 0.26	14.38 ± 0.16	12.31 ± 0.16	448	17.2	781	27.9	2525
Tsaneri northern	Enguri	0[b]	13.30 ± 0.11	11.28 ± 0.11	0	0	574	20.5	3000
Kvitlodi	Enguri	12.23 ± 0.13	11.65 ± 0.12	9.58 ± 0.1	598	23.0	883	31.5	2580
Adishi	Enguri	10.48 ± 0.11	10.34 ± 0.1	9.58 ± 0.1	124	4.8	390	13.9	2480
Chalaati	Enguri	12.71 ± 0.18	12.36 ± 0.19	9.24 ± 0.14	460	17.7	223	8.0	1960
Khalde	Enguri	11.87 ± 0.19	10.65 ± 0.18	8.59 ± 0.13	130	5.0	130	4.6	2545

[a, b]Until the 1980s the southern and northern Tsaneris were merged as one compound-valley type glacier. Their division likely happened in 1980–1985

Mulkhura, Adishura, Khaldechala, Enguri head, Zopkhitura, Chveshura, Chanchakhi, Notsarula, Gharula, Jejora, Jomagi and Keshelta. In these gorges, on the slopes, the lateral moraines (often in two lines) of Würm glaciations are preserved. Especially noteworthy are the Zemo Marghi moraine hillocks and a moraine hillock in the Bali Range in the Dolra River basin. Würm moraines are more visible in the Mulkhura River basin.

A 6–7 km long moraine hill is located on the left side of the gorge, on the northern slope of the Zagari Range (Fig. 9.4), while the moraine hills are relatively smaller in the Mestiachala gorge, at a height of 1900 m, in the site of Gvalda and above the opposite Laghami Village. At this time the Lekhziri Glacier was ~3 km wide and ~350–400 m thick. Mulkhura and Lekhziri glaciers occupied the area of the Latali Village with their joint tongue, indicators of which are the lateral and ground moraines present there. During the Würm age, the Mulkhura Glacier flowed over the Zagari Range and entered the Enguri gorge. An indicator is the Ipari erratic boulder (a tower is built on it) and the moraine hills in the area of the upland. The Mulkhura Glacier flowed over the Ughviri Pass even during the Lower Holocene stadial glaciations.

The Würm moraine hills are well represented in the gorges of the rivers of Zopkhito (the site of Khopito), Chveshura (the site of Jojokheta hill), Gharula (the Kvazha hill), Jejora and Jomagi. The whole spectrum of phase and stadial moraines of Late Pleistocene and Holocene ages is represented in the Bubistskali River gorge.

The large size (over 1000 m^3) erratic boulders brought from the Greater Caucasus watershed range prove the distribution of old glaciers, among them can be noted the blocks of Mestiachala (Fig. 9.5), Mulkhura (Latali), Ipari, Adishi, Khalde, Ushguli, Ghebi, Gona, Glola and Saglolo. Separately can be noted the Oni plageogranite boulders (Fig. 9.6), which were brought as a result of instant pulsation of the Buba-Boko Glacier, during the eruption of the Tsitela volcano (Gobejishvili 2011).

Fluvial (water) processes were active in the formation of the relief of the watershed of the central Greater Caucasus along with the nival-glacial processes. The erosive partition of the relief is especially observed below 2500 m. Along with the glaciers, the underglacier meltwaters (the existence of alluvial material in front of the modern glaciers is a result of the underglacier water activity) were important in the formation of the trough gorges. The erosive

Fig. 9.4 Late Pleistocene moraine hill on the left side of the Mulkhura River gorge on the northern slope of the Zagari Range (*photo by* L. Tielidze)

Fig. 9.5 Oni plageogranite erratic boulder (*photo by* R. Gobejishvili)

action of the waters has been activated after the Würm glaciations (Holocene). The erosive action resulted in the dissecting of slopes on the one hand and overdeepening of the bases of trough gorges on the other. At this time, the 20–50 m deep ravines (Tita, Latali, Mestia, Gona, Kvishi) or narrow V-shaped gorges (Buba, Mulkhura (Mestia-Latali), Adishura, Khalde, Ushguli and Dolra) were developed in the antecedent areas built of solid rocks. In the gorge the river terraces are expressed by floodplain and upper floodplain two or three alluvial steps.

Fig. 9.6 Mestiachala erratic boulder (*photo by* L. Tielidze)

9.3 Gorges of the Enguri River and its Tributaries

The Enguri River basin is located in the western part of the southern slope of the central Greater Caucasus. It is bordered by the watershed range of the Greater Caucasus in the north, by the Svaneti-Egrisi ranges in the south, by the Kharikhra subrange in the west and by the Atkveri subrange in the east. The Zagari (2623 m) and Kalmari (2240 m) passes are located in these subranges, through which pass the roads to the Tskhenistskali and Kodori gorges. Within the Enguri River gorge, the relief is constructed from Cretaceous, Jurassic, Carboniferous and Paleozoic formations. We will talk about them during the discussion of individual sections. In the area above the Khaishi Village the Enguri River basin is often called as Zemo Svaneti depression. Below we will consider the Enguri gorge within certain hypsometric frames, within several hundreds of meters from the bottom of the gorge.

The Enguri gorge is divided into several sections in morphological and morphometric terms: from the head to Ushguli Village, from Ushguli to Latali Village, from Latali to Lakhamula Village and from Lakhamula to Jvari Town.

The Enguri River starts from the Shkhara Glacier. Up to Ushguli the gorge is a trough. In the right tributaries the glacier layers are widespread. During Würm the substream of the Khalde Glacier moved to the Enguri gorge. In the Ushguli neighborhood, on the left side, the two moraine rows of Late Pleistocene age are well expressed; the lower moraine ends at the village of Murkmeli at 1900 m asl. The Kvishara gorge of erosive origin, the left tributary (Dadiashi gorge) of which is of trough, conflows the Enguri River at Ushguli.

The Enguri gorge is erosive from Ushguli to Latali. The morphology of the erosive gorge modifies in Kala and Ipari neighborhood, which is the result of the Würm glacier influence. The gorge is developed in the Middle Jurassic sediments: clay and sandy slates, argillites, sandstones and aleurolites. The gorge extends towards the structure and its morphological look is uniform, a deep erosive gorge with symmetrical slopes.

Due to erosion processes and the nature of structural rocks, in some areas (Ushguli-Kala and Ipari-Latali sections) the profile of the gorge is curved. In the Ipari-Latali section, the Svaneti Range subranges and gorges morphologically descend to the bottom of the valley in the form of an echelon. Below Ipari, the Enguri washes the right slope, due to which the gorge is partially asymmetric in this section. The right slope is hypsometrically lower and is characterized by short and soft forms and is less fragmented. And the left slope is dissected by rivers, which are developed on the northern slope of the Svaneti Range. The relative height is 2000 m. The erosive-accumulative processes are being developed alternately in the bed. The Enguri gorge is symmetrical at a length of 10–12 km of Kala-Ipari, the depth reaches 1000–1200 m, the foot of the gorge is relatively wide and complicated with debris cones. Other accumulative forms are presented as lower terraces. The terraces are relatively well represented in the Ipari and Kala areas. The first terrace of 2–5 m high is developed on both sides of the river in Ipari and the Bogreshi Village is located there. The 25–30 m high terrace is developed on the left side, with a length of 0.8–1.0 km, and with a width of 100–150 m. The Nakipari Village is located there. The III (180–200 m) and IV (300–320 m) terraces are occupied by the villages of Zegani and Tsvirmi. On the edges of these terraces the lateral moraines of Würm period are located towards the gorge. In the Late Pleistocene, one of the subranges of the Mulkhura Glacier moved to the crest of the Zagari Range in the areas of Ughviri and extended in the territory of Ipari in the Enguri River gorge (Maruashvili and Khazaradze 1981; Tielidze 2016). The expansion of the gorge in the vicinity of Ipari is a result of the glacier activity. The glacier was spread in the Enguri gorge, both downward the gorge and upwards of it at 1 km. Here, the Ipari Tower is built on the erratic boulder. The Adishura (Adishchala) and Khaldechala rivers conflow into the Enguri River from the right side near Ipari. Both gorges are trough shaped. Adishura maintains the trough form up to the Adishi Village; below Adishi the trough is modified and is of erosive-tectonic nature mostly, with the development of mudflows on the left side. The I and II terraces were created in the Adishura gorge in Early and Late Holocene. In the vicinity of the Ughviri Pass, in the direction of Ipari, except for the lateral moraines of Würm period the Holocene stadial moraines are also well represented. There was a powerful glacier in the Khalde gorge during the Würm period, the tongue of which moved to the Enguri gorge, in the territory of Kala. With the influence of this glacier tongue and fluvio-glacial processes, the vast terraced terrain was created. The erratic boulder (Perkhuli Stone) is located in the heads of Khalde.

The Enguri gorge is expanded from Latali to Lakhamula; along with the erosion the glaciers are important in the forming of gorge morphology. In this section three large tributaries conflow into the Enguri River: Mulkhura and Dolra from the right side and the relatively small Lailachala River from the left side. From all three gorges, the powerful glacial tongues descended in the Enguri gorge in the Würm and the trace of their impact is highly visible (Gobejishvili 1995). Here the fluvio-glacial terraces are developed with the communities of Svaneti: Becho, Etseri, Latali, Tskhumari, and Pari on their surfaces.

At the Latali Village the Mulkhura River conflows into the Enguri River from the right side at a height of 1210 m. The Mulkhura gorge is of sublatitudinal direction in the Mestia-Latali section and it is latitudinal at Mulakhi community. Morphologically, Mulkhura is a trough. At Mestia it is conflowed by the Mestiachala River, the gorge of which is a trough. The bottom of the trough is covered by debris cones. With the erosive effect of the river, the bottom of the trough is cut at a depth of 50–60 m and the depth and side erosions are underway in its modern bed. In the entrance of Mestia, on the right steep slope of 40 m of Mulkhura, the alternation of alluvial and deluvial sediments are observed (Fig. 9.7). The traces of impact of Late Pleistocene and Holocene

Fig. 9.7 Alluvial and deluvial sediments in the Mulkhura River gorge (*photo by* L. Tielidze)

glaciers in the Mulkhura River gorge are well preserved in the whole valley as lateral moraines. Powerful moraine deposits are particularly notable in the Mestiachala gorge, at the top of the airport; a 6–7 km long moraine hill is important on the northern slope of the Zagari Range. This hill is morphologically more or less prolonged until the Ughviri Pass, joins the Ughviri moraines and stretches up to the Tsvirmi Village. In the territory of Ughviri, the morphology of moraines indicates overflow of the Tviberi Glacier towards the Enguri River gorge not only in Late Pleistocene, but in the Holocene (several times). Due to the glacial morphology of the gorges, only the terraces of the Holocene age are developed, especially the two Late Holocene terraces (the area of the airport and the Mulakhi community). In the Mestiachala River gorge large glacial boulders are common. Distribution of these boulders is related to the stages of glaciation in Late Holocene. In Mesti, the rivers of Mulkhura and Mestiachala develop narrow and deep gorges.

The powerful debris cones have been created by presently active mudflows related to the carbonate flysch sediments. Particularly important are the gorges of Pushkheuri, Lekhtagi, Laghami, Lanchvali, Kherazhi, Cholashi, Zhabeshi, and Tskheki. The Lavradashiskhevi is especially active, at the head of which there is a powerful landslide; its debris cone consists of three steps, of Late Holocene age. Landslides are widespread in the Mulakhi community and snow avalanches are active everywhere.

The Dolra River gorge is a trough for its whole length. In Würm, the tongue of the Dolra Glacier descended in the Enguri River gorge and spread both up the gorge (up to 1.0 km) and downward at a distance of 2–3 km. The Dolra Glacier deposits (of Würm and Holocene) are well preserved on the western slope and crest of the Bali Range as moraine hills. The downstream of the Dolra River gorge is constructed from carbonate flysch (of Jurassic-Cretaceous). The upperstream of the gorge is deeply cut in the

Fig. 9.8 Shtugra Waterfall (*photo by* L. Tielidze)

crystalline rocks. In the gorge the 30 m high waterfall of Shtugra is notable (Fig. 9.8). In the bottom of the Dolra gorge the Mazeri ravine develops via the 3 km long and 1 km width powerful debris cone, while terraces are weakly expressed. The left tributaries of the Enguri River are either trough or erosive. The powerful Lailachala valley glacier descended from the Svaneti Range into the Enguri River bed, creating a valley trough with glacial forms.

The Enguri gorge is erosive from Lakhamula to Jvari. The depth of the river cut is 1500–2000 m. The gorge is V-shaped with steep slopes and the gorge acquires a narrow canyon shape. The drop in the bed ranges between 20–40 m/km. The river sometimes entirely occupies the bottom of the gorge. Creating such anomalous sections is conditioned by the geological structures and their intensive uplifting. In these sections the river of Enguri and its tributaries transversally cross the structures and create antecedent gorges. In some sections (Khaishi, Idliani and Jorkvali) the gorge is slightly widened and the river terraces are highlighted. There are three terraces at Khaishi: I—3–5 m, II—25–30 m and III—80–90 m. Among them, the I-II terraces are of Holocene age, and the third is of Pleistocene age.

The lithology of structural rocks is important in the formation of the morphology of the Enguri River gorge in the Jvari-Lakhamula section. The gorges are narrow when crossing the Triassic-Carboniferous system sediments (silicites, flint slates, andesite and dacite pyroclastolites and marble lenses), and Bajocian (porphyritic series) and Cretaceous sediments. Due to boulder material in the bed, rapids are being developed. The gorges are expanded and V-shaped in the relief built of Low and Middle Jurassic soft suites. Jurassic granitoides outcropped at several places in the gorge above the Khaishi Village, where the gorge is narrow and creates rock-gates.

The rivers of Nakra, Nenskra and Larikvakva conflow into the Enguri River in the right side in the Lakhamula-Jvari section and the rivers of Khumpreri and Khaishura conflow Enguri in the left side. Downstreams of the Nakra and

Khumpreri gorges are erosive, while the middle and upperstreams are of glacial genesis; morphologically, they are modified troughs.

The Nenskra River conflows into the Enguri River at the Khaishi Village. It is the largest transversal gorge on the southern slope of the Greater Caucasus watershed range. The low location of the confluence and the neighboring location of high steep sloped ranges make the Nenskra gorge contrast and diverse. The main part of the gorge is constructed from the Paleozoic crystalline rocks; the gorge is expanded on the territory of the Chuberi community. The gorge is built of soft Jurassic suites and at the confluence; there is a narrow deep gorge in the Bajocian porphyritic suite. In the territory of the Chuberi community, in the bottom of the Nenskra gorge debris cones are developed (the gorges of Tetnashera and Gvashkhara). In Late Pleistocene, in the Nenskra gorge a ~36-km long subranged glacier was developed, which descended to 650 m, the lowest height in the central Greater Caucasus (Tielidze 2017c). The trough gorge is developed in the relief built of crystalline rocks. Two rows of Würm moraines (2–3 km long) are visible in the vicinities of the Zemo Marghi Village. The Nenska River develops a 20–25 m deep ravine in the bottom of the trough at the Tita Village.

The Enguri River gorge is occupied by the Jvari water reservoir below Idliani Village (Fig. 9.9). In general, the gorge is narrow with steep slopes. Below Jvari, the Enguri River flows through the hillock and low terrain of Kolkheti.

9.4 Egrisi (Samegrelo) Range

The semi-inverse Samegrelo Range is distinguished from the central Greater Caucasus folded system. It is of sublatitudinal direction and stretches from the Enguri gorge (from the Jvari water reservoir) to the Tskhenistskali gorge (Rtskhmeluri) (Fig. 9.10). The central section of the range is bent to the north due to the reverse erosive action of the Tekhuri and Khobistskali rivers. The central part of the range is a watershed of the Enguri (Khaishura), Tskhenistskali (Kheledula), Khobistskali, and Tekhuri rivers. The western part is a watershed of the basins of Khaishura and Magana—the left tributaries of the Enguri River, while the eastern section separates the basins Devashi and Jonoula—the right tributaries of the Enguri River. The highest peak in the west is Mount Lakumurash-Dudi (3111 m) and in the east, Mount Tsekuri (3174 m). The length of the range is ~60 km and the maximum width is ~30 km. The Samegrelo Range is connected to the Svaneti Range through the Kheleri subrange; the subranges of the southern direction are Jomarda, Jakhunabu, and Bechuna.

The relief of the Samegrelo Range is constructed from Middle Jurassic sediment and volcanogenic suites. The crest of the range and the south slope are constructed from the rocks of Bajocian porphyritic series—basalts and dacite lavas, lava breccias, tuffites, tuffo-conglomerates, sandstones, and clays. In the central and eastern sections the Bajocian suite is surrounded by Lower and Middle Jurassic age sandstones, aleurolites and clay slates. Each lithological complex is reflected in the morphology of the relief.

Despite the low elevation of the range, modern glaciers are spread on both slopes of the western section of the range. Smaller glaciers are located in the deep corrie forms. The existence of glaciers is determined by the morphology and slopes exposition of the terrain, and the favorable location of the range towards the Black Sea, which stipulates the abundance of atmospheric precipitation.

Together with the nival processes, the Pleistocene glaciation is important in the formation of a crestal part (over 2500 m) of the Samegrelo Range. On both slopes of the range are well-expressed corrie shapes, sharp peaks, karlings and modified troughs. The moraines are preserved in the terrain. The length of the Würm age glaciers reached ~4–6 km and they were valley type glaciers; their tongues descended to the height of ~1800–1900 m. The trace of the

Fig. 9.9 Jvari water reservoir in the Enguri River gorge (*photo by* L. Tielidze)

Fig. 9.10 Egrisi (Samegrelo) Range (*photo by* L. Tielidze)

old glaciers impact is also well preserved around Mount Tsekuri. In the glacial cirques the lakes formed; one of them is Tobavarchkhili Lake (Fig. 9.11).

The relief of the Samegrelo Range is dissected by the erosive action of the rivers of Magana, Khobistskali, Tekhuri, Jonoula, Kheledula, and Khaishura. The depth of the V-shaped gorges cut into porphyrites is 1000–1500 m, and the valleys developed in soft rocks are more widespread and characterized by gentle watersheds. In the relief built of the Cretaceous limestones, the 2–3 km long narrow gorges are developed. The main rivers cross the building structures perpendicularly, and many tributaries are coincided with the directions of structures and fault lines, such as

Fig. 9.11 Tobavarchkhili Lake formed in the glacial cirque (Google Earth imagery 22/11/2015)

the Kheledula River, Kasleti—the tributary of the Khaishura River, the heads of the Magana River and the tributaries of Tekhuri and Khobistskali on its extension.

9.5 Svaneti Range

The Svaneti Range, which is mainly of latitudinal direction, morphologically and hypsometrically is well expressed on the southern slope of the Greater Caucasus. There are sections where the direction of the range is sublatitudinal or submeridional. Its highest peak is Laila (Lahili) 4009 m (Fig. 9.12). The range stretches from the Enguri gorge (the Khaishi Village) to the Tskhenistskali gorge (Tsana Village). From the watershed range of the central Greater Caucasus it is separated by the Zagari (Atkveri) Pass. The range represents the watershed of the Enguri and Tskhenistskali river basins. The length of the Svaneti Range is ~95 km, and the width is 20–25 km (on the Lentekhi meridian). From the subranges of the Svaneti Range the following are notable: Bakildi, Tekrashdudi, and Mepkashi in the north, and Khelerdi, Atsalari, and Goldashi in the south.

According to the hypsometric characteristics and the features of glacier distribution, the Svaneti Range is divided into three parts (Ukleba 1983): the central (1) section is the highest and is located between the Leshnuri (3061 m) and Lasili (3086 m) passes. The highest peaks of the range are located here, such as Laila (4009 m), Gvadarashi (3750 m), Leshnili (3817 m) and Mepkashi (3576 m). The eastern (2) and western (3) sections are lower than the central. The highest peak in the eastern section is Dadiashi (3533 m), and in the western section Uskuri (3320 m). The lowest point in the Svaneti Range is the Latpari Pass (2825 m).

The Svaneti Range is built of the Devonian-Triassic and Mesozoic suites. The structural suites indirectly cross the range (sublatitudinal).

Fig. 9.12 Svaneti Range and the Mount Laila (Lahili) (*photo by* L. Tielidze)

The highest central section of the range is built of the oldest Devonian-Carboniferous-Triassic suites (gravelites, clays, sandstones, silicites, black clay slates, marble and limestone lenses, tuffites, phyllites and conglomerates) (Devdariani 1985). The western and eastern sections are built of Lower Jurassic suites (clay slates, schists, marls, limestones, etc.). In the west the Lower Jurassic sediment are covered by the Bajocian suites (porphyrites) from the top and in the east —by the Upper Jurassic carbonate flysch.

The direction and age of the structural rocks give the range anticlinal character, while the rivers of Enguri and Tskhenistskali antecedently cross the axis of the anticlinal structure.

Lithology, neotectonic movements and impact of various exogenic processes at different hypsometric levels of the structural rocks stipulate the morphological peculiarities of the range. The central section of the Svaneti Range with the Bakildi and Atsalari subranges, built of relatively old and solid rocks, is characterized by rocky, pointed and glacial forms. Gentle and low relief is related to the Lower Jurassic soft rocks. The relief built of carbonate flysch is presented by jagged forms. The sections built of Bajocian porphyrites are characterized by rocky and smoothed relief.

Modern glaciers are allocated in the highest central section of the Svaneti Range. Smaller glaciers are located on the slopes of the Mount Dadiashi. Glaciers are mostly allocated on the northern slopes of the Svaneti Range. Corrie and hanging type glaciers of small size are located on the southern slope.

The large glaciers of the Svaneti Range are formed on the slopes of the Laila Peak. Laila itself is covered with a glacial cap. Among the glaciers the six glaciers are valley type, including the largest glacier of Laila, which is located in the Khumpreri basin. The glacier has a northwest exposition. Lailchala is also a large glacier.

In the morphogenesis of the relief of the Svaneti Range, above 2000 m, the Pleistocene glaciations played a great role. In the morphology of the Svaneti Range the paleoglacial forms are well represented, such as: corries, trough valleys, moraine series, and erratic boulders

(Khazaradze 2004). The trough gorge of Lailchala extends to the Enguri River bed. The large glaciers were formed in Würm in the gorges of the rivers of Khumpreri, Arshira, Namkorvali, Leshkuri, Kvishara, Skiliri, and Laskadura: the old glaciers were 8–15 km long and some of them descended to the height of 1000 m above sea level.

The role of erosive processes is important in the formation of the relief of the Svaneti Range. Relation of erosive processes and structural rocks is reflected in the mesoforms of the terrain. Relatively vast gorges are formed in the direction of alignment of structures (the heads of the Khumpreri, Laskadura, Skiliri, Arshira and Mukhra rivers). The main rivers, which indirectly or perpendicularly cross the structural rocks of the relief, develop deep and narrow V-shaped gorges. In the eastern part of the range there are soft rocks related to the development of mudflows. Particularly powerful mudflow basins are the Nesharistskali, Dabieri, Ashkhashuri, Neshkuri, Mukhra, and Laskadura. The active landslide of Tsvirmiani in the Lasakadura gorge while blocking the Laskadura River stipulates the formation of a powerful mudflow, causing large losses to the population of Lentekhi.

9.6 Gorges of the Tskhenistskali River and its Tributaries

The Tskhenistskali River basin is located on the southern slope of the central Greater Caucasus. It is bordered by the watershed of the Greater Caucasus, Svaneti and Lechkhumi ranges and their subranges (Kheledi, Jazanistavi and Lechkhumi Hill) with the Askhi and the Khvamli limestone massifs.

Morphologically, the Tskhenistskali gorge is divided into several sections as follows: from the head to the Lashkheti depression, the Lashkheti-Choluri section, the erosive gorge of Rtskhmeluri, the Tsageri depression and the erosive gorge of Zubi-Saretskela.

The River Tskhenistskali originates in the vicinity of Mount Pasismta (3779 m). Up to 1600 m asl, the gorge has the shape of a modified trough. Active mudflows create a powerful debris cone (this section of the gorge is called Lapuri) on the bottom of the valley. Development of mudflows is stipulated by structural rocks. The Lapuri gorge is built of Upper Jurassic and Cretaceous carbonate flysch suites (limestones, marls, clays, sandstones and gravelites). The bottom of the gorge is somewhat wider, and the slopes of the main river are dissected by tributaries, among which the Lapuri and Shushara gorges are notable.

In the head, the Tskhenistskali River is conflowed by the Zeskho River gorge from the right side. In the Zeskho gorge, the Würm glaciers descended to about 1600 m asl. Because of this, the gorges of Zeskho and its tributary—Koruldashi are troughs (Fig. 9.13). Below the Trojan trough gorges the Zeskho River flows through a narrow and deep gorge. Traces of impact of snow avalanches are visible on the slopes of the gorges.

In the territories of Lashkheti and Choluri communities, the direction of the Tskhenistskali gorge is western in the area of Lashkheti and then western-northwestern in the area of Choluri. The gorge is expanded and morphologically represents a narrow depression. At the bottom of the gorge, along the erosion accumulation processes develop. There are several accumulative terraces in the gorge (1–3). The interfluves and floodplains appear in the bed. Almost all tributaries have developed debris cones. Kheshkuri, Lamashuri, Dabieri and Choluri are distinguished by mudflow activities. The morphology of the gorge and the activity of erosive-denudation processes are conditioned by the Lower Lias suites (schists, clays, clay slates, aleurolites and sandstones) that are soft, and loose rocks. Of the large tributaries, the Ghobishuri River (the Lechkhumi Range) and the Mukhra River (the Svaneti Range) are notable. Both of the river gorges, especially Ghobishuri, are of trough form in their heads and old moraines are common.

From Sakdari to Babili, the Tskhenistskali gorge is of a southwestern direction, and the river diagonally crosses the Carboniferous and Triassic suites (phillites, flint slates, arkosic sandstones, silicites and gravelites). Morphologically

Fig. 9.13 Koruladashi gorge and Tsana Village (*photo by* L. Tielidze)

the gorge is narrow, V-shaped and the depth of the cut is 2000–2200 m.

The gorge is expanded in the Lentekhi-Kheledi depression and its width is 2–3 km. In the depression, the erosion material is accumulated in the Tskhenistskali River bed. In the formation of the depression's morphology, together with erosive-tectonic processes the lithology of rocks is very important. It is built of the Lower Jurassic loose and soft rocks (turbidites, clay slates, argillites, etc.). At Lentekhi the Tskhenistskali River is conflowed by two large tributaries—Kheledula and Laskadura. Kheledula River flows through the bottom of the depression from the Khacheshi Village to Lentekhi; the downfall of the bed is 18–22 m/km with frequent interfluves (islands) and vast floodplain terraces are developed. Above the Mananuri Village (after the joining of the Skilori River) the gorge is V-shaped with steep slopes in some areas, deeply dissected by numerous tributaries.

The Laskadura River is in a submeridional direction and its upperstream is relatively expanded, the result of action of Würm glaciers. Glacier traces (moraines) are weakly preserved, but instead the corrie forms are completely represented, especially on the northern slope of the Atsalari Range. The Laskadura gorge is narrow and deeply cut into the Carboniferous and Triassic suites below 1800 m asl. The mudflows of the Laskadura River cause great damage to Lentekhi, initiated by the active landslide of Tsviriana developed on the left slope of the gorge at a height 1300 m asl.

The Tskhenistskali gorge is known as the Rtskhmeluri gorge at a distance of ~16 km between the Lentekhi and Tsageri. It is of meridional direction and is deeply cut in the relief. The depth of cut is 1500 m at the village of Ghvimbrala. The eastern slope of the Samegrelo Range and the western slope of the Lechkhumi Range are deeply and strongly dissected by the Ghvimbrala tributaries. From the right tributaries it is worth mentioning the Devashi, Lamashuri and Rtskhmeluri rivers, in the heads of which the traces of Würm and Holocene

Fig. 9.14 Muri gorge (rock-gate) (*photo by* L. Tielidze)

glaciations (the site of Labrakhi) are well preserved. Powerful debris cones are developed on some of the tributaries.

From the left tributaries of the Tskhenistskali River the Khopuri gorge is distinguished by its fan-shaped form and mudflow phenomena. Khopuri brings the whole material into Tskhenistskali and has a visible debris cone. The most part of the gorge is constructed from Lower Jurassic suites in the Rtskhmeluri section. Because of this, erosive steps are observed in the gorge. The villages of Laktoria, Zanakari, and Lagvarashi are located on their surfaces. While crossing the structural rocks (Cretaceous limestones, Bajocian porphyrites) of the northern slope of the Lechkhumi syncline, the Tskhenistskali River forms a narrow gorge of small size—the Muri rock-gate (Fig. 9.14).

The Tskhenistskali River flows through the Tsageri depression after the Muri limestone rock-gate. Tsageri depression is a western limb of the Racha-Lechkhumi syncline and is built of Tertiary suites, Oligocene and Miocene marine malasses (clays, sandstones, conglomerates, and marls). The northern and southern limbs of the syncline are built of the Cretaceous marine limestones, marls, argillites, and Bajocian porphyry series. It is noteworthy that the gorge is of a rock-gate form in the areas of the limestone cross-section. The depth of the cut is 300–500 m. In the Bajocian porphyritic suite, the gorge is narrow, V-shaped and quite deeply cut. In these sections, the river occupies the bottom of the gorge and conducts only erosive action.

The Tsageri depression is bordered by the Lechkhumi Hill from the east, and the western border is conditional and follows the end of the syncline. The depression is oval shape, the width is ~10 km and the length is ~12 km. The width of the bottom of the depression is 1.0–1.5 km (Fig. 9.15). The downfall of the bed is 7 m/km. The Tskhenistskali River flows in bayous through the bottom of the depression and creates numerous interfluves and floodplains; in addition, it conducts the lateral erosion and intensive accumulation. In the floodplain the fractionality of alluvial material changes from top to bottom, from gravelites to sands. Landslide processes are developed on the slopes of the syncline. The slopes, especially the western side,

Fig. 9.15 Tsageri depression (*photo by* L. Tielidze)

are dissected by small rivers, with 3–5 m high lower terraces visible on both sides of the depression. Terraces 70–145 m high are formed on the left side as steps.

After the Tsageri depression, the Tskhenistskali River flows through the Saretskela gorge (Zubi rock-gate) at a distance of 3–4 km in the southwestern direction and crosses the structural Cretaceous sediments transversally (antecedently). This section of the valley is cut at 400–500 m deep and is a gorge. In the middle section of the Saretskela gorge, the Tskhenistskali River is conflowed by the Jonoula River on the right side. Up to the Chkumi Village, the Jonoula gorge is cut into the southern limb of the syncline by steep left slope and inclined right slope. In the middlestream at a distance of 5 km, the foot of the Jonoula gorge is wide. The morphology of Jonoula and Tetrighele, its right tributary, show that the bottom of the gorge is filled with rock avalanche material from the north cliff of the Askhi limestone massif, causing also the creation of Jonoula boulders. In the heads of Akhalchala, the left tributary of Jonoula, a trace (moraines) of Würm glaciation is preserved.

From Zubi to Matkhoji Village, the Tskhenistskali River flows through the Bajocian porphyrites and Upper Jurassic suites. The morphology of the gorge is variable. The right slope of the gorge is a 1000 m cliff of the Askhi limestone massif. The left slope is more extended and is dissected by erosive network. Below Matkhoji the Tskhenistskali River flows through the Kolkheti Lowland.

9.7 Lechkhumi Range

The Lechkhumi Range is located on the southern flooded slope of the central Greater Caucasus. It extends from the Tskhenistskali gorge (from Lentekhi) to Mount Lukhuni (3216 m) (Fig. 9.16), where it connects to the Shoda-Kedela Range; it is separated from the Greater Caucasus watershed range by the passes of Sasvano (2800 m) and the Vatsistsveri (2910 m). From Mount Lukhuni to Chutkharo, the range is in a submeridional direction transversally crossed by the tectonic structures. In the west from Samertskhle the Lechkhumi Range is in a common Caucasian direction and coincides to structures. It is a watershed of the Rioni and Tskhenistskali river basins. The length of the range is ~60 km and its width is about 25 km in the central part. The cross-section of the Lechkhumi Range is asymmetric—its

Fig. 9.16 Mount Lukhunistsveri (*photo by* L. Tielidze)

northern slope is short and steep, and the southern slope is expanded and gentle (Devdariani 1983). The Lechkhumi Range is dissected by the Rioni River tributaries, such as: Lukhunistskali, Ritseuli, Askistskali and Lajanuri. From the southern subranges the following are notable: Nalishi, Tetnari, Gormaghali, Tarigoni, Poria and Chutkharo-Kupri. The Chutkharo-Samertskhle massif located in the central part of the Range is the highest sector of the Letchkhumi Range; the highest peaks are: Samertskhle (3540 m), Chutkharo (3562 m) and Kareta (3354 m). Other peaks are also important: Shushara, (3405 m), Lelashkha (3152 m), Lukhunistsveri (3216 m), and Tekali (3043 m). The lowest point is the Lankori Pass (2453 m).

Geologically the Lechkhumi Range is mainly built of the Lower and Middle Jurassic sedimentary suites: clay slates, and sandstone and aleurolitic turbidites, schists and clays. The central part of the range—the Chutkharo-Samertskhle massif and Chutkharo-Kupri ranges, is characterized with horst-synclinal structure; the core is built of porphyritic series (diabases, basalts, andesites, basaltic and andesite tuffs and lava breccias). The syncline limbs are built of the Lower Jurassic sedimentary rocks. In general, the Lechkhumi Range is a semi-inverse, monoclonal structure.

The morphology of the Lechkhumi Range is determined by the lithology and altitudinal distribution of exodynamic processes (erosion-denudation and nival-glacial) of the structural rocks. The leading role in the formation of high mountainous terrain of the range belongs to modern and Pleistocene glaciations along with nival processes. In the Lechkhumi Range the modern glaciers are distributed over the Samertskhle massif and in the area of Mt. Shushara. The glaciers are hanging type and northeastern exposition.

Pleistocene glaciations forms are well preserved in the Lechkhumi Range over 2500 m in the form of degraded cirques and troughs. The trace of glaciations is best preserved around

Chutkharo-Samertskhle and Shushara. There were quite large valley glaciers distributed in these massifs. In the relief there are still remaining slightly modified glacial troughs, cirques and moraines. The 1–3 km long lateral moraines are visible in the heads of Latashuri, Sokhortuli, Ritseuli, Zhrinavi, and Ghobishuri. In Late Pleistocene the length of the glacier in the Latashuri River gorge was ~8.5 km. Its tongue descended to the height of 1500 m. The glacier completely occupied the Latashuri gorge. On the left side the watershed is followed by hills with large diabase boulders. The 5–7 km long glaciers were in the heads of the Ghobishuri River. There is a glacial "Datusela's boulder" in the Zhrinavi gorge (20 × 12 × 10 m), which is documented in the Red Book (Gobejishvili 2011).

The high sections of the Lechkhumi Range are represented by sharp bare peaks and landforms created by physical weathering. Below them, the mountain slopes are strongly dissected by the erosive action of rivers. The river valley and their watersheds are clearly visible on the southern slope. The depth and morphology of the cut of the gorge reflects the lithology of rocks and the alignment direction of rocks. The depth of the cut of rivers gorges on the southern slopes is 500–1500 m. The rivers mostly cross the structures transversally. The gorges formed on the northern slope of the Lechkhumi Range are characterized by short profiles, and the depth of the cut in the relief is shorter ~200–300 m.

The Lukhunistskali River gorge crosses the structural rocks antecedently (the same applies to the gorges of the Ritseuli, Askistskali and Lajanuri rivers). The slopes of the gorges, generated in the carbonate flysch (heads) and Bajocian porphyrites (the Kupri Range), are deep and steep. The gorges (villages of Uravi and Likheti) generated in the soft Lias suites, are expanded; the bottom of the gorge is wide and several low terraces are developed there. In some places the landslides are common in soft rocks. The western slope of the Lechkhumi Range is dissected by the Khopuri River, the left tributary of the Tskhenistskali River.

9.8 Gorges of the Rioni River and its Tributaries

The Rioni River gorge extends to the Tvishi Village on the southern slope of the Greater Caucasus. The complex geological and tectonic structure, the nature of erosive-accumulative processes and the modern vertical differential movement of terrain lead to the existence of different morphological units. In the Rioni River gorge from its head to Tvishi the following morphological units are distinguished: the Rioni gorge in mountainous Racha depression, then Rioni crosses the Shoda-Kedela Range, then from the Nakieti Village to the Ghvardia Village the Rioni gorge extends in the Racha depression, which is divided into the Zemo (Upper) Racha and Kvemo (Lower) Racha depressions by the Tsesi-Khidikari gorges, and finally, the Ghvardia-Tvishi section cut in the limestones. We will consider the Rioni River gorge only within the height of 300–400 m from the bed, because this section is formed by the erosive action of the Rioni River and clearly depicts the morphological look of the gorge.

In the head, the Rioni River flows through the mountainous Racha depression. The depression is located between the higher uplifted Greater Caucasus and the Shoda-Kedela ranges. The depression is stretched to the sublatitudinal direction and tectonically it is of a graben nature. The surface of the depression is dissected by the tributaries of the Rioni River. The dissection depth of the depression is 2000–2500 m from the bottom of the Rioni River, while the depth of cut of the individual tributaries depends on the river size (up to 500–1500 m long). The Rioni River flows through the bottom of the mountainous Racha and is often referred to as the Rioni-Chanchakhi graben (depression) in the literature. The Rioni River gorge up to Saglolo is built of the Upper Jurassic and Cretaceous carbonate sediments as follows: the shallow marine limestones, sandstones, marls, clays, gravelites, conglomerates and others (carbonate flysch).

Fig. 9.17 Place of blocking the Rioni River by the Gizheshuri River mudflow in 2010 (*photo by* L. Tielidze)

This region is tectonically a tension region. The Lower Jurassic structural sediments of the mountainous Racha are separated from axial crystalline sediments of the Greater Caucasus by a deep fault line. In the south, the live fault line passing between the Cretaceous and Jurassic sediments, divides the bottom of the depression into two parts. The bottom of the gorge is wider above the site of Shkhilori and intensive accumulation (10–15 cm/year) and lateral erosion took place in the whole Late Pleistocene and Holocene. The thickness of alluvium deposits is 50–60 m. The downfall of the bed is 10–15 m/km. The depth erosive processes are developed in the bottoms of the gorges of the Rioni and Chanchakhi rivers below the site of Shkhilori. The downfall of the bed is 40–60 m/km. There are four terraces visible in the gorge in the areas of Brili, while in other areas the floodplain and upper floodplain terraces are visible. Powerful debris cones are formed in the bottom of the gorge. On the slopes of the gorge, powerful active mudflow basins are aligned, which often block the Rioni River and cause floods. The active mudflow basins are Gadareula, Khmaura, Gizheshuri (Fig. 9.17), Didghele, Shkhilori, Mukhameshi, Dghviora and others. Among the tributaries of the Rioni River the gorges of Zopkhitura, Chveshura, and Chanchakhi are notable, which are more or less modified troughs as a result of erosion. Pleistocene glaciers—Zopkhito and Kirtisho showed in the bottom of the Rioni River gorge, the indicator of which is the existing moraine deposits there. The Ghebi Village is constructed on the terrace created by a glacier (Fig. 9.18). The Chanchakhi River gorge is a strongly modified trough due to erosion processes. In late

Fig. 9.18 Ghebi Village constructed on the terrace formed by Late Pleistocene glacier (*photo by* L. Tielidze)

Pleistocene the strong tongue of the Buba-Boko Glacier descended to the height of 1050 m and morphologically was a subranged glacier.

Below Saglolo (up to the Oni depression) the Rioni River crosses the horst-synclinal Shoda-Kedela Range built of carbonate flysch and develops a deep, V-shaped, antecedent gorge. The depth of the cut is 2000 m on average. Depending on the structure of the range, the slopes of the gorge are dissected by tributaries. Mudflows develop powerful debris cones, while erosion and accumulation processes alter the bottom of the gorge.

In the Zemo Racha depression the Rioni River gorge is expanded. It crosses the Lower Jurassic, soft sedimentary suites that build the Sori anticlinal structure. The gorge is asymmetrical below Oni; its right slope is steep and the left is expanded. The Rioni River mainly flows through the bottom of the right slope. Such morphology of the gorge is caused by the activity of landslides and mudflows occurring on the left slope throughout the Quaternary period.

The left slope of the gorge in the Zemo Racha depression is dissected by the erosive action of the left tributaries (Gharula, Jejora, Kheori, Shardometiskhevi, and Barula) of the Rioni River. And the right slope is relatively weakly but deeply fragmented. In the Zemo Racha depression the Rioni River terraces are well represented, especially on the left slope. The first and the higher terraces are well represented in the form of individual fragments. The terraces are the main areas of settlement of the population.

The entire spectrum of the Rioni River terraces are represented in the areas of Oni; the town itself is constructed on the first terrace (Fig. 9.19). The altitudinal synchronic distribution of terraces allow us to relate the time of their formation to the phases and stages of the Pleistocene glaciation (Gobejishvili 2011).

The landslide processes and the associated mudflows are important in the formation of morphology of the Zemo Racha depression. Landslides were active throughout the Pleistocene and are still active (Tskhmori, Chordi, Somitso, Ankara, Korta, Skhieri, Khirkhonisi, Parakheti, Seva, Shardometi, etc.). The higher terraces of the Rioni River are fragmented by landslide processes.

In the Zemo Racha depression the Rioni River tributaries are from the right—Sakaura, Chalula, and Sontarula, and from the left—Gomila,

Fig. 9.19 Oni town constructed on the first terrace of the Rioni River (*photo by* R. Gobejishvili)

Fig. 9.20 Khidikari rock-gate built of the Cretaceous limestones (*photo by* R. Gobejishvili)

Gharula, Jejora and Kheori. Especially notable are Jejora, Gharula, and Sakaura. The thalweg of the Sakaura River is a southwestern border of the Shoda-Kedela Range along a certain section. The Jejora River starts from the Shoda-Kedela Range and a trace of the glacier action is visible in its head. And below, the morphology of the gorge is determined by the lithology of structural rocks. In the soft rocks the gorge is expanded (the areas of Kvaisa) and cut into the Bajocian porphyrites the gorge is deep with steep slopes. The gorge is asymmetric in the Bajikhevi-Oni section. The River Gharula gorge is erosive from the village of Kvazha to Oni, and it is glacial in its heads.

From the Sori Village to the Kvatskhuti Village the Rioni River crosses the Bajocian porphyrites; the gorge is narrow with inclined slopes. At Khidikari the gorge built of Cretaceous limestones (northern limb of the Racha syncline) is antecedent and is a rock-gate (Fig. 9.20). The width of the bottom of the rock-gate is 15–20 m. In the relief built of Bajocian porphyrites the Rioni and Lukhunistskali rivers (Sori, a section of Tsesi) develop rocky, deep and V-shaped gorges. In the left side of the gorge a curve landslide is clearly expressed in the relief.

The Rioni River gorge flows through the bottom of the Racha syncline from Kvatskhuti to Ghvardia. The gorge has a latitudinal direction and its morphology is determined by the direction of the Racha syncline. The lower landmark of the bottom of the gorge is located at a height of 400 m and the upper landmark at 580 m. The downfall of the bed gradually decreases from 9 to 2 m/km. This section of the gorge is known as the Kvemo Racha depression. Accumulative processes mainly prevail here. The alluvion capacity is over 10–15 m and width of the bottom of the gorge (floodplain) is 200–400 m, but its maximum width is about 1 km at Ghvardia. The Rioni River flows through the bed in bayous and forms interfluves.

The bottom and lower sections of the slope of the Racha syncline are built of Eocene, Oligocene and Miocene marine molasses with clays, sandstones and conglomerates (Maikopian series). The syncline limbs are built of the Cretaceous shallow marine limestone, marls, carbonate clays, arkosic and glauconite sandstones. The higher sections of the syncline limbs are constructed from the porphyritic series of Bajocian age. On the right side of the syncline the Saelio Range and its parallel and higher Sochibi Range built of porphyrites are visible in the relief. These ranges are well represented up to the Ritseuli gorge, and in the west they are dissected by the right tributaries of the Rioni River. The fragmented relief is represented in the forms of hills and hillocks.

The landslide processes are important in forming the Rioni gorge. Landslides are widely developed on the left and right sides of the gorge. Landslides are active and their tongues come down to the bottom of the gorge. At the right side of the gorge the landslides are located at the heads of the small ravines. The impact of landslides on the terrain was conducted throughout the Pleistocene.

On the bottom of the Rioni River gorge, old and new debris cones are common. The old debris cones are located on the first terrace on the both sides of the gorge. Such cones are present in the Kvatskhuti, Dzirageuli, Ghviara, Bostana, Chorjo, Chkvishi, and Znakva. The old cones and the first terrace are the main areas of settlement and farming for population.

The terraces are well developed in the Kvemo Racha depression. The high terraces are located in fragments in the small watersheds of tributaries. Nine terraces are distinguished here. The first terrace is at a height of 10–15 m and the 9th at 340–380 m (Nemanishvili 1962, 1982).

Both limbs of the Racha syncline are fragmented by erosive processes. The main rivers are Ritseuli, Askistskali, Shareula, and Krikhula. They start from the Lechkhumi and Racha ranges and cross the limbs of the syncline perpendicularly. The gorges are narrow and deep in the sections built of porphirites with a V-shaped profile. The sectors built of limestones are often rock-gates; examples include the Kldekari, Kldisubani and Shareula sectors. The depth of erosive cut is 300–500 m.

The Rioni River gorge is cut into the Cretaceous limestones between Alpana and Ghvardia.

Fig. 9.21 Natural columns of Sairme in the Rioni River gorge (*photo by* R. Gobejishvili)

The Rioni gorge separates the Khvamli limestone massif and Labechina Range on the right side and the western end of the Racha Range on the left side. The gorge is quite deeply cut (200–500 m) in the limestones. The height of the limestone cliffs on the right side (400–600 m) is higher than the left one (150–300 m). The gorge has a canyon form on Tvishi and Alpana-Ghvardia section. On the right side of the gorge, at the Sairme Village, the gravitational-block landslide is developed and in its detaching area there are fascinating rock columns of Sairme (Fig. 9.21).

The bottom of the gorge is narrow, while erosive and accumulative (Alpana surroundings) processes alternate in the bed. The downfall of the bed is 7–8 m/km. The River Rioni does not have tributaries on the left side, while from the right side it is conflowed by the rivers of Lajanuri, Ghvirishisghele, Minatskarosghele, and Lakhepisghele.

The Lajanuri River originates at the Lechkhumi Range. The lower part of its gorge flows through the Orbeli depression. The gorge is wide with landslides (Orbeli, Lailashi) developed on the both sides. After the Orbeli depression, The Lajanuri River perpendicularly crosses the Labechina anticlinal hill of latitudinal direction. The hill is stretched from the Gvirishisghele to the Zemo Chkvishi Village and is elevated by 600–800 m from the Rioni gorge. While crossing the Labechina Hill, the Lajanuri River forms the 3 km long and 200–400 m deep gorge. The water reservoir is built on the Lajanuri River in the Orbeli depression.

9.9 Shoda-Kedela Range

The Shoda-Kedela Range is parallel to the Greater Caucasus Range. It stretches from Mt. Lukhunistsveri (3216 m) to the Zekari Pass (3184 m), a distance of about 60 km. The northern slope of its eastern section does not enter Georgia. In this section, the Kedela Range morphologically covers the Greater Caucasus watershed. With the antecedent gorge of the Rioni River the range is divided into two parts—the Shoda and Kedela sections. The highest peak of Kedela is Mt. Khalatsa (3938 m), and of

Fig. 9.22 Kedela Range (*photo by* L. Tielidze)

Shoda—the Mt. Shoda (3609 m) (Fig. 9.22). The lowest point of the range is in thalweg (1050 m) of the Rioni River. As a result of active reverse erosion of the rivers developed on the slopes, the crest of the Shoda-Kedela Range is winding. The range represents a morphological and climatic barrier for mountainous Racha.

Morphologically, the Shoda-Kedela Range is a Horst-synclinorium built of Lower Cretaceous carbonate flysch. The terrain is constructed from sandstone-aleurolitic flysch rocks, such as sandstone and aleurolitic turbidates, pelagic clays, argillites and marls. It is distinguished by its complex internal structure, which is more complicated by faults and overthrusts. In the relief's morphology the terraces and the peculiarities of modern erosive-accumulative processes in the Rioni River bed are related to the live fault line passing through the northern slope of the Shoda-Kedela Range. The bed has a longitudinal profile, namely, at shorter distances the alternation of anomalous sections (Tsereteli 1966) (the downfall of the bed varies within 18–44 m/km). In the Rioni River gorge the greatest intensity of erosion processes are observed at the Tsidrotsikhe-Shkhilori section. There is a powerful layer of alluvion accumulated on the bottom of the Rioni River at the Chiora-Ghebi section, above Shkhilori.

The nival-glacial and erosive processes have a leading role in the formation of the relief of the Shoda-Kedela Range. In high mountainous areas (above 3000 m), small corrie-type glaciers are developed in deep cirque forms. Only in the heads of the Jejora River are there two large glaciers of the corrie-gorge type. Glaciers can be found in the basins of the rivers of Shodura, Geske, Khamijauri, Gomila, Gharula, and Jejora. Above 3000 m, physical weathering takes place in the bare corrie forms and debris cones, rock-fills, sand flows, pointed crests and peaks are created.

In the late Pleistocene and Holocene periods the trace of action of powerful glaciers is well preserved on both slopes of the range in the form of trough gorges, cirques and morains. Particularly noteworthy are the troughs and lateral moraines developed in the gorges of Gharula, Jejora, Khamijauri ,and Shodura, which determine the distribution of tongues of the valley glaciers (Fig. 9.23). In the Gharula basin, on the right slope, the Würm moraine is visible as a hill, which bending at a height of 1300 m moves to the terminal moraine. The large Jejora Glacier descended in the gorge to 1500 m above sea level, indicated by the lateral moraine remaining in its left side. Glaciers in the Shodura and Khamijauri basins descended to the bottom of the main gorge. On the northern slope of the Kedela Range, almost every gorge has a hanging trough shape. Glaciers here were the tributaries of the Chanchakhi River basin glaciers. In some valleys, the stadial moraines are visible into the Würm trough. The base of the trough slopes is covered by continuous debris cones. Lakes are formed in some corrie forms.

Fig. 9.23 Trough of the Khamijauri River on the northern slope of the Kedela Range (*photo by* L. Tielidze)

A trace of erosive action is obvious in the Shoda-Kedela Range. The northern slope is dissected by ravines with short profiles and almost with no tributaries. By morphology and erosive activity they are divided into two groups: 1. The rivers that originate from the Shoda-Kedela crest, have a well-developed trough gorges, erosive processes are weakened, have no debris cones and are distinguished by the constant stream of water; 2. The ravines starting from the middle part of the slope and are distinguished with intense erosive action. The relief of their basins is

Fig. 9.24 Sakaura River captured due to reverse erosion (Google Earth imagery 11/10/2017)

bare, which is the result of strong physical weathering. The slopes are steep and powerful mudflows are formed during the heavy rains. They often block the main rivers and cause flooding, and great damage to the population. Such ravines have large debris cones. The mudflow ravines expand their area by regressive erosion and often capture neighboring rivers. The remnants of the gorges are preserved in the Skodorula and the Beghlori basins. The classic example of such capture is the Sakaura River, which intruded on the northern slope as a result of the regressive erosion and captured the heads of the Rubodzali River. This action is thoroughly reflected in the relief: the presence of the sharp curve in the heads of the Sakaura River, deep and narrow gorge in the area of capture and the lowest section of the Shoda watershed range (Fig. 9.24). The southern slope of Shoda-Kedela is deeply dissected by the active action of the Sakaura, Gomila, Gharula, Jejora rivers and their tributaries. The depth of the cut of the main gorges is 1000–1500 m, and that of their tributaries 400–600 m.

In the interfluve of the Jejora River and its right tributary—Kvedrula and in the vicinities of the villages of Skhanari, Tsedisi, and Kudaro, the isolated exits of the limestones are found, where the karst forms are developed—wells, sinkholes, caves, etc. The waters leak into the Skhanari karst relief, and flow out in the Kvedrula River gorge as vaucluses. The town of Oni is supplied with this water. The Kudaro caves were dwellings of the Pleistocene human.

In 1891, due to earthquake, a powerful rock avalanche descended from the limestone massif of Mount Kdebodzali located on the left side of the Kvedrula River, blocked the gorge and Kvedi Lake emerged; (the Racha song "Kvedrula modidebula" is related to it). There stand three rock columns in the area from where the rock avalanche was cut off.

9.10 Racha Range

The Racha Range is located in the southern part of the Greater Caucasian folded system. It extends from the Tvishi Kldekari (the Rioni gorge) to Ertso Lake in the east. The Racha Range is a watershed of the Rioni and Kvirila rivers. Its length is ~110 km along the crest and width is ~25 km in the Sachkhere-Oni section. The western and central sections are characterized by a monoclinal structure, for which the northern slope is better developed than the southern slope. The highest peak of the Racha Range is Lebeuri (2863 m). The significant peaks include Leknari (1746 m), Tskhrajvari (1569 m), Satsalike (1996 m), Khikhata

Fig. 9.25 Rocky southern slope (monocline) of the central section of the Racha Range (*photo by* R. Gobejishvili)

(2239 m), Potskhvrevi (2402 m), and Daghverila (2724 m). Highways pass through the lower passes (Nakerala 1217 m, Ertso 1785 m) of the Racha Range.

On the southern slopes of the Racha Range built of limestones, sharp cliffs are developed. The height of the cornice (cliff) within the Nakerala Range is 300–600 m. The cliff overlooks the Okriba depression from the north (Fig. 9.25). The cliffs are developed in the Ertso-Tsona depressions built of limestones. Depending on the morphology of the Racha Range, it has the subranges only on the northern slopes in the central and western sections. The Khikhata and Potskhvrevi subranges, built of limestones, develop cliffs in the north. It is worth mentioning the cliff—The Tamardedopali Rocks, developed in the Potskhvrevi Range.

In the eastern section built of Bajocian porphyrites, the subranges are developed on the both slopes and they are the watersheds of the tributaries of the rivers of Kvirila and Jejora.

The Ertso-Tsona depression is located in the easternmost part (over 1700 m asl) of the Racha Range, which is located between the hills of Valkhokhi and Lebeuri. The Shaori, Khotevi and Shkmeri depressions, formed in the relief built of limestones, are located on the northern slope of the Racha Range. The Shkmeri depression is a synclinal structure, bordered by the Khikhati, Racha and Potskhvrevi ranges.

The central and western sections of the Racha Range (to the west of the Mount Tsiteli Klde) are built of the Cretaceous age limestones, and the eastern section of the Bajocian porphyritic suites. Between the latter suites and the limestones, the narrow strip of the range is constructed from the Oligocene and Miocene marine molasses: sandstones, clays and conglomerates. The easternmost part of the Racha Range is built of Upper Cretaceous limestones. Geological structures are complicated by tectonic fault lines. The crossing (seismo-tectonic knot) of the fault lines of different direction is related to a strong earthquake of April 29 1991, which caused significant changes in relief. It is important to note the powerful rock avalanches of Khakhieti and Patsa, which blocked the gorges and created lakes. The

Fig. 9.26 Rarest speleothems of the Muradi Cave (*photo by* G. Lominadze)

rockslides were developed in the relief built of Bajocian porphyrites, the rock avalanches were developed in the limestone precipices of Potskvrevi and the landslides were activated in the Tertiary sediments. The influence of lithological and tectonic conditions is sharply reflected in the relief of the Racha Range. The karst relief is represented to the west, the sharp rocky shapes in the section built of Bajocian porphyrites, and in the Tertiary sediments the relatively soft forms develop and the relief is also relatively lower.

The limestone section of the Racha Range is a classical example of western Georgia's mountain karst, with sharply expressed karst terrain, diverse surface, and underground forms. Due to the monoclinal structure of the range, the karst forms are only represented on the northern slope to the west of the Nakerela Pass. Among them the Muradi Cave is distinguished by speleothems unique in the world. Several years ago this cave was discovered several kilometers away from the Nakerala Pass by the members of "Tetnuldi"—the group of mountain climbers and travelers. Rare stalactites can become the subject of serious study of world scientists (Fig. 9.26).

The central section of the Racha Range (Nakerala-Tsiteliklde) is distinguished by the abundance of karst forms. Among them are the sinkholes, which are numerous in the Shaori depression, heads of the Dzusa River, Didi Satsalike, Patara Satsalike, Khikhata, etc. In the central part of the Racha Range bare and closed karsts are developed. The number of sinkholes is 20–30 units per km^2. Sinkholes with tiers are common. In Satsalike the karst sinkholes are symmetrically located in the direction of rock alignment. There is an outlet of vaucluses on the southern slope of Khikhata, at the foot of the cliffs, and the Krechia vaucluse (the river Jruchula basin) is notable. The Shaori depression is the largest closed karst poljes

Fig. 9.27 Shaori water reservoir located on the northern slope of the Racha Range (*photo by* L. Tielidze)

in the Caucasus, located on the northern slope of the Nakerala Range. Its flat bottom height is 1150 m asl. The length of the depression is ~6 km and the width is ~3.5 km. The Shaori depression is connected to the Shareula River gorge by underground channels. In the past century, the underground water passages in the Shaori depression were closed and the reservoir was created (Fig. 9.27). There are numerous karst forms on the northern slope, in the canyon-shaped gorges of the Shareula, Velevura, Khotevura, Krikhula and Kheori rivers, such as shafts, wells, sinkholes, and poljes. There are also many caves: Tskhrajvari, Nikortsminda, Skhvava, Usholta, and Shkmeri. In some caves (Nikortsminda and Skhvava) ice is formed. These "Freezers" are described by Vakhushti (1941).

Karst can be found in the Potskhvrevi and Khikhata Ranges—the wings of the Shkmeri depression. The Kheori River develops a narrow and deep gorge (local population call it Kvagakhetkila) while crossing the Potskhvrevi Range. The river disappears in the karst cavities and flows out in the end part of the gorge in the form of a vaucluse. The Buja River has similar morphological conditions in Satsalike polje, which disappears here and flows out as vaucluse after 0.5 km. A waterfall is interesting in the Shareula River gorge, which is formed by the river flowing out from the cave.

The erosive action of rivers is remarkable in the modeling of the Racha Range. The southern slope of the range is dissected by the tributaries of the Khakhietistskali, Dzusa, Tkibula, Tskaltsitela, and Lekhidori rivers. The main gorges are developed in Bajocian porphyrites; the depth of the gorges is 200–400 m. On the northern slopes, the valleys cut in the limestone relief are canyon-shaped (Kheori, Krikhula, Khotevura, and Shareula).

The northeastern slope of the nonlimestone Racha Range is dissected by the left tributaries of the Jejora River. The gorges are stretched and characterized by short profile. Landslide processes (Chordi, Tskhmori, Somitso, Korta, Parakheti and Bari) are developed in the relief built of Oligocene marine molasses. The landslides have affected the relief throughout the entire Pleistocene. The trace of old glaciations can only be seen on high peaks (Daghverila-Lebeuri) in strongly deformed corrie forms.

9.11 Gorges of the Didi Liakhvi and Patara Liakhvi Rivers

The Liakhvi River basin is located on the southern slope of the central Greater Caucasus. In the west it is bordered by the Likhi Range and in the east by the volcanic terrain of the Keli upland. The gorges of the Didi Liakvi and Patara Liakhvi are separated from each other by the Germukhi Range and its subrange-Gudisi.

The Liakhvi River basin is mainly constructed from Upper Jurassic and Cretaceous carbonate flysch suites: limestone and sandstone turbidites, marls, clay slates, limestones, and argillites. The vicinities of the Java depression are built of Tertiary sediments, such as the Oligocene and Miocene marine molasses: sandstones, clays, gravelites and gypsum clays, and the Kekhvi-Patsa section of Bajocian porphyritic series.

The Didi Liakhvi gorge can be divided into five parts by morphological, morphometric and orographic characteristics as follows: (1) From the head to Kvemo Roka, (2) the Paneli gorge (from Kvemo Roka to the mouth of the Gudisi River), (3) The Java depression, (4) Patsa-Kekhvi, and (5) Shida Kartli section.

The head of the Liakhvi River is called the Maghran-Dvaleti depression. The gorge was formed through erosive-tectonic, glacial and volcanic processes. The gorge has a latitudinal direction, it is V-shaped and symmetrical, which indicates the even development of erosive processes. The gorges of right tributaries (Sbastts-kali, Cheliatistskali, Rokistskali, and Jomagistskali) are located on the southern slope of the watershed range of the Greater Caucasus. The main part of the gorges is erosive and deeply cut (600–900 m) in the relief. Morphologically, the upper parts of the gorges are quite modified troughs. In Würm, the 7–9 km long glaciers descended to the height of 1600–1800 m. The large glaciers were located in the heads of the Jomagistskali and Didi Liakhvi (Kalasani) rivers, where small size corrie and corrie-valley glaciers are still developed today.

Above the Akhubati Village the Liakhvi River gorge is represented by volcanic relief. The Akhubati volcanic streams in the Late Pleistocene, the Khozi volcanic streams in Holocene, which descended in the paleo Liakhvi gorge at 1600 and 2100 m, substantially changed the bed of the Liakhvi gorge. The surfaces of lavas are wavy, and small size depressions are common. The Liakhvi River and its tributary—Ermanistskali developed their gorges in the lava edges in contact with the base rocks. The morphology of newly cut gorges is asymmetric; slopes on the side of lavas are steep. The depth of the lavas from their surfaces is 150–200 m.

The Didi Liakhvi gorge is of submeridional direction from the Jomagi gorge to the Kvemo Khtse Village. The river diagonally dissects the relief building structures. The relief is erosive, 900–1000 m deep and its slopes are dissected by smaller ravines.

The Didi Liakhvi River flows through the Java depression from Khtse to Patsa. The depression is developed in the Java syncline built of Tertiary suites. The Didi Liakhvi diagonally crosses the syncline and the gorge is expanded. The width of the bottom is up to 1 km and the Didi Liakhvi bayous flow through the expanded floodplain. The first terrace is relatively visible in the depression, while the other terraces and steps are represented in fragments, with villages constructed on them. The Gudisistskali River conflows the Didi Liakhvi River from the left in the Java depression. It is erosive and takes its origin in the Germukhi Range. From Patsa to Kekhvi the Didi Liakhvi River develops deep gorge in the relief built of Bajocian porphyrites. Lower terraces are weakly developed.

The Patsa River conflows the Didi Liakhvi River from the right side, the gorge of which is mostly submeridional and starts in the vicinities of the Ertso depression. At the Kemulta Village, the Keshelta River conflows from the left side, which originates from the Greater Caucasus (Mount Zekari) and represents the modified trough in its head (at a distance of 6–7 km) and downward it is erosive. The lower part of the gorge is formed in

Fig. 9.28 Patsa Lake formed as a result of earthquake in 1991 (*photo by* R. Gobejishvili)

the Bajocian suite and morphologically is narrow and deep. As a result of the 1991 earthquake a rock avalanche occurred, which blocked the Patsa River gorge and a ~1 km long lake appeared (Fig. 9.28). By 2007 the Patsa River filled the lake with solid deposits.

The Patara Liakhvi River conflows the Didi Liakhvi River at the Shertuli Village. Its basin is located on the southern slopes of the Germukhi Range and its subrange. It takes its origin from the top of Mount Mangavi (3339 m). The gorge changes its direction several times. The upper stream relief of the gorge is built of Cretaceous sediments, and the downstream relief of Tertiary and Quaternary sediments. The valley gorge in its head is trough shape. Generally the Patara Liakhvi flows in the V-shaped gorge, which is especially narrow on its meridional section (between the villages of Atseriskhevi and Knogho). The gorge is extensive and relatively wide in its middlestream. Both slopes of the gorges are dissected by tributaries, such as Pachuristskali, and Chaparukhistskali. The lava stream flows down from the Knogho volcano from the heads of the Patara Liakhvi.

The Liakhvi River flows through the Shida Kartli plain below Kekhvi Village in a wide bed. Low terraces are well represented in the gorge. Near Gori the Liakhvi River antecedently crosses the Kvernakebi Range and on its slopes the terraces are more completely represented.

9.12 Gerumukhi Range

The eastern part of the central Greater Caucasus is morphologically interesting. This section of the watershed range is the lowest in the whole central Greater Caucasus morphometrically. Morphologically, are distinguished the Germukhi Range, the gorges of the Lekhura and Ksani rivers and the Kharuli and Lomisi ranges. All these orographic units begin from the Keli volcanic upland (plateau).

The Germukhi Range is a watershed of the Didi Liakhvi and Patara Liakhvi. The gorge is erosive-tectonic, and the watershed is heavily curved. It extends ~30 km from the Mount Knogho to the southwest up to the Didi Liakhvi River gorge. From the Germukhi subranges, on the northern slope, the submeridional Dareni Range is separated from the highest peak (Mangavi, 3339 m), as is similarly the 25 km long Gudisi Range in a southwest direction from Mount Germukhi. The peaks of the Germukhi Range include the following: Mangavi (3339 m), Germukhi (3196 m), Dareni, and Chaparukhi. Passes are located higher, at 2700–2900 m asl in the Gerumukhi Range.

The Gerumukhi Range is constructed from carbonate flysch suites: sandstones, slates, limestones, and marls. Young volcanic lavas are common in the eastern part.

The morphological look of the Germukhi Range is created by glacial and erosive processes. On the crests and subranges of the range, with height exceeding 2000–2200 m, the Pleistocene glacial forms are preserved, such as cirques and modified troughs. The cirques, troughs and moraines are well preserved on the slopes of northern exposition of the range. The highest peaks of the mountain range are serrated. In the easternmost part, in the vicinities of Knogho (in the heads of the Patara Liakhvi River), the Holocene volcano of Knogho (2875 m) exhibits a 3 km long volcanic stream to the south in the gorge. The slopes of the Germukhi Range are deeply dissected by the right tributaries of the Didi Liakhvi and Patara Liakhvi rivers. The slopes of the gorges are steeply inclined. In general, the form of the valley is V-shaped, but the shape of individual sections is defined by the petrography and fault lines of the structural rocks. The nival, gravitational erosive processes and related morphosculptural forms are characteristic to the different sections of the Germukhi Range.

9.13 Kharuli Range

The Kharuli Range is in a meridian direction, starting from the peak of Knogho (3284 m), in the south of the Keli volcanogenic upland. It is divided into three sub ranges at the peak of Dziri

(2594 m) in the south and is the watershed of the Ksani, Lekhura and Patara Liakhvi rivers. In the north of the Kharuli Range is the Pachuri subrange in a latitudinal direction, with the highest peaks of the range in Mount Tsitelkhati (3026 m) and Gelavduri (3230 m). On the northern slope of the Pachuri Range, the volcanic cone of Tsitelkhati is located in the old glacial cirque, while the lava stream is spread at the bottom of the trough first, and then in the Ksani gorge (the Zhamuri gorge). The Tsitelkhati cone closed the cirque exit and created two lakes at a height of 2760 m (Gobejishvili 2011).

The Kharuli Range is built of the Upper Jurassic and Lower Cretaceous carbonate flysch suites: the sandstones, clay slates, marls, and limestones. There is a red colored volcanic cone located in the range.

In the high sections of the Kharuli Range, the trace of Late Pleistocene (Würm) glaciations is preserved; especially in the northern slopes of the peaks of Knogho, Tsitelkhati and Gelavduri, the old glacial formations are widespread, such as cirques, modified troughs, and erratic blocks. The western and eastern slopes of the Kharuli Range are common and deeply dissected by erosive processes below the glacial terrain. The morphology of the gorges reflects the lithology of the structural rocks. In the areas of the limestone distribution, debris cones are developed. The cross-section of the river gorges is V-shaped.

9.14 Limestone Massifs of the Central Greater Caucasus

A Limestone strip is located in the southern part of the Egrisi (Samegrelo) Range, which is divided into the limestone massifs by the gorges of the Magana, Chanistskali, Khobi, and Tekhuri rivers, these massifs are as follows: Tsulishi, Kvira, Gaucha, Migaria, and Askhi. Spatial distribution of limestone rocks determines the form and size of these massifs. The Gaucha massif is the lowest and smallest.

The Tsulishi massif is located between the Enguri and Magana river gorges. It is a steeply inclined slope and surrounded by forest. Due to its morphological signs, the karst forms are weakly expressed. The area of the massif is ~ 7 km^2.

The Kvira massif is located between the Magana and Chanistskali gorges. In the north it descends to the Magana gorge, in the east to the Tsintskali gorge, and in the south it gradually descends and the vaucluse rivers of Intsra and Morozha begin on its edges. The karst forms, such as sinkholes and dry gorges are well developed in the treeless part of Kvira (above 1700 m asl). The Tsotne Dadiani (depth 77 m) and Albov (45 m) cave and abysses are also located there adorned with speleothems.

The Gaucha massif is located between the Chanistskali and Khobistskali river gorges. It extends in a sublatitudinal direction at a distance of 3–4 km long and 1–2 km wide. It is built of the Lower Cretaceous limestones. The karst forms are weakly expressed as dry ravines and sinkholes in the Zinda cavity.

The Migaria massif is located between the Khobistskali and Tekhuri river gorges. It extends from the west to the east at a distance of ~ 17 km. The highest place is Mount Migaria (2025 m). The Migaria massif is bordered in the north (the Khobistskali River and its tributary Chitastskali) and east (the Tekhuri River and its tributary Khipagali) by a 200–400 m high cliff. The upper part of the latitudinal massif is constructed from Urgonian limestones; its length is ~ 8 km and width—3–4 km. It is depression of synclinal structure; the height of the bottom of the depression is 1000–1100 m. Mount Migaria is located in the east of the depression and Mount Jvari (1838 m) in the west. The unity of the depression forms the Migaria complex polje. The Tsipuria abyss is located in the western part of the polje. The micro-relief of the abyss is represented by karst sinkholes and erosive ravines. There are shafts formed on the southern slope of the abyss and ice can be found in some of them. The bottom and slopes of the Migaria depression are perforated by karst sinkholes. Extensive areas of corrie surfaces are developed there covered by Colchis Forest; the southern part is built of the monoclonal suites of Upper Cretaceous age. Tselami Hill is distinguished in the relief, dissected by the Ochkhamuri River and its tributaries. Karst is also common in Tselami Hill.

The Lekhari rocks are located in the north of the depression; their northern slopes form a cliff directed towards the Khobistskali gorge (height 150–200 m). The southern slope is inclined and is distinguished by the abundance of karst forms.

The Askhi massif is the largest and most complex massif of Georgia. The high and medium mountainous karst regions are distinguished within this massif. It is stretched both meridionally and latitudionally at 20–21 km between the gorges of the Tekhuri and Tskhenistskali rivers. In the north and east the Askhi limestone massif is separated from the adjacent areas by 300–500 m high cliffs. Its highest peak is Gadrekili (2520 m). The Askhi massif has a complex morphological look with locked cavities and high ranges, plateaus and canyon gorges. These include: Abasha, Okatse, and Tsachkhuri. In addition to canyons, the relief forms are complicated by karst formations. The following large morphological units are distinguished within the Askhi massif: the high mountainous karst region is built of Urgonian limestones. Morphologically, here is distinguished the oval-shaped Kvibia syncline cavity, with length of ~12 km and width 5–6 km. Kvibia's bottom is inclined westward from 1600 to 850 m. In the bottom and on the slopes of Kvibia, there are numerous karst sinkholes and corrie surfaces that give peculiarity to the Kvibia relief. The karst waters absorbed in the Kvibia cavity find outlets in the western edges of the cavity as numerous vaucluses. All tributaries of the Tsachkhuri River are fed from its vaucluses. The Kvibia depression is bordered by the Opitsare Range in the north and by Tsekilda Hill in the south. The highest peak of Gadrekili is located in the Opitsare Range. It is separated from Jonoula and Chkhorotsku gorges by cliffs in the north. Its southern slope is inclined and weakly dissected by rivers. The karst forms are developed in the relief built of limestones. The Maidani Plateau is an interesting morphological unit in the Askhi massif. It is treeless and the limestone layers lie horizontally, which ensures the abundance and depth of karst forms. The waters absorbed into the karst cavities find their outlets at the edges of the Maidani or the Askhi massif in the form of vaucluses.

The Turchu polje is distinguished by its size and abundant of karst forms in the Askhi medium mountainous massif. The flatland extends from the northeast to the southwest (width 2 km, length 3 km) and is characterized with sharply distinguished flat bottom. There are larger streams flowing through them creating the Turchu River, which disappears in the karst wells in the southwestern part of the polje and flows out after 3 km as Toba, the left tributary of the Abasha River. The polje is 80–150 m deep. To the south-west of the Turchu Polje is located the Didghali cavity, with its karst forms.

Together with the above-listed morphological units, it is worth mentioning the following karst plateaus: Tabakela, Ghulukuna, Sachikvano and the canyons of the Tsachkhuri, Abasha, and Okatse rivers. The Tsachkhuri River flows out to the west of the Kvibia polje and its watershed basin includes the Maidani Plateau. The Tsachkhuri River forms the 2 km long canyon above Salkhino. The Abasha River originates with numerous and watery vaucluses, which are located in the site of Rachkha. Toba, the left tributary of the Abasha River is interesting, a continuation of the underground flow of Turchu and creates large waterfalls in the heads. The gorge of Okatse, the tributary of the Tskhenistskali River is located in the south of the Askhi massif. At one place, the width of the gorge is 2–5 m, and the depth is 60 m.

The Khvamli massif is located between the gorges of the Tskhenistskali and Rioni river gorges. The length of the massif is 11–12 km from Saretskela to the Tvishi gorge and the width is ~7 km. Its highest peak is Khvamli (2001 m). The Khvamli massif is of monoclinal structure.

Fig. 9.29 Khvamli limestone massif (*photo by* L. Tielidze)

Its northern slope is the southern limb of Lechkhumi syncline; the inclined slope runs from 2000 to 1000 m. The Lakhepisghele gorge divides the slope of the cuesta hill of Khvamli. The ravine is cut quite deeply to 200–250 m and is asymmetrical. The left slope is steep and the right one maintains the morphology of the monoclinal series. The KhvamIi massif is constructed from Urgonian limestones. Its southern slope is separated from the Okriba massif built of Bajocian porphyrites by a 300–500 m high cliff. There are several historical caves in the cliff, where the kings of Georgia kept the treasure. These cliffs of the Khvamli massif are visible from entire Imereti and form an iconic benchmark (Fig. 9.29). The southern part (the right slope of the Lakhepisghele) of the northern slope of the Khvamli massif is strongly karstified, with numerous sinkholes, deep cracks and shafts. The Boga icy cave is notable. The northern section of the Khvamli massif (cuesta) is constructed from Upper Cretaceous thin-layered limestones, and is weakly karstified.

References

Devdariani G (1983) Qartuli Sabchota Enciklopedia [Georgian Soviet Encyclopedia], vol 6. Tbilisi (in Georgian), pp 208

Devdariani G (1985) Qartuli Sabchota Enciklopedia [Georgian Soviet Encyclopedia], vol 9. Tbilisi (in Georgian), pp 310–311

Geology of the USSR (1964) Vol. X. Georgian SSR. Part I—M. Nedra, Moscow (in Russian)

Gobejishvili RG (1995) Saqartvelos tanamedrove mkinvarebi da Evraziis mtebshi gamkinvarebis evolucia gvian Pleistocensa da Holocenshi [Present day glaciers of Georgia and evolution of glaciation in the mountains of Eurasia in late Pleistocene and Holocene]. sadoqtoro disertacia, Tbilisi (in Georgian), 320 pp

Gobejishvili R (2011) Relief of Georgia, Monograph. Universali, Tbilisi (in Georgian)

Gobejishvili R, Lomidze N, Tielidze L (2011) Late Pleistocene (Wurmian) glaciations of the Caucasus. In: Ehlers J, Gibbard PL, Hughes PD (eds) Quaternary glaciations: extent and chronology. Elsevier, Amsterdam, pp 141–147. https://doi.org/10.1016/b978-0-444-53447-7.00012-X

Khazaradze R (2004) Kavkasiis Pleistocenuri Gamkinvareba [Pleistocene glaciation in the Caucasus]. Metsniereba, Tbilisi (in Georgian), pp 319

Maruashvili L (1981) Kavkasiis fizikuri geografia [Physical geography of the Caucasus], Monograph. Metsniereba, Tbilisi (in Georgian)
Maruashvili LI, Khazaradze RD (1981) Ancient glaciation of the Caucasus (The USSR). Perspectives in Geomorphology, Concept's International Series in Geography, No 2, vol 1, pp 281–300. New-Delhi, India
Nemanishvili S (1962) For the geomorphology of the northern part of Kvemo Racha. Works of Vakhushti Institute of Geography, vol. 13. Tbilisi (in Georgian)
Nemanishvili S (1982) River terraces of mountainous areas, vol 2. Tbilisi (in Georgian)
Tielidze L (2016) Modern and old glaciers of Georgia, monograph. Samshoblo, Tbilisi, pp 216
Tielidze L (2017a) The morphological types, exposition, snow, and firn line location of the glaciers of Georgia. In: Chapter in Glaciers of Georgia, Springer, Geography of the Physical Environment. https://doi.org/10.1007/978-3-319-50571-8-4
Tielidze L (2017b) Dynamics of the glaciers of Georgia. In: Chapter in glaciers of Georgia. Springer, Geography of the Physical Environment. https://doi.org/10.1007/978-3-319-50571-8-5
Tielidze LG (2017c) Late Pleistocene and Holocene glacier extent in the Georgian Caucasus. Open J Geol. https://doi.org/10.4236/ojg.2017.74036.7.517-532
Tielidze LG, Wheate RD (2018) The Greater Caucasus glacier inventory (Russia, Georgia and Azerbaijan). Cryosphere 12:81–94. https://doi.org/10.5194/tc-12-81-2018
Tsereteli ED (1966) Geomorphology of the southern slope of the central Caucasus (within Racha). Author's abstract of the candidate's thesis, Tbilisi (in Georgian)
Ukleba D (1983) Qartuli Sabchota Enciklopedia [Georgian Soviet Encyclopedia], T. 6. Tbilisi (in Georgian), pp 151
Vakhushti B (1941) Agcera samefosa Saqartvelosa [Description of Georgian Kingdom], Geography of Georgia. Tbilisi State University Press, Tbilisi (in Georgian)

Levan Tielidze is a Ph.D. of Ivane Javakhishvili Tbilisi State University in the field of geomorphology-glaciology-cartography and a senior research scientist at Vakhushti Bagrationi Institute of Geography. He is also a researcher-scholarship holder of the Department of Earth Sciences of the Georgian National Academy of Sciences. The field of his research is mountain cartography and glacial-geomorphological study of the mountainous areas in the Quaternary (Late Pleistocene and). He is a member of several international organizations. In 2014/2015, he conducted a research stay at the Climate Change Institute of the University of Maine, USA, and in 2015/2016–2017 at the University of Northern British Columbia, Canada. He is the author of about 40 scientific papers, maps and eight monographs.

Ramin Gobejishvili (1941–2014) was a Doctor of Geographical Sciences, a geomorphologist-glaciologist. Until his last days he was Professor of the Faculty of Exact and Natural Sciences, Ivane Javakhishvili Tbilisi State University and the chair of Geomorphology and Cartography. In 1996–2008 he was the president of the Geographical Society of Georgia, and in 2005–2007—the director of the Institute of Geography. He was the author and co-author of about 150 scientific works and ten monographs.

Levan Maruashvili (1912–1992) was a prominent Georgian geographer. Doctor of Geographical Sciences (1954), Professor (1965). Honored Scientist of the Georgian Soviet Socialist Republic (1966). Honorary member of the Russian Geographical Society (1985). He produced over the 500 scientific and popular scientific works in Georgian, Russian, and English languages.

Nikoloz Astakhov (1911–1993) was a Doctor of Geographical Sciences, Professor (1965). He was researcher at the Department of Geomorphology of the Vakhushti Bagrationi Institute of Geography. His research area was structural geomorphology.

10 Eastern Greater Caucasus

Levan Tielidze, Ramin Gobejishvili and Alexander Javakhishvili

Abstract

This chapter presents the geomorphological characterization of the sub-region of the eastern Greater Caucasus. The orographical units are considered in detail, including: the main watershed range of the eastern Greater Caucasus, Khokhi Range, the gorges of the Tergi River and its tributaries, Kuro, Shavana and Kidegani ranges, the Asa River gorge, Khevsureti Range, the Arghuni River gorge, Atsunta and Pirikita ranges, the Tusheti depression, Keli volcanic upland, Lomisi (Alevi) Range, the Ksani, and Aragvi river gorges, Gudamakari and Kartli ranges, the Iori River gorge, Kakheti Range and Kakheti Caucasus. The text also provides the geological structure, tectonic conditions and current exogenic processes of the study region. Also the rates of modern and Late Pleistocene glaciations of the eastern Greater Caucaus are considered. Each subchapter is accompanied by appropriate images and maps.

Keywords

Eastern Greater Caucasus · Khokhi Range Tergi River and its tributaries · Kuro Range Shavana Range · Kidegani Range · Asa River gorge · Khevsureti Range · Arghuni River gorge · Atsunta Range · Pirikita Range Tusheti depression · Keli volcanic upland Lomisi (Alevi) Range · Ksani River gorge Aragvi River gorge · Gudamakari Range Kartli Range · Iori River gorge · Kakheti Range · Kakheti Caucasus

L. Tielidze (✉) · R. Gobejishvili · A. Javakhishvili
Department of Geomorphology and Geoecology, Vakhushti Bagrationi Institute of Geography, Ivane Javakhishvili Tbilisi State University, Tbilisi, Georgia
e-mail: levan.tielidze@tsu.ge

L. Tielidze · A. Javakhishvili
Department of Earth Sciences, Georgian National Academy of Sciences, Tbilisi, Georgia

R. Gobejishvili · A. Javakhishvili
Department of Geography, Faculty of Exact and Natural Sciences, Ivane Javakhishvili Tbilisi State University, Tbilisi, Georgia

10.1 Research Area

The eastern Greater Caucasus is located in the east of Mt. Vatsisparsi (3573 m). Only its ~200 km long section lies within the territory of Georgia up to Mt. Tinovroso (3374 m). The northern slope of the Greater Caucasus also belongs to Georgia from Vatsisparsi up to Mt. Shaviklde (3578 m). The main watershed range reaches its maximum width of 60 km on the

L. Tielidze (ed.), *Geomorphology of Georgia*, Geography of the Physical Environment,
https://doi.org/10.1007/978-3-319-77764-1_10

meridian of the Arghuni and Iori rivers, and the minimum width of 15 km at the Azerbaijan border. Hypsometrically, the eastern Greater Caucasus is far behind the central Greater Caucasus. The highest peaks are Mt. Mkinvartsveri (Kazbegi), 5047 m and Mt. Tebulosmta 4492 m. Many peaks exceed the height of 4000 m (e.g. Komito, Diklosmta, Kideganismaghali, Kurostsveri, Shavana) (Fig. 10.1). Despite such heights, the glaciers are less developed in the eastern Greater Caucasus than in the central and western Greater Caucasus due to the climate conditions.

The eastern Greater Caucasus is built of Jurassic age shale and sandstone suites. In the southern part, the relief is built of the Cretaceous carbonate flysch. In the construction of the terrain the Paleozoic metamorphic rocks (in Kakheti), intrusives (Dariali gorge) and diabases (in Chaukhi and Chimghisklde) are involved. Effusive volcanic sediments also participate in the terrain structure. Tectonically, it is by isoclinals folds; the Jurassic folds are turned over to the south. A special role in the formation of structures is given to the latitudinal fault lines. Overthrusts and faultings are directed from the north towards the south. The carbonate flysch is distinguished by the amount of faulting in the Pshavi area. Tectonic fault lines—the lineaments are well expressed on both slopes of the Greater Caucasus (especially in Tusheti and Pshavi).

Modern glaciers and nival processes are important in the modeling of the high mountainous terrain of the eastern Greater Caucasus. Modern glaciers are mainly located in the old corrie forms located at the higher altitudes. Main centers of Pleistocene and modern glaciations are associated with the Kazbegi massif. Relatively smaller glaciers are distributed in the Chaukhi massif and hypsometrically high lateral ranges (Pirikita, Atsunta, Khevsureti, Kidegani, Shani and Kuro). The glaciers are small sized and morphologically belong to hanging, corrie and corrie-valley types (Tielidze 2017a). Lateral and terminal moraines and corries are well developed in the basins of the glaciers. It is noteworthy that the deposits of modern and old rock glaciers are widespread in the high mountainous region. Slopes are covered by debris fall cones and rockfills. Over 3500 m above sea level, the relief is under total denudation.

The major role in the formation of the topographic surface of the eastern Greater Caucasus belongs to the Würm glaciation, though due to lithological conditions of structural rocks of the relief, the glacial forms are incompletely preserved. The low morphometry of the eastern Greater Caucasus and the continental climatic conditions defined the lesser development of glaciers in Late Pleistocene and at present as well, than in the central Greater Caucasus. The traces of the Late Pleistocene glaciation are preserved in the high mountainous zone over 2600 m in the forms of degraded cirques, troughs and moraines. The Chaukhi massif was a powerful center of glaciations, where the length of the glaciers exceeded ~11.5 km. The Roshka Glacier descended to the lowest altitude of 1600 m asl. The larger glaciers (Kistura, Chaukhi, Akhieli, Shatili, Tebulo, Didkhevi, and Diklo) were developed in the Khevsureti, Kidegani, Shani, and Pirikita ranges (Gobejishvili et al. 2011; Tielidze 2017b).

The nival-glacial zone of the eastern Greater Caucasus includes the relief located above 3500–3600 m. Due to total denudation the soft and rounded shapes are related to soft rocks. In solid rocks pointed forms, karlings, deblis fall cones, and rockfills dominate.

In the eastern Greater Caucasus, the volcanic relief is presented only in its western part. Among the volcanic land forms, the volcanic upland is particularly remarkable, the characterization of which is presented in the individual subchapter below. The central volcanoes are as follows: Sakokhe (3268 m), Sadzele (3080 m), Miliona, Tsiteldziri, Kalko, and Kabarjina (3135 m). The lava, erupted from Sakokhe, extends through the paleo Tetri Aragvi gorge up to the Kvesheti Village (1370 m). The volcanic lava is spread upward in the Khadistskali gorge, which is reflected in the morphology of the gorges. The Tetri Aragvi and Khadistskali rivers cut the gorges at the contact of base rocks and volcanic lavas. Rocky slopes are formed in the area of volcanic lava, contrasting with gentle slopes in the base rocks. The surface of the lava

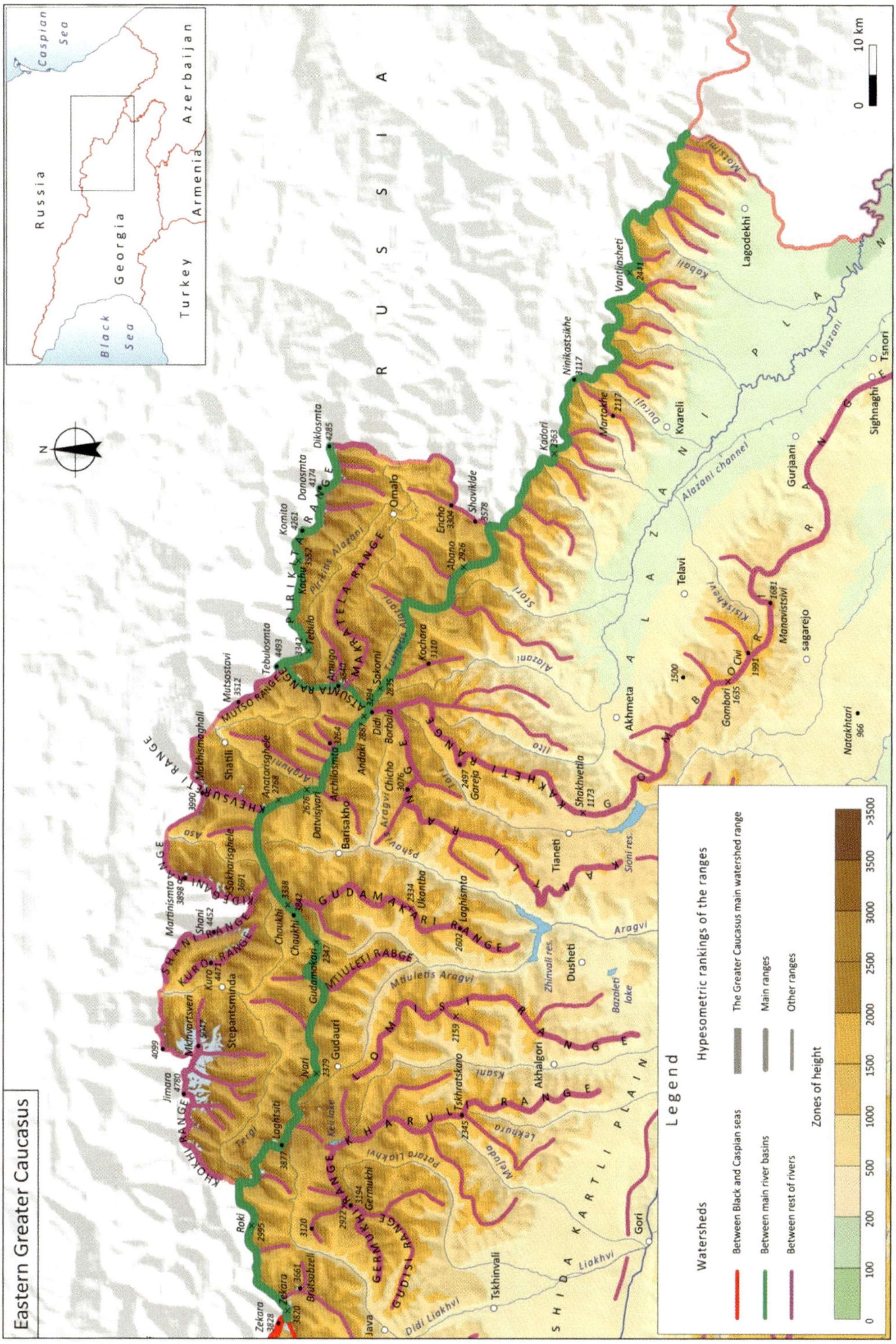

Fig. 10.1 The orography of the eastern Greater Caucasus

flow is terraced. The villages of Kaishauri and Seturebi and a mountain ski resort Gudauri (Fig. 10.2) are situated on the plain surfaces. Erupted lava from the volcanoes of Miliona, Sadzele, and Tsiteldziri extends on the northern slope. The Kabarjina volcanic cone and its surroundings are interesting. The bottom of the volcano is spread along the road at a distance of ~7 km from the Kobi Village up to the Sno Village. The volcano slopes are eroded with erosion-denudation, resembling badlands. The columns occurring within tuffo-breccias due to denudation, are common in the vicinities of the Sno Village. Volcanic lavas and loose efflusives are composed of andesite-dacites, dacites, and andesite-basalts (Gobejishvili 2011).

Erosion forms and processes are well developed on both slopes of the eastern Greater Caucasus. The northern slope of the main range and the slopes of its subranges of Khokhi, Kuro, Shavana, Kidegani, Khevsureti, Komagi, Mutso, Atsunta, and Pirkiti are dissected by erosive action of the rivers of Chkheri, Mna, Suatisi, Snostskali, Khdestskali, Asa, Arghuni, Andaki, Pirikita, and Tusheti Alazani. The gorges are deeply cut in the relief to 1000–1500 m in depth. The gorges are formed in Jurassic sediments. The slopes are characterized by different inclinations, developing narrow cuts in some sections. The main gorges (except the rivers of Tusheti) crosscut the geological structures; therefore, the morphology of the gorges is determined by the resistance of structural rocks to exodynamic processes. The peaks of the ranges of Kuro, Shani, Kidegani, Khevsureti, Atsunta, and Pirikita, exceed 4000 m. Their high hypsometry is favorable for distribution of small-sized modern glaciers. The tributaries of the main rivers develop debris cones on the bottom of the gorge. The old debris cones are associated with the existence of settlements and arable lands.

The morphology of the southern slope of the eastern Greater Caucasus is defined by the main gorges and ranges of meridional direction. The gorges of the Ksani, Mtiuletis Aragvi, Pshavis Aragvi, Iori, Alazani rivers and its numerous left tributaries are morphologically distinguished. Their watersheds are the Lomisi, Gudamakari,

Fig. 10.2 A mountainous-ski resort of Gudauri built on the volcanic lava (*photo by* L. Tielidze)

Kartli, and Kakheti ranges. The watershed ranges decrease in heights from the north to the south (from 3000–3200 to 1000–1200 m). In general, these ranges are of average height. Their crests are located in the alpine zone and are characterized with morphologically soft peaks and slopes. The slopes are dissected by river erosion actions. The rivers cross the structures transversally. The main gorges are deeply cut in the relief, at a depth of 1000–1500 m. Their morphology is determined by the lithology of the structural rocks together with the erosive processes. Due to the low hypsometry, the Pleistocene glaciations were weakly represented; modern nival processes are developed only in the area of some peaks.

The erosive dissection of the terrain was mainly undergone perpendicular to the tectonic structures; therefore, the direction of the main gorges and their watersheds are meridional and submeridional. The gorges and ranges of latitudinal direction are developed in the Tusheti and Ukana Pshavi depressions.

The exogenic processes are characterized by vertical distribution of the processes starting from nival-glacial and ending with erosive, denudative, and accumulative processes.

From a geomorphological point of view the eastern Greater Caucasus is divided into the following orographical units/regions: (1) the main watershed range; (2) the main ranges and gores of the northern slopes of the eastern Caucasus (Khokhi, Kuro, Shavana, Kidegani, Atsunta and Pirikita ranges; the Tergi, Asa and Arghuni river gorge and Tusheti depression); (3) main orographic units of the southern slope of the eastern Greater Caucasus (the Keli volcanic Upland, Lomisi, Kartli and Kakheti ranges, the Ksani, Aragvi and Iori river gorges); and (4) the Kakheti Caucasus.

10.2 Eastern Greater Caucasus Watershed Range

The leading orographic unit in the eastern Greater Caucasus is a main watershed range, which separates the Mtkvari River basin from the Tergi and Sulak river basins. It is far below the central Greater Caucasus with morphometric indicators.

The watershed range stretches from Mt. Vatsisparsi (3573 m) to Mt. Tinovroso (3374 m). The orographic appearance of the range is heavily curved due to river erosive actions, particularly in the Khevsureti section. The highest peak of the main watershed range is Chaukhi (3842 m). Only two peaks are above 3500 m, the Roshka-Khokhi and Shavi Klde (3578 m). There are low passes in the eastern Greater Caucasus (except the Abano Pass at 2926 m)—Jvari (2379 m), Bursachiri (Gudamakari) (2347 m), Datvisjvari (2676 m), and Kadori (2366 m). The watershed range is built of the Jurassic suites, Upper Lias shales and sandstones. The Chaukhi massif is constructed from diabases, while the metamorphic sedimentary rocks of the Upper Paleozoic and Triassic age are found in the Kakheti section. The volcanic formations are prevalent distributed in the Mtiuleti Range at the heads of the Aragvi and Tergi gorges.

The nival-glacial, ancient glacial and erosive processes are important in the formation of the relief of the watershed range. The nival landscape extends mostly to high mountainous sections—Chaukhi, Roshka-Khokhi and other peaks that exceed 3400 m.

The modern glaciers are distributed only on the northern and southern slopes of the Chaukhi massif. The moraines of the LIA maximum are located adjacent and in front of the glacial tongue. The snow avalanches have a certain role in the modeling of the relief of the watershed range. There are several degraded and one well-defined rock glaciers around the Chaukhi massif (in the head of the Chaukhi River). The Chaukhi rock glacier is 40–50 m thick with a straight, slightly angled surface (table-shaped) (Fig. 10.3). The steep slopes of the Chaukhi massif are intensely weathered and their surroundings are covered with debris fall cones. Plain areas are covered with rockfills.

The main role in modeling the relief of the watershed range of the eastern Greater Caucasus belongs to the old glaciation. Traces of the Late Pleistocene glaciations are preserved along the entire length of the range over 2600 m asl in the

Fig. 10.3 Chaukhi rock glacier (*photo by* L. Tielidze)

forms of the degraded cirques, troughs, and moraines in some areas. The Chaukhi massif was a powerful center of glaciation there. In Late Pleistocene glaciers descended from this massif in all directions. On the southern slope of the watershed the largest glaciers were located in the Bakurkhevi and especially in the Roshka gorges. The glacier of the Bakurkhevi gorge overflowed into the Kvenamtistskali gorge through the Bursachiri Pass. The glacier developed in the Roshka gorge descended to the Khevsuretis Aragvi River gorge. The length of the glacier was ~11.5 km and its tongue ended at ~1600 m (Gobejishvili 1995; Tielidze 2017b). The Holocene stadial (rock glaciers) moraines are deposited in the bottom of the Würm trough above the Roshka Village. The glacier developed on the northern slope of the Chaukhi gorge, occupied the area of the Juta Village and blocked the Juta gorge. The moraine deposits are found on the both slopes of the Juta gorge, and the Chaukhi gorge is trough-shaped, the slopes of which is covered by the debris falls.

The young volcanic terrain is represented only in the western section of the watershed range, both on the crest and on the northern and southern slopes. Among the volcanic centers Sakokhe, Sadzele, Tsiteldziri, and Miliona should be noted including their lava flows: the Sadzele-Tsiteldziri and andesite-basalt lava was erupted from Sakokhe and extends on the southern slope up to the Kvesheti Village. The moto road passes through it (Mleta-Gudauri) and the mountainous-ski resort Gudauri is also located on it. The Tsiteldziri dacite cone is located in the Khokhi River gorge.

Slopes of the watershed range are characterized by gentle shape and dissected by erosion processes forming gores with depth 100–300 m.

The Jvari and Bursachiri passes are interesting from a genetic point of view, formed firstly due to erosive actions of the rivers oriented from south to north, and then glaciers of the same direction developing the trough shape, an indication of which are the glacial deposits found there (the presence of diabases unfamiliar for those areas). In the following period both rivers were overtaken by the reverse erosion of the rivers of Mtiuletis Aragvi and Bakurkhievi (Gobejishvili 2011). Both passes, along with Kadori (2367 m), are the lowest in the central and eastern Greater Caucasus. The Georgian Military Road crosses the Jvari Pass (2379 m).

10.3 Khokhi Range

The Khokhi Range comprises the eastern part of the lateral range of the Gretaer Caucasus. The lateral range (the Tepli Range) stretches ~55 km from the Ardon River up to the Tergi River in the latitudinal direction; only the eastern and southern slopes of the Khokhi Range lie within the territory of Georgia. The Khokhi Range is the highest in eastern Georgia (Fig. 10.4). Its highest and most beautiful peak Mkinvartsveri (Kazbegi) (5047 m) is located in the eastern part of the range (Fig. 10.5). The second highest peak is Mt. Jimara (4780 m). Together they are known as the Kazbegi-Jimara massif (Kutuzov et al. 2016). The valley type glaciers of Mydagrabyn, Maili, Kolka (Russia), Suatisi, Mna, Gergeti, Abano, Chachi, Devdoraki (Georgia) are located on the slopes of the Kazbegi-Jimara massif and characterized by strong glacial hazards (Tielidze and Wheate 2018). Other high peaks of the Khokhi Range are Mt. Maili (4598 m), Suatasi (4552 m) and Ortsveri (4222 m).

The Khokhi Range is built of the Lower Jurassic suites: black clay slates, sandstones, dacite lavas and lenses of conglomerates. In the Mkinvartsveri massif, the Quaternary cones and lavas of effusives are common: andesite-dacites and andesites. The Quaternary period is represented by glacial and colluvial deposits. An active role in the formation of the Khokhi Range belongs to the glacial, volcanic, ancient glacial (Late Pleistocene) and erosive processes.

Volcanic formations are widely represented in the surroundings of the Kazbegi massif. Kazbegi is a dormant volcano with a two-headed cone and several side cones. The main cone erupted about 185.000 ± 30.000 years ago, while the side cone of New Tkarsheti erupted about 6000 years ago

Fig. 10.4 Khokhi Range (*photo by* L. Tielidze)

Fig. 10.5 Mt. Mkinvartsveri (Kazbegi), 5047 m asl (*photo by* L. Tielidze)

(Chernyshev et al. 2002). The lava flows descended from the volcanic massif were radially distributed in the old gorges. Among the volcanic cones and lavas, the following are notable: in the south of the massif, in the Mna River gorge, the Mna lava flow descended from the Sherkhota (Sirkhisari) volcano; in the east, from the volcano of Didi Tkarsheti the lava flow overflowed into the Tergi River bed through the Kesia gorge and blocked it. The narrow, deep, and almost impassable gorges were developed in the edges of volcanic lava and at the contact of base rocks. In Middle Holocene (6000 years ago), the lava erupted from Patara Tkarsheti blocked the Tergi River bed, the dammed Tergi River cut the lava flow and formed the 30–40 m deep canyon (Gobejishvili 2011) (Fig. 10.6). The Arsha, Chkheri, Tsdo, and Gveleti lava flows descended from the eastern slope of the Mkinvartsveri massif. They hardly reached the bottom of the Tergi gorge. Among them the longest were the Chkheri and Tsdo flows (8–9 km).

A great role in the modeling of the Khokhi Range belongs to the modern and Pleistocene glaciation. In the late Pleistocene there was a powerful glacial cover in the Khokhi Range. The Mkinvartsveri-Jimara massif was covered with an ice cap and glaciers of 14–17 km long descended in all directions of the massif and their tongues overflowed into the Tergi River bed. The glaciers created jams in the gorge and often supported the creation of powerful glacial mudflows. The Devdoraki Glacier descended at the lowest altitude—~1200 m, even passing the Dariali gorge. The glacier brought the huge boulder—Ermolov's rock. The Gergeti Glacier ended at the height of ~1550 m asl, the glaciers of Mna and Suatisi at ~1950 and ~2150 m respectively. Valley type glaciers were located in the other river basins (Jimara, Tepistskali, Resistskali, and Siverauti). Their lengths were 5–8 km, and some of them reached the Trusso depression (Gobejishvili 1995; Tielidze 2017b).

In the Khokhi Range the modern glaciers are mostly spread in the Kazbegi-Jimara massif. The large glaciers descend from the massif in the radial direction, the firn basins of which are connected to each other. The large glaciers are: Chachi, Devdoraki, Gergeti, Mna, and Suatisi. The tongue of the Devdoraki Glacier descends to the lowest altitude asl. Morphologically, these glaciers are of hanging-valley type glaciers. The Gergeti Glacier is the longest and the Eastern

Fig. 10.6 Patara Tkarsheti lava (canyon) dissected by the Tergi River (*photo by* L. Tielidze)

Suatisi (8.89 ± 0.09 km^2) is the largest in area (Tielidze and Wheate 2018).

The relief of the Khokhi Range is dissected by the gorges, which are modified troughs. The depth of the cut of the gorges formed by the fluvio-glacial processes is 800–1000 m. The geomorphological landscape in the nival zone consists of corries, debris falls, karlings, rockfills, and rock flows.

10.4 Tergi River and its Tributaries

The Tergi gorge is located on the northern slope of the Greater Caucasus and covers the ethnic region of Khevi. The lowest point of the gorge is at a height of 1210 m asl in the Dariali gorge (Kldekari), while the highest point is Mt. Mkinvartsveri 5047 m. The main morphometric and morphological units of Tergi gorge relief are formed by interaction of tectonic and erosive-accumulative processes. The main background of tectonic-erosive terrain is complicated by the forms of glacial, volcanic, gravitational, and karst origin.

The Tergi River basin is built of Jurassic suites: black clay slates, aleurolites, and in some places limestones, conglomerates and sandstones. In the Dariali gorge the Paleozoic gray granitoides are outcropped. The Quaternary and Holocene volcanic lavas, alluvial, proluvial, glacial and colluvial layers are common in the relief (Tsereteli 1966). The Tergi gorge is characterized by complex tectonic nature. Structures are separated from each other by fault lines and overthrusts. Against the background of differential movement of the relief, the lateral and main watershed ranges are notable with high rates of uplifting (10–15 mm/year) (Gobejishvili 2011).

Morphologically, the Tergi gorge is divided into three parts: 1. from the head until the Okrokana Village is called the Truso depression and is of sublatitudinal direction (Fig. 10.7); 2. the Okrokana—Stepantsminda section of submeridional direction; and 3. the most narrow and rocky Dariali gorge of meridional direction.

The Truso depression is formed mainly by tectonic and erosive processes. Truso depression length is about 20 km and the width of the bottom is 1–2 km. The depression is located above

Fig. 10.7 Truso depression, the Tergi River upstream (*photo by* L. Tielidze)

2000 m. The Tergi River flows in bayous through the bed covered with gravellites. The lateral tributaries create powerful debris cones. The formation of the Truso depression was caused by the erosive-tectonic processes together with the Khorisari volcanic lava descended from the right side of Tergi. The volcanic cone of the Patara Khorisari and the volcanic lava are developed in the bottom of the trough gorge. The lava flow is three-stepped. There are several small lakes located in the cirques between the volcanic cone and slope. The sides of the lava flow are elevated by 80 m from the surface of the lava, and are separated from the main slopes by dingles. In Middle Holocene (6000 years ago) the lava flow closed the gorge and a powerful water reservoir was created. After damming, the Tergi River formed a narrow and deep asymmetric gorge or "Kasriskelis vitsrobi" (the Kasriskeli narrowness) by the depth erosion at the contact of base rocks and lava flow (Fig. 10.8). The length of the Kasriskeli narrowness is 3–4 km. The fragmented columns of the lava cliff of 300 m in height borders the Tergi River from the right side, while the left side is built of shale and is less steep. The Kasriskeli narrowness is a lava flow configuration.

The subranges of the main and lateral ranges and the modified trough gorges located between them descend to the bottom of the Truso depression. In Würm the most powerful was the Suatisi Glacier located in the Khokhi Range, and its tongue outflowed in the Tergi gorge. In the gorges of the Truso depression, the traces of Würm and Holocene glaciations are well preserved in the forms of corries and troughs. Lateral moraines are preserved only in some gorges.

In the Truso depression the Tergi River bed is cut into the old (when there was a lake) alluvial-prolovial layers. The downfall of the bed is in the bottom of the depression is 10–15 m/km, and in the Kasriskeli narrowness is 30–35 m/km.

On the right slope of the Truso depression, in the areas of limestones distribution, the karst sinkholes are developed, namely in the Desistskali gorge. In the relief, in the different areas, a

Fig. 10.8 Khorisari volcanic flow and the Kasriskeli volcano (Google Earth imagery, 12/07/2014)

Fig. 10.9 Travertine fields in the Tergi River gorge (the Truso depression) (*photo by* L. Tielidze)

wide cover of travertines is visible. Above the Kasriskeli narrowness, at a height of 220 m, the fascinating "Travertine fields" (Fig. 10.9) are in the process of intensive development.

The Okrokana-Stepantsminda section is morphologically interesting. The bottom of the Tergi River gorge is quite wide ~0.5–1 km, and is filled with a powerful alluvion. The thickness of the alluvion at Stepantsminda is over 250 m. The origination of the alluvial plain is caused by many factors (the old landslide and the Kuro mudflow gorge descended from Kuro at Stepantsminda, and descending of the tongue of the Gergeti Glacier in the Tergi gorge in Würm). The leading factor is the live tectonic fault line in the common Caucasian direction, which passes through the town of Stepantsminda. Differential movements of the Earth's surface and other factors conditioned the accumulation of a strong alluvion along the entire length of the valley. The river flows in the form of bayous. The downfall of the bed is 8–12 m/km. In the bottom of the gorge the old and new debris cones are formed (Kanobi, Khurtisi, Goristsikhe and others) (Fig. 10.10).

In this part of the Tergi gorge, the volcanic relief is interesting, which is formed by the lavas erupted from the volcanic cones of Kabarjina, Didi Tkarsheti and Patara Tkarsheti. In the Late Pleistocene, the Khurtisi lava erupted from Didi Tkarsheti, extended through the paleo-Kesia gorge and blocked the Tergi gorge. The lava erupted from the Patara Tkarsheti parasitic cone in Middle Holocene, blocked Tergi and the lake was formed. In the lacustrine sediments the tree remnants were found, which were dated as 6000 years (Gobejishvili 2011). In this section the Mna, Kesia and Chkheri rivers conflow the Tergi River from the left side, while the Snostskali and Bidara rivers conflow it from the right side.

The leading role in the formation of the Mna River gorge belongs to the glacial, erosive and volcanic processes. In the Pleistocene, the lava flow erupted from Sherkhota (3625 m) in the paleo-Mna River gorge, and extended up to the Kobi Village. The glacial river and the Würm glacier of Mna formed a 400–600 m deep gorge at the contact of the base rocks and lava with their joint action. In the relief, on the left side of the gorge, a 6–7 km long and 300–500 m wide step covered with lava flow is preserved, on the surface of which the lakes are located. By the action of the left tributaries of the Mna River, the surface of the old lava flow is divided into the separate sections—mesas.

The Chkheri River gorge was formed by joint action of erosive, volcanic and glacial processes along its entire length.

Fig. 10.10 Tergi gorge, the Okrokana-Stepantsminda section (*photo by* L. Tielidze)

The Snostskali (Juta) gorge is the alongside depression. It is mainly built of the Middle and Lower Jurassic suites: black clay slates and aleurolites. Its shape is formed by actions of erosion and accumulation. The Late Pleistocene glaciers played a significant role in the heads of the Snostskali gorge, which resulted in forming cirques, troughs and moraines. The Würm glaciers descended to the height of 2000 m. The Chaukhi Glacier descended to the lowest altitude and blocked the gorge. The signs of this are the glacial deposits with large-sized angular diabase boulders (Fig. 10.11) on the right slope of the gorge at the Juta Village. A modern table-shaped rocky glacier is formed in the head of the Chaukhi gorge. The entire lower half of the Snostskali gorge is wide and built of gravelites, and the old and newer debris cones are deposited there.

The Bidara gorge is erosive; here, on the left slope of the gorge the travertines are beautifully expressed.

Below Stepantsminda the Tergi gorge is antecedent; it is cut in the axial zone of the Greater Caucasus and forms a rocky corridor. This section of the gorge is called Dariali gorge (Fig. 10.12). The most narrow, rocky and severe part of Dariali is cut in the Paleozoic granitoides, between the Amali and Khdestskali conflows, where our ancestors constructed a fortress (the Tamari Fortress) on the erosive witness. Terraces are developed in the gorge in fragments. Volcanism plays an active role in the formation of the relief.

The Amali River conflows the Tergi River from the left side and the Khdestskali River conflows from the right side. The Amali River gorge is of a glacial and erosive origin. The large glaciers of the Kazbegi mssif, the Devdoraki and Chachi glaciers descend in the head of the Amali gorge. The upper parts of the Amali gorge are of trough type; in the lower part of the gorge, the trough bottom is deeply cut by the river in granites. In this section of the gorge, a 2 km long and 300–500 m wide Würm step is found in the right side, which is covered by the moraine hills of the Holocene stadial glaciation. The lakes are formed among the moraines. The Devdoraki Glacier is characterized by glacial avalanches, often blocking the Tergi River gorge and causing destructive floods and mudflows. The last glacial avalanche was observed in May 17, 2014, when

Fig. 10.11 Glacial diabase boulders in the Juta River gorge (*photo by* L. Tielidze)

Fig. 10.12 Tergi River gorge (Dariali gorge) (*photo by* L. Tielidze)

the rocky and icy mass collapsed at about 4500 m asl on the heavily inclined north-eastern steep slope of Mkinvartsveri. The collapsed mass was transformed into a rocky-ice avalanche, which was split into three flows and with high speed descended on the surface of the sloping glacier tongue. The total volume of mudflow mass brought out from the detaching wall up to the Tergi River confluence was estimated to be 5–6 million m^3 (Fig. 10.13). As a result of the disaster nine people died and the road of international importance connecting Georgia-Russia was damaged. The disaster breached the international (Transcaucasian) gas pipelines, brought down the high voltage-power transmission mast, separated the border guards' base, the customs checkpoint, and the residence of the Georgian Patriarchate from the outside world and created great problems for their staff (Tielidze 2016).

The upper part of the Khdestskali (Brolistskali/Kistura) gorge is trough-shaped; it is developed from the Shavana and Kuro ranges by actions of the glaciers descended in the Late Pleistocene. The Khde gorge is an isolated kingdom of rocks, detritus, alpine meadows, beautiful, and small-sized hanging glaciers and flows with rapids and waterfalls. The Kdestskali River enters the erosive gorge from its middle section, and its ending section is cut in the granites creating a continuous stairway of foamed rapids. The length of this section with rapids is about 4 km and the height is 300 m. In the Khde gorge rare but powerful mudflows are developed. The Kibesha Glacier is located in the head of the Khde gorge.

10.5 Kuro Range

The Kuro Range is a watershed between the Tergi and Khdestskali river gorges. It is of submeridional direction, stretching a distance of ~16 km from Mt. Kora (3631 m) up to the confluence of the Khdestskali and Tergi rivers. The highest peaks are Kurostsveri (4089 m) (Fig. 10.14), overlooking the Stepantsminda settlement, and Mt. Shino (4047 m). The relief of the range is built of the Lower Jurassic clay slates and sandstones; in some areas the diabases are outcropped. Small sized glaciers are located on the northern slope of the Kuro Range. Tectonic, nival-glacial, erosive, and weathering processes

Fig. 10.13 Consequences of the Devdoraki glacier disaster in the Dariali Gorge in May 17, 2014

participate in the creation of the look of the range. Glacial forms are predominantly presented in high mountainous zone, where debris fall cones are common and below, the slopes are dissected by erosion processes and snow action. It is noteworthy that a powerful debris cone has formed by the mudflow occurring on the western slope of the Kuro Range on which the Stepantsminda settlement is built.

10.6 Shavana Range

The Shavana Range is a watershed of the Armkhistskali and Khde gorges. Some peaks are over 4000 m high, such as Bachakhi (4291 m), Rustavi Peak, Gori, and Japaridze. The highest is Mt. Shani (4452 m). The relief is built of the Lower Jurassic clay slates and sandstones. There

Fig. 10.14 Kuro Range and the Mt. Kurostsveri (4089 m) (*photo by* L. Tielidze)

are no passes in the range due to the morphological and morphometric character. The small-sized modern glaciers are of hanging type. The leading role in the formation of the relief belongs to the glacial and nival processes. Due to the active weathering processes, debris fall cones are developed on the slopes.

10.7 Kidegani Range

Only the eastern slope of the Kidegani Range lies within the territory of Georgia. The range is of meridional direction and is a watershed of the basins of the Asa and Armkhi rivers. Hypsometrically the range is high including the peaks of Chimghismaghali (3970 m), Kideganismaghali (4262 m), Akhielismaghali (4274 m), and Sakharismta (3963 m). There are several passes in the range, the heights of which are over 3400 m (Samtekhlosghele, Sakharisghele, Sajikhvesghele and Salashkrosghele). The relief is built of the Lower Jurassic clay slates and sandstones. There are four glaciers located in the highest central part of the range (in the Akhielistskali River basin), the largest of which is the Akhieli Glacier of corrie-valley type with eastern exposition. The nival-glacial processes, especially the Late Pleistocene glaciation, played an important role in the formation of the relief. The glacial forms and the forms created due to weathering are well represented: cirques, karlings, debris fall cones, modified troughs and small sized rock glaciers. The lithological softness of the structural rocks contributed to the strong dissection of the slopes.

10.8 Asa River Gorge

Only the heads of the Asa River get within the territory of Georgia at a distance of 20–22 km. Its lower point is located at 1220 m. In the south it is bordered by the Greater Caucasus watershed range, in the west by the Kidegani Range, and in the east by the Khevsureti Range (Fig. 10.15). The Asa River Basin is built of the Lower Jurassic (Lias) schists and slates with the middle

Fig. 10.15 Asa River gorge (*photo by* R. Kumladze)

layers of aleurolites and fine-grained sandstones. The Asa River tributaries are Chimghistskali, Akhielistskali and Chkhotani from the left, and Tskarostavi, Taniestskali (Kogra) and Bisna from the right. The Asa gorge is mainly of meridional direction, morphologically V-shaped and deeply cut in the relief. The depth of the cut is 2000–2500 m below Akhieli, and 1500–2000 m above it. The bottom of the gorge is inclined, and the common downfall is 80 m/km. In the heads of the tributaries of the gorges, modified troughs are formed. The other sections of the gorges are erosive. Tanie and Tskarostsavi gorges are interesting: in the heads, in the bottom of the gorges, the rock glacier's material has spread, which is overlapped by powerful debris fall cones from the sides. Smaller lakes are common in the gorges. The river terraces are weakly developed.

10.9 Khevsureti Range

The Khevsureti Range connects the Greater Caucasus watershed range at Mt. Tanie (3499 m), in a northeastern direction and extends to Mt. Makhismaghali (3990 m). The Khevsureti Range is a watershed of the Asa and Arghuni river basins. The lowest pass is the Isartghele (3486 m). The relief is built of the Middle and Lower Jurassic shales, clay slates, and sandstones. Glaciers are spread on both sides of the range, but are small in size. In front of them the terminal stadial moraines are well preserved. The old glaciations played a great role in the transformation of the relief. In Würm the glaciers descended much below, where their effect is preserved as modified troughs. Old and modern rock glaciers are common in the higher zone of the range. Due to the active weathering processes debris fall cones are formed. The slopes of the range are dissected by the erosion of the Shatilistskali, Gurostskali, Giorgitsminda, Bisna, and Kogra rivers.

10.10 Arghuni River Gorge

The Arghuni River is represented by two main rivers in Georgia: Arghuni itself and its right tributary the Andaki River, which is longer than the Arghuni River (Fig. 10.16). The Arghuni and

Fig. 10.16 Andaki River gorge (*photo by* G. Dvalashvili)

Andaki gorges are bordered by the Greater Caucasus watershed range in the south, by the Atsunta and Mutso ranges in the east and north-east. The Arghuni and Andaki river basins are separated from each other by the Ardoti (Khomi) Range. These two rivers joins each other at the site of Anatori (1130 m asl). The gorges are deeply cut in the relief (1500–2000 m). Morphologically they are V-shaped, and in some areas they are narrow and antecedent. The relief is constructed from Lower Jurassic schists, slates, and sandstones. The direction of the main gorges are commonly northward and their main tributaries—Gurostskali, Giorgitsminda, Shatili, and Chanchakistskali are characterized with the latitudinal direction that coincides with structures. The creation of the heads of the gorges are caused by the actions of modern and old glaciers. In the relief there is a trace of the Würm glaciations preserved in the forms of the modified toughs. In the Late Pleistocene the largest glacier of ~8.5 km long was located in the Shatili gorge. Its tongue descended to the height of ~1800 m. In the surrounding areas of Shatili the lacustrine layers are preserved. Old glacial forms are preserved in the Ardotistavi massif. The Alerdoi River gorge has a trough form at a distance of 3–4 km, and in the head the old rock glacial layers are common. In the heads of Arghuni, on the slopes of the gorge landslides are developed (Khakhabo and Bakhao), while debris cones are formed in the bottom the gorge.

10.11 Atsunta Range

The Atsunta Range is of meridional direction. It is a watershed of the river basins of Tusheti and Pirikita Khevsureti. The range is stretched from the peak of Tebulo (4492 m) to the peak of Patara Borbalo (3134 m) joining the main watershed. The highest peak of the range is Amugho (3839 m). Hypsometrically the range is high, with small glaciers widespread. Tusheti is connected to Pirikita Khevsureti by the Atsunta Pass (3431 m). The Late Pleistocene glaciation played the main role in modeling relief, with modified troughs, corries, moraines, and karlings preserved. Nival forms are common. The slopes of the range are dissected by the erosive action of the rivers.

10.12 Pirikita (Tusheti) Range

The Pirikita (Tusheti) Range is the watershed of the rivers of Arghuni and Andi Koisu in a latitudinal direction. Only the southern slope of the range lies within the territory of Georgia. The eastern and western parts of the range are high with the peaks more than 4000 m asl, such as Mt. Tebulosmta (4492 m), Diklosmta (4285 m), Komito (4261 m), and Danosmta (4174 m). The crest of the range is the lowest between the peaks of Tebulosmta and Komito. Tusheti is connected with Chechnya by lower passes.

Glaciers in the Pirkita Range are located around the Tebulosmta and between the peaks of Komito and Diklo. According to Kurdghelaidze (1983), the Tebulo Glacier was the largest valley type of glacier in area with southern exposition by the 1960s, situated in the deep cirque of the southern slope of the Tebulosmta. According to his data, in the years of 1953–1968 the glacier retreated by ~ 107 m.

The Late Pleistocene glaciation made a large contribution to the formation of relief. Valley glaciers 5–7 km long descended to the height of 2100–2400 m in the Pirikita Alazani River tributaries: Khaoskhevi, Cheroskhevi, Didkhevi, Kachu, Parsmaskhevi, and Kvakhidi. The glaciers impact trace is preserved in the relief in the forms of corries, modified troughs, and moraines.

The southern slope of the Pirikita Range is dissected by the left tributaries of the Pirikita Alazani River below 2600–2700 m. The depth of erosive dissection is 500–1000 m. The relief in the upper part of the gorge is of erosive-glacial genesis, and below it is erosive with V-shaped form (Kurdghelaidze 1983). The deluvial aprons are developed on the slopes of the gorges.

On the southern slope of the Pirikita Range, the most beautiful villages of Tusheti are located, such as Parsma, Chesho, Dartlo (Fig. 10.17), Dano and other villages.

10.13 Tusheti Depression

An extensive mountainous depression of latitudinal direction is located on the northern slope of the Greater Caucasus, which includes the historical-ethnographic region of Tusheti. In the south it is bordered by the southern Greater Caucasus watershed range, in the north by the Pirikita and in the west by the Atsunta high ranges.

The Tusheti depression is divided into two alongside gorges of latitudinal direction. In the north the Pirikita Alazani basin is located and in the south the Tushetis Alazani basin (Gometsari gorge). Both basins are asymmetric. The Pirikita Alazani River mainly has left tributaries: Parsmaskhevi, Cheshoskhevi, Didkhevi, Chigoskhevi, Dikloskhevi, and Cheroskhevi. From the right tributaries, Larovanistskali is worth mentioning. The right tributaries of the Tushetis Alazani River are Ortskali and Khisos Alazani, while the Tsovatistskali conflows from the left. Pirikita and Tushetis Alazani join at the Shenako Village at 1600 m asl and form the Andi Koisu River. The lowest point of Tusheti at 1500 m asl is marked in its bed. It is noteworthy that in the Tushetis Alazani gorge, the highest in Georgia and one of the highest settlements in Europe, is located the Bochorna Village (2345 m asl). Until the general census of 2014, the status of the highest village of Georgia belonged to the Ushguli Village (Svaneti).

The Tusheti relief is of erosive-tectonic origin and the depth of erosive cut is 1500–2500 m. The direction of the main ravines are determined by the fault lines—the lineaments and geological structures. The relief is constructed from Lower Jurassic soft rocks—clays, argillites, clay shales, and sandstones. There are powerful heaps of travertines in the surroundings of the Dartlo Village.

The gorges of the Pirikita Alazani and the Tushetis Alazani (Gometsari) are separated from each other by the Makratela Range. Its highest peaks are Mt. Samkhulismta (3480 m) and Tsiva (3360 m). It is a curved range descending to the east and going into the Omalo-Shenako erosive-denudative plain (Fig. 10.18). The main villages

Fig. 10.17 Dartlo—one of the most beautiful village of Tusheti, located on the southern slope of the Pirikita Range (*photo by* L. Tielidze)

Fig. 10.18 Omalo Village plain-depression (*photo by* L. Tielidze)

of Tusheti are built in these plains, such as Omalo, Shenako, and Diklo. The slopes of the Pirikita and Gometsari gorges are dissected by erosive actions of tributaries. Fortress-towers are built in the erosion remnants. The debris cones are formed in the bottoms of the rivers and the deluvial aprons on the slopes. In the heads of the main rivers, the trace of the Würm glaciations is preserved in the form of moraines, cirques and modified troughs. In Würm, the Tebulo glacier length was ~ 11 km and descended to the height of ~ 2400 m asl.

10.14 Keli Volcanic Upland

The Keli volcanic upland is a distinct morphological unit in the Greater Caucasus, located in the heads of the rivers of Aragvi, Ksani, and Liakhvi at a height of 2600–3694 m asl (Fig. 10.19).

The plateau-shaped surface itself extends from 2600 to 3300 m. There are many volcanic cones located in its surface. Among them the lowest is Mt. Narvan-Khokhi (3247 m), and the highest are Mt. Sherkhota (3694 m) and Keli (3627 m).

The length of the Keli Plateau or the Upland is ~ 11 km, and the width is 12–15 km. The relief is built of the Quaternary and Holocene volcanic cones and lava composition rocks, such as andesites and andesite-dacites. The areas free of volcanic rocks and around the Keli Upland are built of carbonate flysch of the Upper Jurassic and the Crescent age. In addition to volcanic processes, nival, glacial and erosive processes take place in the formation of the Keli volcanic Upland.

The relief of Keli Upland is divided into the southern and northern sections by morphological signs. The northern section is wider and is characterized by numerous volcanic cones; among them can be distinguished young volcanoes such as the Sharkhokhi (3303 m), Shadilkhokhi (3250 m), Keli, Sherkhota, Pidarkhokhi (2975 m) and Levinson-Lesing (3374 m). The Mepiskalo volcanic cones are relatively older. The surface of the terrain built of volcanic lavas is wavy-hollow; up to 20 different sizes of lakes can be found in the hollows. The largest are the Archvebi Lakes located at 3060 m above sea level.

The lava erupted from the Shadilkhokhi volcanic cone in Holocene formed the Khodzi lava plateau in the heads of the Didi Liakhvi River,

Fig. 10.19 Keli volcanic upland (*photo by* L. Tielidze)

Fig. 10.20 Keli Lake, located in the volcanic plateau at 2920 m asl (Gobejishvili 2011)

and the lava flow erupted from the Pidarkhokhi and Sharkhokhi formed the Akhubati lava plateau (Gobejishvili 2011). The Keli, Sherkhota and Levnison-Lesing volcanic cones are distinguished in the relief; their slopes are covered with barrancos and no trace of Würm glaciation is observed.

In the Mepiskalo volcanoes and heads of the Tetri Aragvi River, the trace of the Late Pleistocene glaciation is well preserved in the forms of corries and morains.

The volcanic plateau of the southern section was formed by the lavas erupted from the Mepiskalo and Narvan-Khokhi volcanoes in different periods. Narvan-Khokhi is a relatively young cone; the lava erupted from it was spread out in all directions. In the west the lava flow created a vast depression in contact with the Kharuli Range, where the Keli Lake (Fig. 10.20) is located at 2920 m asl. From the lake the Ksani River originates, the gorge of which is cut in its beginning at the contact of Lava and flysch sediments; then it crosses the Arkhi Range and flows through the Zhamuri gorge. The main laval flow was spread to the south, and after the contact with the Arkhi Range it extended to the east and above the Ereto Village in the Aragvistavi River gorge, it ends at 2200 m asl in the form of a narrow tongue. The plateau surface is a wavy-hollow. Lakes are found in the hollows of different sizes. Often the distribution of lakes is related to the contact of main rocks and lava.

In the Keli Upland, due to the nival processes the polygonal forms, rocks, and rock flows are formed in the relief. Overall, there are widespread biogenic forms and round hills ("Kudurebi").

The gorges of the Tetri Aragvi and Liakhvi rivers and their tributaries, developed at the contact of the main rocks and lava flow, are of asymmetric forms, and while crossing the lava flows, they develop narrow canyon-shaped forms. Rapids and waterfalls are common in the Tetri Aragvi River gorge.

10.15 Lomisi (Alevi) Range

The Lomisi (Alevi) Range is a watershed of the Mtiuletis Aragvi and Ksani rivers. The range begins at the Mt. Miketi (2937 m) located at the Keli volcanic plateau. It has a south-eastern direction (2681 m) up to Mt. Northern Kurkuta, then taking a southern direction. The high peaks are Lomisi (2452 m), Mujukhi (2681 m) and Alevisklde (1973 m). At Mt. Miketi the Arkhi Range joins the Lomisi Range. The Arkhi Range extends to the Kharuli Range. The peaks of the Arkhi Range raise above the Keli Upland at 50–150 m, and at 800–1100 m above the Zhamuri depression.

The terrain is built of Cretaceous flysch suites: sandstones, marls, limestones, and clays. The slopes of the range are greatly dissected by the gorges of the Ksani and Aragvi river tributaries. The crest and slopes of the range are soft-shaped due to the structural rocks; the Alevisklde is exceptional, built of the limestones. Outcropped surfaces in the heads of the ravines promote intensive weathering and development of mudflows, especially characteristic of the Mleta-Kvesheti and Chokhi ravines (Fig. 10.21).

10.16 Ksani River Gorge

In the west the Ksani River is bordered by the Kharuli Range, in the east by the Lomisi Range, and in the north by the Arkhi Range. The Ksani River begins from Keli Lake (2914 m) located in the volcanic upland, crosses the lava flow of Keli and denudative-erosive Arkhi Range, develops rapids in the bed and flows through the Zhamuri gorge in a latitudinal direction; then it flows southward below the villages of Pavliani and Chighoiani and in the last section it flows through the Mukhrani Plain first and then, antecedently crossing the Kvernakebi Range monocline, joins the Mtkvari River.

The Ksani gorge is constructed from the Cretaceous carbonate flysch suites (clay-slates, sandstones, slates, limestones, marls and argillites). Further down, the hillock zone is built of continental molasses of the Paleogene period (conglomerates and clays). Young volcanic cones and lavas are widespread on the northern slope of the Pachuri Range in the upstream of the Ksani River.

The direction and morphology of the Ksani gorge largely depend on the lithology of the structural rocks. The gorge has a sublatitudinal direction and follows the direction of structure. The rest of the sections are of meridional direction and the morphology of the gorge, along the small sections, is determined by the strength of the structural rocks.

The landforms of the Ksani River basin are formed with the combination of erosive-tectonic, old-glacial, volcanic and accumulation processes. The leading role in the modeling of the gorge belongs to erosive processes. The intensity and

Fig. 10.21 Mleta mudflow (*photo by* L. Tielidze)

morphology of the developed landforms are determined by the solidity and direction of the rocks along with the river flood. Its tributaries dissect the slopes of the Kharuli and Lomisi ranges quite deeply. The tributary gorges of sublatitudinal direction are widespread, while the gorges of meridional directions are narrow and deeply cut in the relief.

In its head the Ksani River flows through the sublatitudinal Zhamuri gorge and is deeply incised. The river thalweg is located at 2100 m asl at the Bagina Village, and the crest of the Pachuri Range is 2800–2900 m high (some peaks exceed 3000 m). The volcanic relief is formed in the Zhamuri gorge with the Tsitelkhati volcano and its lava flow that extends across the Ksani gorge. The tongue of the volcano flow is dissected into mesas by the tributaries of the Ksani River flowing from the Armkhi Range. Tsitelkhati itself is located in the bottom of the Würm glacial cirque that indicates its Holocene age, and the lava flow extends in the bottom of the trough. Old glacial forms—cirques and modified troughs are preserved on the northern slope of the Kharuli Range (Pachuri Range). Two lakes are formed in the cirque blocked by the volcanic cone.

The Ksani River crosses the subranges of the Kharuli and Lomisi ranges in the sublatitudinal direction below the Zhamuri gorge and develops a deep and narrow gorge, while the gorge is relatively enlarged below the Largvisi Village and the lower terraces are observed. In the downstream of the Ksani gorge, in the hillock relief and above it, there are smoothed surfaces well represented at different heights. Their number is up to four and is used by the population (villages, farm lands). The Ksani River has an accumulative and lateral erosion effect on the Mukhrani Plain and transversally crosses the Kvernakebi Range.

10.17 Aragvi River Gorge

The Aragvi River gorge is located on the southern slope of the Greater Caucasus. In the west it is bordered by the Lomisi (Alevi) Range and in the east by the Kartli Range. Two main gorges are distinguished here: the rightward is called the Mtiuletis (White) Aragvi and starts in the Keli Upland; at Pasanauri it is conflowed by the Gudamakari (Black) Aragvi. At the Zhinvali Village the river is conflowed by the Pshavis Aragvi (there is a Zhinvali water reservoir there today), after which they are called Aragvi (Fig. 10.22).

The relief of the Aragvi River gorge will be considered in two parts: the Mtiuletis Aragvi

Fig. 10.22 Zhinvali water reservoir in the Aragvi River gorge (*photo by* L. Tielidze)

Fig. 10.23 Tetri (Mtiuletis) Aragvi Gorge (*photo by* L. Tielidze)

gorge and Pshavi-Khevsuretis Aragvi gorge. Genetically they are close, but they differ from a geological point of view.

The Mtiuletis Aragvi gorge is developed at a contact of lavas and Upper Jurassic flysch from the head to Kvesheti. In the head the bottom of the gorge is filled up with the Nepiskalo volcanic lavas. The river cuts it in half and develops the canyon-shaped gorge (Fig. 10.23). Further below, the Aragvi gorge is developed at the contact of lava flow and base rocks. The gorge has an asymmetric shape here; the left side is coral, and the right side generated in the Jurassic sediments is inclined. The Mleta (Kaishauri) lava flow is located on the left side of the valley. The slopes built of the deluvial aprons are gullied, and active mudflow processes are developed that produce the debris cones at the bottom of the gorge.

The Shavi Aragvi gorge is V-shaped; only in the upper stream of the Bursachiri River is the gorge trough-shaped. It is noteworthy that the trough gorges built of diabases are well preserved in the heads of the gorge. In Würm, the glacier tongue, descended there, crossing the Bursachriri Pass. Relief is constructed from black clay-shales and slates, and therefore, the river gets black in the rainy period.

From Pasanauri to Ananuri, the Mtiuletis Aragvi gorge is formed in the Cretaceous system's terrigenous flysch suites, perpendicularly crosses tectonic structures and is characterized by narrowness. While passing through the Tertiary suites below Ananuri, the gorge is relatively expanded and forms alluvial terraces. The Mtiuletis Aragvi conflows the Zhinvali water reservoir at a height of 740 m asl. Below the Zhinvali water reservoir, the Aragvi River flows through the Mukhrani-Saguramo Plain and joins the Mtkvari River at Mtskheta (Fig. 10.24).

The gorges of the Tetri Aragvi and Shavi Aragvi rivers are divided by the Mtiuleti Range because of the features of structural rocks and hypsometric conditions. The highest peaks are Mt. Sabadlo (2831 m), Chelistavi (2739 m) and Tsipuri (2574 m). Traces of glaciations are weakly preserved in the relief.

The relief of the Pshavi-Khevsuretis Aragvi River is strongly dissected by rivers. The Khevsuretis Aragvi basin built of the strongly

Fig. 10.24 The confluence of the rivers of Mtkvari and Aragvi (*photo by* L. Tielidze)

dislocated clay-shales and sandstones is distinguished by the high rate of dissection. The Pshavis Aragvi basin is less dissected. The Pshavis Aragvi upstream is wider and dissected; it is located between the Kartli Range and subrange of the Greater Caucasus, which is formed on the border of the flysch suites of the Jurassic and Cretaceous ages. The gorge has a common Caucasian direction and is characterized by a complex folded structure and a whole series of overthrusts. The Pshavis Aragvi gorge is distinguished by strong mudflows and landslides. In the bottom of the gorge there are extensive debris cones, both modern and old ones.

The Khevsureti Aragvi River gorge is deeply cut in the relief and is characterized by V-shaped morphology and strongly inclined slopes. The gorge is basically of erosive-tectonic origin. In the heads of the gorge, namely, in the Roshka gorge, the trace of Würm glaciations is preserved. In Late Pleistocene, there was a powerful glacier in the Roshka gorge, the tongue of which blocked the Khevsureti's Aragvi gorge. In the gorge, on the slopes, the Würm age moraines are well-preserved, and the bottom of the gorge (above the Roshka Village) is covered by the deposits of the Holocene period rock glaciers. The huge boulders of Roshka, brought down from the Chaukhebi, are notable (Fig. 10.25). There is a small glacier in the head of the Roshka River. The lakes are formed between the neoglacial moraines in front of the glacier tongue.

After joining the Pshavis Aragvi and Khevsuretis Aragvi, the Pshavis Aragvi gorge is developed in the carbonate flysch suites. The tectonic structures are dissected by the river and the form of the gorge depends on rock lithology. In soft rocks it develops a relatively wide bottomed gorge, and in solid rocks a narrow bottom and inclined slopes.

10.18 Gudamakari Range

The Gudamakari Range is a watershed of Mtiuletis Aragvi and Pshavis Aragvi river basins. It is a range of medium height separating from the Greater Caucasus watershed range at the eastern Chaukhi (3644 m) and has a meridian direction. From the north to the Zhinvali water reservoir

Fig. 10.25 Roshka erratic boulders (*photo by* G. Lominadze)

(1000 m) the height of the range gradually decreases. The high peaks are Mt. Sachali (2910 m), Laghismta (2602 m), Saorbe (2518 m), Sakharonosmta (2344 m), and Khmali (2141 m).

The northern part of the Gudamakari Range is built of Middle and Upper Jurassic clay slates, sandstones, clay limestones and argillites, and the southern part with a carbonate flysch. It is stretched perpendicularly to the range structures.

The morphological appearance of the range was formed by interaction of erosive-tectonic processes. In the northern part, in the areas over 3000 m high, the trace of the Late Pleistocene glaciations is preserved.

The western and eastern slopes of the range are dissected by 1000–1500 m deep gorges. The main direction of the ravines coincides with the direction of the structural suites.

10.19 Kartli Range

The Kartli Range is a watershed of the Aragvi (Pshavis Aragvi) and Iori rivers. It is a medium height range with the highest peak of the Chichosmta (3076 m). The Range is of submeridional (southwestern) direction from Mt. Chargalistavi (2781 m) to its southern ending. The rest of the range is of sublatitudinal direction. This section is sometimes called the Tianeti Range. It joins the Kakheti Range to the south of Mt. Didi Borbalo. The main section of the Kartli Range is constructed from the Cretaceous sediments (carbonate flysch) and the northern part with the Upper Jurassic sediments. The Kartli Range becomes lower from north to south to about 1500 m and is represented as individual hills in the vicinities of the Tianeti depression. The slopes of the range are dissected by the left tributary of the Aragvi River and the right tributary of the Iori River. The traces of the Würm glaciation are weakly preserved around the high peaks.

10.20 Iori River Gorge

The Iori River gorge is situated between the Kartli and Kakheti ranges, flows to the Artana Village to the south-west, and below is of meridional direction. The gorge slopes are dissected by the tributaries. The gorge is erosive, V-shaped and crosses the relief-building flysch suite antecedently. The bottoms of the gorge is connected to lithology of structural soft rocks. The main role in forming relief forms is related to erosive processes. The debris cones are spread at the bottom of the gorge. There are accumulative plains of small size among them, where the floodplain and over floodplain terraces are formed.

10.21 Kakheti Range

The Kakheti Range is separated from the Greater Caucasus watershed range in the south of Mt. Didi Borbalo at a height of 3276 m asl and extends in the meridional direction up to the Shakhvetila Pass (1130 m). It is a watershed of the Iori and Alazani River basins. The highest peaks of the range are Mtsvetagora (1661 m), Mukhata (2042 m) and Laghismta (2597 m). The range is constructed from carbonate flysch. The range is of erosion origin and is dissected by the small tributaries of the Iori and Ilto Rivers, which are cut in the direction of carbonate sediments.

10.22 Kakheti Caucasus

The southern slope of the eastern Greater Caucasus from Mt. Borbalo (3294 m) to Mt. Tinovroso (3374 m) is called the Kakheti Caucasus.

The peaks over 3000 m are located on the crest of the Kakheti Caucasus, they are as follows: Gvelismta (3089 m), Natsidristsveri (3102 m), Ninikastsikhe (3116 m), and Kachalismta (3428 m). The highest is Mt. Shaviklde (3578 m) and the lowest is the Kadori Pass (2363 m). The Tusheti road passes through the Abano Pass (2926 m). It should be noted that Abano is the motor road highest pass in the whole Caucasus region (Fig. 10.26).

Together with the tectonic movement in the formation of the relief of the Kakheti Caucasus, the nival-glacial and erosive-denudative processes were actively involved (Ukleba 1980).

The basins of the Alazani River and its tributary Ilto are located in the western part of Kakheti Caucasus. The Ilto River gorge is erosive with V-shaped profile, while the upper stream of the Alazani gorge (the Pankisi gorge) is characterized by erosive terrain and a wide base (box-shaped gorge). The trace of old glaciations is preserved in the heads of the Alazani River. The Pankisi gorge is built of alluvial material. Here are visible river terraces (with relative height of 5–10, 10–15, 40–50, and 120–140 m).

In the east of the Pankisi gorge, the Kakheti Caucasus is almost a straight line of mountainous wall ($\sim$95 km long), which is formed by a grandiose fault. The width of the slope is 15–25 km, and the relative height is 2000–2500 m on average. Subranges descend from the short and very steep slope of the main range like noses, and located among them the river basins intensively dissect this slope: Stori, Didkhevi, Lopota, Intsoba, Chelti, Duruji, Avaniskhevi, Kabali, Shorokhevi,

Fig. 10.26 Abano Pass through Kakheti Caucasus (*photo by* L. Tielidze)

Fig. 10.27 Duruji River mudflow (*photo by* G. Lominadze)

Lagodekhiskhevi, and Matsimistskali river basins. All the rivers are of active mudflow nature and form powerful debris cones in the Alazani Valley, which are reflected in the relief.

A crest of the Kakheti Caucasus main range and the main part of the southward steep slope are built of the Lower and Middle Jurassic age intensely folded clay slate suites. In the deep gorges of the Stori and Didkhevi Rivers, as well as in the Speroza rocky massif erected in the watershed of Tsilovaniskhevi and Samkuristskali, the Paleozoic metamorphic rocks are outcropped, such as arkozic sandstones, marbles and crystalline shales. Upper Jurassic and Cretaceous sediments are represented in the heads of the Alazani River, while in the east the Upper Jurassic sediments (clay shales, marls, limestones and argillites) are distributed in the form of a narrow strip and build the hillock terrain in the foothill.

The geomorphological character of the Kakheti Caucasus is formed by river erosion, which was related to tectonic processes. Against the background of Alazani-Agrichai intensive sinking from the Middle Pliocene, the active uplifting of the Greater Caucasus was in progress, which caused erosion processes. Because of this the relief is intensely fragmented. The gorges found here are morphologically rocky narrows with many waterfalls. The local population call "Torne" the rocky section of the gorge the upper part of which is wider and the lower part is narrow. In high mountainous terrain (in high peaks), the traces of action of the Würm glaciers are well preserved, such as cirques, corries, moraines, and modified troughs. In some gorges the Pleistocene glaciers descended to 1800 m asl.

Active mudflow ravines are developed in the Kakheti Caucasus creating powerful and broader debris cones. Mudflow ravines cause great damage to the population and economy. The Duruji (Fig. 10.27), Kabali, Chelti, and Shorokhevi rivers are distinguished by mudflow activity. The lower terraces of Alazani are overlapped by debris cones. The river terraces are relatively visible in the heads of the Alazani River and in the Ilto gorge.

References

Chernyshev IV, Lebedev VA, Bubnov SN, Arakelyant MM, Goltsman YV (2002) Isotopic geochronology of Quaternary volcanic eruptions in the Greater Caucasus Range. Geochem Int 40:1–16

Gobejishvili RG (1995) Saqartvelos tanamedrove mkinvarebi da Evraziis mtebshi gamkinvarebis evolucia gvian Pleistocensa da Holocenshi (Present day glaciers of Georgia and evolution of glaciation in the mountains of Eurasia in late Pleistocene and Holocene). Sadoqtoro Disertacia, Tbilisi, 320 pp (in Georgian)

Gobejishvili R (2011) Relief of Georgia. Monograph, Publishing House "Universali", Tbilisi (in Georgian)

Gobejishvili R, Lomidze N, Tielidze L (2011) Late Pleistocene (Wurmian) glaciations of the Caucasus. In: Ehlers J, Gibbard PL, Hughes PD (eds) Quaternary glaciations: extent and chronology. Elsevier, Amsterdam, pp 141–147. https://doi.org/10.1016/b978-0-444-53447-7.00012-X

Kurdghelaidze G (1983) Tusheti. The Vakhushti Institute of Geography, Publishing House "Metsniereba", Tbilisi (in Georgian)

Kutuzov SS, Mikhalenko VN, Grachev AM, Ginot P, Lavrentiev II, Kozachek AV, Krupskaya VV, Ekaykin AA, Tielidze LG, Toropov PA (2016) First geophysical and shallow ice core investigation of the Kazbek plateau glacier, Caucasus Mountains. Environ Earth Sci 75:1488. https://doi.org/10.1007/s12665-016-6295-9

Tielidze L (2016) Glaciers catalog of Georgia. Publishing House "Samshoblo", Tbilisi, p 116

Tielidze L (2017a) The new inventory and general description of the modern glaciers. In: Glaciers of Georgia. Geography of the physical environment. Springer. https://doi.org/10.1007/978-3-319-50571-8-3

Tielidze L (2017b) Late Pleistocene and Holocene glaciation. In: Glaciers of Georgia. Geography of the physical environment. Springer. https://doi.org/10.1007/978-3-319-50571-8-6

Tielidze LG, Wheate RD (2018) The Greater Caucasus Glacier Inventory (Russia, Georgia and Azerbaijan). Cryosphere 12:81–94. https://doi.org/10.5194/tc-12-81-2018

Tsereteli DV (1966) Pleistocenovie otlojenii Gruzii (Pleistocene deposits of Georgia). Publ. Metsniereba, Tbilisi, pp 1–583 (in Russian)

Ukleba D (1980) Qartuli Sabchota Enciklopedia (Georgian Soviet Encyclopedia). T. 5, Tbilisi, pp 447–448 (in Georgian)

Levan Tielidze is a Ph.D. of Ivane Javakhishvili Tbilisi State University in the field of geomorphology-glaciology-cartography and a senior research scientist at Vakhushti Bagrationi Institute of Geography. He is also a researcher-scholarship holder of the Department of Earth Sciences of the Georgian National Academy of Sciences. The field of his research is mountain cartography and glacial-geomorphological study of the mountainous areas in the Quaternary (Late Pleistocene and Holocene). He is a member of several international organizations. In 2014/2015 he conducted a research stay at the Climate Change Institute of the University of Maine, USA, and in 2015/2016–2017 at the University of Northern British Columbia, Canada. He is the author of about 40 scientific papers, maps and eight monographs.

Ramin Gobejishvili (1941–2014) was a Doctor of Geographical Sciences, a geomorphologist-glaciologist. Until his last days he was Professor of the Faculty of Exact and Natural Sciences, Ivane Javakhishvili Tbilisi State University and the chair of Geomorphology and Cartography. In 1996–2008 he was the president of the Geographical Society of Georgia, and in 2005–2007—the director of the Institute of Geography. He was the author and co-author of about 150 scientific works and ten monographs.

Alexander Javakhishvili (1875–1973) was one of the oldest geographer and anthropologist in the Soviet Union. He was the founder of Georgian geographical school. His qualifications and awards include Doctor of Geographical Sciences (1937), Academician of the Georgian National Academy of Sciences (1944), an honorary member of the Soviet Union Geographical Society, and winner of a State Prize of the Georgian Soviet Socialist Republic.

11 Western Georgia (Kolkheti) Intermountain Plain

Levan Tielidze, Ramin Gobejishvili, Levan Maruashvili and Nino Chikhradze

Abstract

This chapter covers the geomorphological characterization of the subzone of Western Georgia (Kolkheti) intermountain Plain. The detailed description of the following orographic units are considered: the Kolkheti Lowland, Northern Kolkheti foothill hillock relief, and Leselidze-Gantiadi hilly relief, Akhali Gagra—Akhali Atoni and Sokhumi-Merkheuli subregions, Samurzakano region, Odishi Plateau, Okriba or Northern Imereti hilly subregion, Southern Kolkheti foothill hillock relief, Imereti Highland, and Likhi Range. The text also provides the geological structure, tectonic conditions, and current exogenic processes of the study region. Each subchapter is accompanied by relevant images and maps.

Keywords

Western Georgia (Kolkheti) Plain · Odishi Plateau · Imereti highland · Likhi Range

11.1 Study Area

Western Georgia Plain is inclined as an amphitheater toward the Black Sea; its hydrographic network belongs to the Black Sea (Fig. 11.1). The relief of the Western Georgia Plain was formed against the background of differential tectonic motion. The central plain is the sinking area and was formed by the accumulative action of rivers in Pleistocene. The Kolkheti Lowland is surrounded by the Tertiary sediments and in some places by the foothill hillock strip built of Cretaceous and Jurassic suites. This strip is divided into two parts by the Imereti crystalline massif, the eastern wing (Likhi Range) of which is the watershed of the Black and Caspian seas.

The Western Georgia Plain is divided into several regions, according to the orographic and morphogenetic signs:

1. Kolkheti Lowland with a height of 0–200 m extends from Sokhumi to Sarpi.

L. Tielidze (✉) · R. Gobejishvili · L. Maruashvili · N. Chikhradze
Department of Geomorphology and Geoecology, Vakhushti Bagrationi Institute of Geography, Ivane Javakhishvili Tbilisi State University, Tbilisi, Georgia
e-mail: levan.tielidze@tsu.ge

L. Tielidze
Department of Earth Sciences, Georgian National Academy of Sciences, Tbilisi, Georgia

R. Gobejishvili
Department of Geography, Faculty of Exact and Natural Sciences, Ivane Javakhishvili Tbilisi State University, Tbilisi, Georgia

L. Tielidze (ed.), *Geomorphology of Georgia*, Geography of the Physical Environment,
https://doi.org/10.1007/978-3-319-77764-1_11

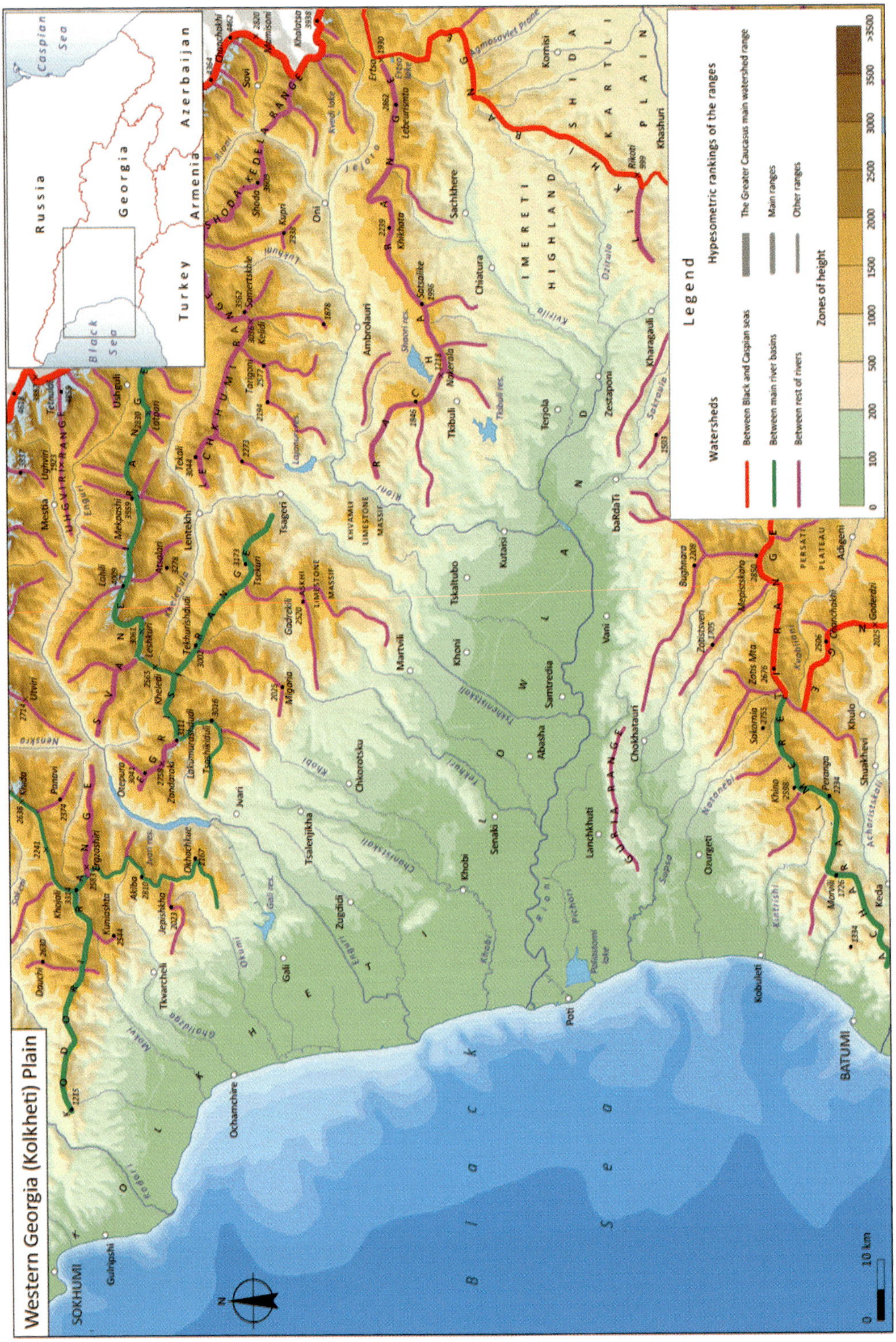

Fig. 11.1 The orography of the Western Georgia Plain

2. Foothill hillock relief (strip) of the Northern Kolkheti, stretches as a narrow strip from the Psou River to the Imereti Highland (to the Dzusa River).
3. Southern Kolkheti foothill hillock relief, including Achara-Guria-Imereti up to 600–800 m height.
4. Imereti Highland (also called the Dzirula massif), is located in the east of Kolkheti; it is morphologically visible with smoothed surface. The highland extends to the height of 1200–1400 m asl.
5. Morphologically the Likhi Range is an eastern part of the Imereti Highland and orographically, it is a connecting bridge of the Greater Caucasus Range and Achara-Imereti Range. It is also the morphological barrier dividing the eastern and western parts of Georgia in terms of geomorphological, climatic, hydrological, and biogeographical conditions. Its height ranges among 1000–2500 m. It should be noted that the northern part of the Likhi Range belongs to the Greater Caucasus fold system.

11.2 Kolkheti Lowland

The Kolkheti Lowland extends from the Sokhumi City to Sarpi Village. It lies at 100 m elevation, sometimes reaching 150 m asl. From the Black Sea, the plain intrudes deep through the Rioni River valley and its tributary Kvirila River and then spreads up to Zestaponi. Some places are located below sea level. The Kolkheti Lowland is created by the river deposits in the sea gulf. The eustatic variability of sea level led to the layering of the Kolkheti building sediments and presence of organic layer (peat) in these layers. The materials of drilling, conducted in Kolkheti show that the surface of the lowland was formed in Late Pleistocene and Holocene. The thickness of the layers of this age is 65–80 m above the modern surface, out of which 40–50 m thick layers are of Holocene age. There are three horizons of peat visible in the Holocene layers at depths of 10, 37, and 41 m. As a result of drilling, a number of anticlinal (Samtredia, Pichora, Kvalovani, Kulevi, and Maltakva) and synclinal (Poti, Grigoleti, Khobi, and Nabada) sunken structures were identified (Gobejishvili, 2011). These structures are of longitudinal distribution and are reflected in the modern relief of the lowland.

In the southwestern sector of the Kolkheti Lowland (Achara territory), two morphological units—the Kakhaberi and Kobuleti lowlands are distinguished.

The Kakhaberi Lowland encompasses the entire delta of the Chorokhi River, the foothill plain of Anaria and a narrow strip that follows the Black Seashore from Sarpi to Chorokhi confluence.

The Kakhaberi Lowland has a form of a triangle; its top is at the Erge Village and the base leans against the Black Sea. Its absolute height varies between 0 and 20 m. In the west, it is slightly inclined toward the Black Sea (Fig. 11.2). The Chorokhi River cross-sects this plain along the entire length in its southern part, where the river is divided into several branches and creates islands. These branches join each other and conflow the sea as a strong stream (nowadays, due to the dams built on the Chorokhi River in the Turkish territory, the stream of the river is significantly weakened). In the middle part of the lowland, a small Mejina River flows. The surface of this lowland is less swamped than in the Kobuleti Lowland, but long ago the marshes covered its important areas.

Kobuleti Lowland is located in the downstream of the Kintrishi and Choloki rivers. Compared to the Kakhaberi Lowland, it is lower and swamped. The sand dunes follow the coast almost its entire length, masked by artificially planted pine forest-park in the resort zone of the Kobuleti Town. The old dunes in the area of the pine forest are 200 m away from the sea shore; the dunes are bare and are located in three rows. They are one of the main reasons for swamping of this lowland; they obstruct the surface runoff from reaching the sea and cause water ponding. Currently, the irrigation channels are constructed and a significant part of the wetlands are drained.

Fig. 11.2 Kakhaberi Lowland (*photo by* G. Lominadze)

At a glance, the Kolkheti Lowland is characterized by a uniform plain surface. But in fact, there are three types of relief: coastal accumulative, river accumulative, and accumulative-erosion (Maruashvili 1981). Each type of relief corresponds to the area of its spread (sea narrow strip, central, and elevated periphery zone).

The river accumulative type is mainly spread along the Rioni River at a distance of 54 km from Poti to Samtredia in a triangular shape and is formed due to accumulative actions of the Kodori, Enguri, Khobi, Rioni, Supsa, and Chorokhi rivers. Erosion cutting does not take place. The Rioni and Khobi rivers flow through the beds elevated from the surface of the lowland, the origin of which is related to the artificial dams constructed along the river bed. The surface of the plain is horizontal at a glance. The lowered areas are distinguished by vast marshes. In the central part of Kolkheti, the Paliastomi, Imnati, Bebesiri, and other small lakes are located. This type of relief originated in the Upper Pleistocene and Holocene.

The width of the peripheral line of the lowland varies from 2–3 to 30–35 km. The peripheral line is stretched along the foothill hillock relief. Its height varies from 0–20 to 100–150 m. The riverbeds are cut down to 10–12 m. There are nearly no wetlands and lakes. This type of relief was formed in the Pleistocene.

The coastal accumulation relief spreads as a narrow strip along the sea from Kobuleti to Sokhumi and in some places as individual areas, for example, the Kakhaberi Lowland and the surroundings of the tributaries of the Bzipi and Psou rivers. This type of relief is mainly represented by sea coastal dunes, which in the center of Kolkheti detain the water filtration process and facilitate the creation of wetlands and lakes. Some rivers (Supsa, Churia, Kaparcha, and the rivers of Samurzakano) are forced to flow along the coast until they empty into the sea.

The relief of the Kolkheti Lowland has been modified due to human influence. Development of coastline mainly for the purposes of recreation, ports, rivers bank dams, roads, and populated areas. The Black Sea level fluctuation and sinking of the central part of Kolkheti (6 mm/s) are of great importance in the shifting of the coastline.

11.3 Northern Kolkheti Foothill Hillock Relief

The Northern Kolkheti foothill hillock relief stretches ~265 km from Zestaponi. The following subregions are morphologically distinguished: Leselidze-Gantiadi, Akhali Gagra-Akhali Atoni, Sokhumi-Merkheuli, Samurzakano, Odishi Plateau, and Okriba.

Leselidze-Gantiadi hillock relief (the left bank of the Psou River). Its northern border is separated from the Gagra limestone Range (southwestern slope) by cliffs. The relief is dissected by the Lapsta, Mekhadiri, and Khashupse erosive gorges, constructed from Cretaceous and Tertiary suites, which have a common Caucasian direction. The coastline is composed of Pleistocene marine and river deposits (Devdariani 1980). In the formation of hillock relief, erosive, karst, and gravitational processes have an active role together with tectonic processes. This zone is tectonically sinking and in the north, it is separated from the Katsirkhva limestone range by the fault line. The marine terraces, formed due to the abrasive actions of the sea, are visible in the relief. The first terrace is larger and is 5 m high, the second terrace is 15–25 m high, and the third is 40–45 m high. These terraces are represented as fragments. The landslide forms are locally distributed in Tertiary suites (clays, clay shales).

From Gantiadi to Bzipi delta, the Gagra Range (Arabica massif) descends near the sea concealing the hillock relief. The coastline is narrow and terraced on the left bank of the Psou River.

Akhali Gagra—Akhali Atoni subregion (between the Bzipi and Gumista rivers). In the east of Akhali Gagra, the seacoast is expanding and is represented as a 10–15 km wide hillock relief up to Akhali Atoni. The 500–600 m high foothill hillock strip is sharply visible in the relief. In the north, it is separated from the Bzipi Range by a limestone cliff.

The relief is constructed from Oligocene and Miocene marine and continental molasses (conglomerates, sandstones, clays, gravelites, and sands). The coastline and main gorges are built of gravels, sands, and clays.

The western part of the hillock subregion is occupied by the 100–200 m high Miusera Hill (Highland). It is dissected by the numerous tributaries of the Miusera River and independent short and shallow river gorges. Their origin is associated with the relief builder Pliocene conglomerates, which are inclined south, indicated by the presence of the Bzipi River paleodelta. The Miusera Highland's distinctive microrelief is identified by the lithological peculiarities of the building rocks (Gobejishvili 2011). Soft forms are associated with clays, and sharper and higher relief with strongly cemented Pliocene conglomerates. Landslides are common in the relief built of clays.

In the hillock relief, vicinities of the villages of Jirkhva and Duripshi occupy a special place, known as the Duripshi Plateau. In the north and west it is surrounded by a deep and narrow gorge of the Tetritskali River (at Khicha), and in the east by the gorges of the tributaries of the Dokhurti River. The plateau gradually lowers toward the sea in the south without a noticeable step. The Duripshi Plateau is built of limestone conglomerates, covered by a powerful (5–7 m) layer of soil.

The nature of the building rocks, the abundance of atmospheric precipitation and peculiarities of the relief facilitate the formation of the karst (clustokarst) processes and forms including sinkholes, caves, and karst lakes. The drainage of karst waters is directed mainly toward the deep gorge of the Khipsta River conditioning the exits of vaucluses and large number of caves.

There is a hillock relief on both sides of the downstream of the Bzipi River. Absolute height of the hillocks is 250–300 m. The river gorges are V-shaped and are characterized by low depths in the upper streams, but in the downstream, they are large and box-shaped. The watersheds are low hills with plain crest and stepped terraces toward the sea. The relief is constructed from Miocene–Pliocene marine and continental molasses—conglomerates, clays, sandstones, and sands, as well as Oligocene carbonate clays. The character of the building rocks is visible in the hypsometry of the relief—

the areas built of clays are lower than the ones built of conglomerates and sandstones.

Exogenous processes play a leading role in the formation of micro landforms. Erosion forms are represented by gorges, ravines, and cliffs, and landslide phenomena are mainly developed in the clay substrate. The landslide relief is developed in the surroundings of the Kaldakhvara Village. There are several active landslides on the left side of the Bzipi River below the bridge, the genesis of which is related to lateral erosion of Bzipi. Accumulative relief is developed on the seashore covering the delta of the Bzipi River and surroundings of Bichvinta. Their height is up to 100 m. Landforms are emerged due to riverine–marine accumulation processes and shapes (terraces and coastal dunes).

The Bichvinta Peninsula (Cape) is interesting, a delta of the Bzipi River. It is bordered by the Miusera hillock hill in the north. Dunes 5 m high are developed along the coastline. The central part of the delta is lower with wetlands and Inkiti Lake. Today the coastal and hillock reliefs are modified due to anthropogenic processes. There are five marine and riverine terraces along the Bzipi River: 5–7, 12–15, 35–40, 120–130, and 250–300 m in height.

The Sokhumi-Merkheuli subregion stretches as a 28–30 km long hillock strip, and 15–20 km wide between the Gumista and Kodori rivers. The limestone sectors include the canyon-shaped gorges of the Gumista and Kelasuri rivers. Relief built of limestones is tectogenic with anticlinal hills (Chaama, Akhupacha, Akhbiuki, Iashtudi, Ghvardia, and Apiancha), and small synclinal depressions. Asymmetric anticlines are constructed from Lower Paleogene and Cretaceous limestones, and the synclinal depressions from Oligocene clays. Landslide relief is developed due to the erosion processes in the depressions, creating wavy shapes in the bottoms of the depressions. The karst relief is represented by numerous sinkholes in the limestone areas. There are karst caves in the following villages: Shroma, Lindau, Baghmarani, Tsebelda, Shakurani, and Pali. The underground sector of the Amtkeli River is situated in the same region. The dwellings of the Paleolithic, Mesolithic, and Neolithic humans have been discovered in many caves (Tatashidze 1997). There are dry gorges in the relief as well, a result of catch up of rivers.

From Sokhumi to Kodori, from karst relief to the Kolkheti Lowland, coastal low hillock relief is widespread. The genesis of the relief is mainly related to the erosive action of the rivers, which cross-sect the Neogene suites. The look of mesorelief is formed by the landslide forms and marine terraces, visible in the area of Sokhumi. Here, the six terraces of 2–4, 6–8, 18–20, 40–45, 60–70, and 100–120 m in height are distinctive (Gobejishvili 2011). The relief of this region is anthropogenically transformed.

The Kodori River delta is characterized by interesting relief located for ~15 km between the villages of Naa and Dranda. There are six river terraces well represented (4–6, 12–15, 20–25, 40–45, 65–70, and 150–160 m), constructed from conglomerates and gravels.

Samurzakano covers the northern Kolkheti region from the Kodori River to the Enguri River. It is bordered by the Kolkheti Lowland in the south and by the narrow limestone strip of the southern slope of the Kodori Range, separated from Tertiary sedimentary rocks by the fault line. The hillock relief stretches from the west to the east over 60 km, and from the north to the south at 12–15 km. The relief of Samurzakano is a uniform body both geomorphologically and geologically. The relief is constructed from Oligocene and Miocene ages (Maikop suite) carbonate and gypsum clays, conglomerates, and gravelites, while the low hillocks of the Kolkheti Lowland from Cretaceous gravels, conglomerates, and clays. The Duabi, Mokvi, Ghalidzga, and Okumi rivers cross-sect the relief building rocks with their deep gorges. Landslides are well-developed, especially along the river gorges. The clustokarst shapes are well expressed in relief. In the high hillocks, caves are developed, such as Otipistavi, Jgerde, and Chkhortali. From the karst massifs Panavi, Mokvi and Lashkenderi are notable. One of the longest Abrskili Cave (3 km) is located in the Mokvi massif. In the hillock strip, the six terraces of 5 m, 15–20, 40–45, 67–70, 110–130, and 150–160 m height are distinguished. They are developed in the gorges and watersheds of the main rivers (Enguri, Ghalidzga, Mokvi, and Okumi) in the form of steps.

The Odishi Plateau is a name of the foothill hillock strip from the Enguri River up to the Tskhenistskali River. It stretches for 53–55 km from the northwest to the southeast. The Northern Kolkheti hillock strip is the widest in this sector and extends for 45 km. The Odishi Plateau is separated from the southern slopes of the Samegrelo Range by the Terzeni stair. The plateau has a sharp morphological border with the limestone massifs of Kvira, Gaucha, Migaria, and Askhi. It is separated from the Kolkheti Lowland by the Urta-Abedati low-mountainous hills, which are well visible brachyanticlinal microstructures, distinguished hypsometrically and morphologically, and are attached bow-shaped to the synclinal plateau of central Odishi. Stretching from the southwest to the northeast, the hill is divided into three parts: the massifs of the Eki, Nokalakevi, and Abedati. The length of the hill from the Abasha canyon to the resort of Menji is 38 km and the length of the Urta-Bia Hill does not exceed 15 km.

The Odishi Plateau is constructed from Neogene and Quaternary sediments. Upper Cretaceous shallow sea sediments—the layered limestones, marl limestones, sandstones, and marls build the hearts of the brachysynclinal hills (Urta, Eki Mountain, and Nokalakevi Mountain). The central Odishi synclinal plateau is constructed from Miocene marine and continental molasses, sandstones, clays, conglomerates, and marls. In the eastern and northern edges of the Odishi Plateau, the Eocene and Oligocene shallow sea limestone, marl limestones and marls are outcropped. The Eocene molasses are covered from above by the Lower Quaternary gravelites, conglomerates and clays. They are located along the Enguri, Khobistskali and Tekuri rivers, and morphologically represent the debris cones of these rivers, dissected by the erosive network.

The relief of the Odishi Plateau is uplifted by 200–400 m above the nearby Kolkheti Lowland due to tectonic movements. The morphostructural look of the Odishi syncline was formed by differential movements after Pliocene. The syncline, uplifted in the north at the altitude of 300–400 m, formed a morphologically inclined plain. This plain is surrounded by the anticlinal hills in the south and east, the speed of vertical movement of which exceeded that of the Odishi Plateau relief (Gobejishvili 2011).

The central Odishi Plateau's morphological look is created by the erosive gorges and terraces, landslides, and clustokarst forms along with the old debris cones.

The erosive gorges of the submeridional direction and their watershed plain hills are important in the morphology of the relief. The depth of the erosive section is 140–160 m; the height of the watersheds increase from south to north from 130–150 to 300–400 m. These plain watersheds are located higher than 100–120 m from the bottoms of the river gorges. In the gorges of the rivers of Khobistskali, Chanistskali, Ochkhamuri, Tekhuri, and Abashistskali are smooth surfaces of low altitude, which are the young terraces of these rivers. The terraces are fully visible in the Tekhuri gorge, in the Didichkoni depression (12–14 km long, 6–8 km wide). The relief of the central Odishi Plateau is dissected by soft, hilly, and wavy watersheds averaged inclined slopes and sometimes, by narrow gorges and ravines.

The gorges of the rivers of the plateau were formed at the end of the Lower Quaternary. There basically 2–3 terraces formed in the river gorges and their age is Middle and Upper Pleistocene. They are located at the height of 6–8, 14–18, and 30–40 m above the river level (Nemanishvili 1982).

The clustokarst formations are widely distributed in the central Odishi Plateau in the conglomerates and sandstones of Neogene age. The large sub-horizontal caves are spread across the villages of Kortskheli, Garakha, Nazodelavo, Savekuo, and Nogha. The longest is the Nazodelavo Cave (up to 700 m), formed in the conglomerates and sandstones. Kortskheli Cave is formed in the same sediments (up to 800 m).

The Urta-Bia Hill is divided into two parts by the Khobistskali River gorge—the Bia Hill (214 m) in the east and higher Urta Hill (466 m) in the west.

The Urta Hill is low mountainous, the crest of which is slightly dissected by erosive processes. Karst forms are widely developed. Among the karst caves, the caves of the villages of Urta and Tsaishi are notable, located in two levels and have an appearance of a long horizontal tunnel. The height of the levels is 30 m. During heavy rains, the temporary streams flow from the caves. The monoclinal hills are spread along both slopes of the Mount Urta.

Bia is a low hill, which is partitioned into the individual hillocks due to erosion. In the morphology of the Bia Hill, the alongside gorges and ravines are notable, which are formed by the left tributaries of the Khobistskali River.

The Unagira Hill occupies the submeridional 4–6 km wide and 36–38 km long strip, exhibiting brachyanticlinal folding. Morphologically, it is divided into individual sectors: Eki Mountain, Nokalakevi Mountain, Abedati Mountain, and Dadiani's Takhta.

The slopes of the Eki Mountain overlook the Senaki town. The mountain is uplifted at 400 m above the Kolkheti Lowland. The limestone part of the Eki Mountain is rich in karst formations, such as sinkholes and caves.

The Abasha (Martvili) canyon is interesting in the Odishi Plateau. There is a zigzag canyon with steep slopes cut into the wider part of the Abasha gorge at 3–4 km distance between the villages of Skurda and Inchkhuri. Its depth is 40–50 m, the width of the upper part is 15–25 m, but in some places 5–10 m, with many bridges of natural limestones. The canyon is a former cave with a collapsed ceiling (Fig. 11.3). In some places, the river creates 12–15 m high waterfalls. One sector distinguished with special beauty is Oputskhole known as "Dadianis' Bath". At present, this place is included in the state protected areas of Georgia (Martvili canyon nature monument) due to its picturesque nature and visited by thousands of tourists annually. The landslides are common in many places in the Odishi Plateau; the surroundings of the villages of Dojrabeni-Khuntsa are especially distinguished.

Fig. 11.3 Abasha canyons (Martvili canyon nature monument) (*photo by* L. Tielidze)

Okriba or the Northern Imereti hillock subregion is bordered by the Tskhenistskali River gorge in the west, by the Dzusa River—in the east, the southern boundary passes through the Matkhoji-Kutaisi-Ghvankiti (at a height of 120–150 m) and in the north—through the southern slopes of the Racha Range and Khvamli massif. The highest peak is Mount Nadeshturi with a height of 1612 m (in the south of Khvamli) asl.

The relief of Okriba is constructed from Bajocian porphyritic series (basalt and andesite lavas, lava breccias, tuffites, etc.) that are encompassed by Cretaceous limestones. Middle Jurassic sheet slates and Upper Jurassic sediments of color suite can be found there. The relief built of porphyries is characterized by erosive, hillock forms. Peripheral limestone suites are represented by karst relief. Landslide processes develop in the Bathonian and Upper Jurassic loose deposits. Tectonically Okriba is the inclined anticlinal dome, the ark of which is washed due to erosive processes. Sediments of different ages are separated from each other by faults, overthrusts, and faultings.

According to the morphological signs, geological structure, and exogenic processes, Okriba is divided into the individual areas: (1) Porphyritic Okriba with erosive hillock relief; and (2) Limestone Okriba represented by hills and hillock plains.

The western part of the porphyritic Okriba includes the Rioni-Tskhenistskali watershed and is represented by tectonic-erosive relief of medium and low mountainous mountain-gorges. It extends 10–11 km to the south of Khvamli and 8–9 km between the rivers of Rioni and Tskhenistskali.

The main orographic units of relief are the hills of Nadeshturi, Gormaghali, and Kanari, which are the mountainous massifs of a whole watershed range. The highest peaks of the Gormaghali Range are Mount Nadeshturi (1612 m) and Gormaghali (1471 m). From the Gormaghali massif originate the right tributaries of the Rioni River: Sakalmakheghele, Korenishistskali, Mekvenastskali, Tsiplarastskali, and Semistskali (head of the Gubistskali River). In the relief built of porphyries, they develop narrow, deep, V-shaped gorges characterized by a 500–800 m high cliffs. From modern geomorphological processes, the depth of erosion and rockfalls are notable.

The erosion-landslide relief of Shida Okriba is located in the east of the Rioni River and south of the Racha Range. Tectonically it is the Okriba anticline outcropped due to erosion and built of porphyritic volcanogenic suites, Bathonian sheet slates and continental sediments of carbon suite. On the top, they are covered by Upper Jurassic sandstones, clays and marls (color suite). Shida Okriba consists of small depressions and narrow watersheds; small depressions are mostly originated in the synclinal folds and the watersheds in the anticlinal folds.

The northern part of Shida Okriba is dissected by the tributaries of the Lekhdari and Tskaltsitela rivers. The look of the relief built of porphyries is created by a sharp mountain-gorge relief. The Lekhdali River originates on the western slope of the Nakerala Range, with a developed profile. In the middle and upper parts, the valley is asymmetric. The right slope is more steep and high and deeply dissected. In the lower part of the gorge, the left slope is steep and weakly dissected. Here, it crosses the Namakhvani porphyritic anticline and develops the cut in a meandering narrow gorge that lacks terraces and floodplains.

In the last part of the Lekhidari gorge, there is a morphologically visible erosive witness of "Kvamatro", built of porphyries. The average depth of the dissection of the relief of Northern Okriba is 250–300 m, and the maximum is 500–600 m. The Tskaltsitela River basin is opened as a fan in the central part of Shida Okriba; its right tributary is the Chala River and the left is the Mtischala River, which begins on the western slope of Nakerala, from karst vaucluses. The relief of the gorges and depressions is complicated by erosive-landslide forms. The Tskaltsitela River flows through the Gelati-Kursebi gorge-depression below the Orpiri Village (Fig. 11.4). Low mountainous relief is dissected by dense erosive ravines of small streams. The rise of morphological sharpness in the southwest,

Fig. 11.4 Tskaltsitela River gorge (*photo by* L. Tielidze)

the formation of isolated hills and highlands with dome crests are related to the teschenite layer-lodes or dykes.

The low mountainous tectonic-erosion relief is located in the south of Shida Okriba. Its main morphological peculiarity is a continuation of erosive-tectonic flatlands along the 5–6 km wide latitudinal strip, surrounded by the low mountainous relief. The bedrocks in their foot are covered with alluvial–lacustrine and proluvial–deluvial sediments; such areas are the flatlands of Akhalsopeli, Tsutskhvati and above discussed Kursebi and Gelati. The Tkibuli water reservoir lies in the bottom of the Akhalsopeli flatland (Fig. 11.5). Prior to this, the thickness of Pleistocene sediments was 80 m in the flatland (Apkhazava 1986); creation of sediments was stipulated by the accumulation of the Tkibula River deposits. This process continues today. The Tkibula River has an underground passage in the Argveta limestone Hill flowing as a vaucluse in the other side and then originates the river of Dzevrula. The landslide processes are common in the eastern part of the flatland.

The Tsutskhvati flatland is located in the west of the Tkibuli flatland (reservoir), where the

Fig. 11.5 Tkibuli water reservoir located in the Akhalsopeli flatland (*photo by* L. Tielidze)

Tsutskhvati Village is located. The flatland is a quadrangle shape with a hilly surface. The Maghara River flows across the flatland joining the Shabatghele River, then it flows through the underground karst gorge under the Okriba-Argveta Hill and originates the Chishura River in the form of vaucluse.

In the south, the porphyritic Okriba is surrounded by the Sataplia-Samgurali and Okriba-Argveta hills and Tskaltubo-Simoneti hilly Plains built of limestones. The Tskaltubo limestone massif is spread over the Rioni-Tskhenistskali interfluves. The Didvake hilly relief and the Samgurali-Sataplia limestone highland are distinctive here.

The relief of Didvake is constructed from Cretaceous thick-layered, monoclinally inclined, and fissured limestones with karst waters exits. Karst sinkholes, wells, perforated cirque forms, denudative remnants, dry gorges, and caves are common in the Didvake relief. The unique cave of Kumistavi (Prometheus), which attracts thousands of tourists every year is worth mentioning (Fig. 11.6). The new cave halls were discovered in 15 July, 1984 by the members of the speleological expedition of Vakhushti Bagrationi Institute of Geography of Georgian Academy of Sciences.

The main orographic units of the Samgurali-Sataplia limestone massif are the Samgurali Hill and the Sataplia hilly relief itself, constructed from Cretaceous thick layered limestones. On the eastern slope of Samgurali Hill, a cliff of 160–200 m high stretches along the Rioni River, in the bottom of which there are deluvial aprons composed of limestone material with karst waters exits. The Rioni River crosses the limestones in the area of Kutaisi and creates the Jachvi Bridge canyon with the cliff overlooking from the top. In the crest of the Sataplia highland, the karst sinkholes, wells, and serum fields are formed. In the Sataplia Cave, about 200 footprints of predator and herbivorous dinosaurs were discovered in the marl limestones (Mamisashvili

Fig. 11.6 Kumistavi (Prometheus) Cave (*photo by* L. Asanidze)

1984). This is of great paleogeographical importance.

The Okriba-Argveti Hill is a well-defined morphological unit, built of Urgonian limestones. Its crest is higher by 150–300 m above nearby areas. The highest point is 871 m. In Neo-Pleistocene and Upper Quaternary the rivers crossed the crest, but now they flow in the underground karst gorges (Tkibula, Chishura, etc.). The Maghara River was able to make a depth cut of the Okriba-Argveta Hill by the erosive-karst action creating the 13-storey height through Tsutskhvati Cave system. The length of the main karst tunnel is 250 m and is a modern river bed. After leaving the cave the Chishura River begins. Water flows out from the water reservoir through the underground bed of the Tkibula River and a hydropower plant is built on it. There are denudative remnants, debris and karst shapes (wells, sinkholes, and caves) formed in the low mountainous relief of the hill.

The Simoneti wavy Plain is located in the south of Okriba-Argveti, inclined toward the Kvirila River. In the relief built of limestone, the karst shapes are formed. The Navenakevi well-improved cave, Bodogani, and Nagharevi tunnels are notable. Erosive terraces are also common in the Simoneti Plain.

11.4 Foothill Hillock Relief of the Southern Kolkheti

The foothill hillock region of the Southern Kolkheti extends for a distance of 225 km from Zestaponi to Sarpi. It surrounds the Achara-Imereti Range from the north and west in the form of a strip. Its width is 5–10 km, only in Guria it reaches 20–22 km. It is built more simply than the Northern Kolkheti region. The weak spread of limestone stipulates the absence of karst forms. In the Southern Kolkheti region, there is a thick red weathering rind, which has a great influence on the microrelief. Here are distinguished several subregions: Southern Imereti, Guria, and Achara.

The foothill hillock belt of the southern Kolkheti located within Imereti, covers the lower zone of the northern slope of the Achara-Imereti Range, stretching between Zestaponi and Samtredia with the width of 5–7 km. From the north, it is surrounded by the Kolkheti Lowland.

The foothill relief of southern Imereti is built of Tertiary (Oligocene, Miocene) sedimentary rocks: clays, sand, conglomerates, and marls. The above-mentioned suites are eroded and along with erosive processes, stipulate the occurrence of landslide phenomena. The Eocene volcanogenic suite and the Upper Cretaceous sediments in the form of tiny islands take part in the construction of the belt near the Achara-Imereti Range.

The relief of the southern Imereti foothill is mostly represented by hillocks and low mountains with a height from 200 to 700–800 m asl. The relief is very fragmented, but it has a wavy shape. Characters of meso and microreliefs are identified mainly by the erosion and accumulation in the depressions. The transverse and longitudinal gorges, represented there are related to tectonics and erosion.

There is an alongside lowered area morphologically distinguished in the foothill relief, which begins from the surroundings of the village of Ukhuti and stretches to the west up to the Khevistskali River. More to the west it coincides with the transversal gorge of the Supsa River. The fault lines follow the lowered area from the north and south. In the north, the longitudinal lowered area is followed by the foothill low mountainous hill, which is known as Guria "Mountains" in the west. In some sectors, the lowered area is expanding and acquires the form of a small depression. Their watersheds have a relative height of 60–70 m. River terraces (2–4, 8–10, 18–20, and 40–50 m high) and landslide shapes are formed on the slopes of the depressions and in the northern sector of the hill. In the foothill belt, the series of the Rioni terraces (height 120 m) are particularly visible between Bagdati and Vani.

Within Guria, the hilly belt of the southern Kolkheti Lowland is relatively wider (20–22 km). The morphological look of the relief is created by the depressions formed along the Supsa and Natanebi river gorges and two hills of

Nasakirali and Guria. The height of the crest of the Guria Hill increases to the east and reaches 150–200 m in the area of the gorge cross-secting the Supsa River and 600–700 m in the eastern sector. The slopes of the Guria Hill are dissected by the ravines-gorges of the smaller depths. The bottoms of the Supsa and Natanebi gorges are vast and built of alluvial deposits. On the slopes of the gorges, low accumulative and pedimental terraces are represented. Nasakirali Hill is located between the Nasakirali and Natanebi depressions.

Within Achjara, the foothill belt of southern Kolkheti Lowland has a small width (3–4 km). The belt is dissected by the Chorokhi, Chakvistskali, and Kintrishi sublatitudinal gorges. The foothill belt can be divided into the relatively high inland and the lower hilly parts of the coast. The first is characterized by deep narrow gorges and strongly dissected relief, and the other is represented by low, rounded hills, in the forming of which the red weathering rind is very important.

11.5 Imereti Highland

The Imereti Highland is located in the middle part of the intermount zone. In its east, in the Likhi Range, the watershed of the Black and Caspian seas are located, which is also the connecting bridge of the Greater Caucasus and southern highland. In the north, the Imereti Highland is bordered by the Racha Range and in the south by the Achara-Imereti Range and extends for a distance of ~60 km. The eastern border goes along the bottom of the eastern slope of the Likhi Range, while the western one along the Zestaponi meridian. The modern relief of the Imereti Highland was formed with the interaction of endogenous and exogenous processes.

Hilly, low, and medium mountainous relief of the Imereti Highland (Plateau) is dissected by a dense network of erosive gorges. The highland is constructed from Paleozoic crystalline rocks and in the literature is often known as the Dzirula crystalline massif (Maruashvili 1970). The dissected and denuded old peneplain is overlaid by horizontal-structured Mesozoic–Cenozoic sedimentary rocks. Due to the geological conditions and relief peculiarities, two morphostructural and geotectonic steps are distinguished. The lower step is constructed from old Paleozoic and Cambrian crystalline rocks, crystalline slates of granites, etc. Metamorphic rocks of the Cambrian age are outcropped in the watershed of Dzirula-Chkerimela, and cross-sected by the pink granites. The Paleozoic suites are represented by porphyritic tuffs, breccias, and phyllites.

Morphometric indicators of the relief of the Imereti Highland are diverse. The watersheds of Dzirula-Chkherimela and Likhi Range are the highest (1300–1500 m). The western and northern parts of the highland are 600–800 m high. Individual sectors of the watershed of the Dumala-Dzirula are higher than 900 m, namely, the gently sloping Korbouli Plateau.

Due to the differences in morphological, geotectonical, and morphometric indicators of the relief of the highland, two genetic regions are distinguished: the Chiatura structural Plateau and the denudative, strongly dissected peneplain.

The Chiatura Structural Plateau extends over the watershed of Dzirula-Kvirila and on the right bank of the Kvirila River. The relief of the plateau is wavy and dissected. On the right bank of the Kvirila gorge, the structural plateau is built of Upper Cretaceous limestones and marl limestones, covered by Oligocene and Miocene clays, sandstones and carbonate clays. In the areas of limestone spreading, the Kvirila River and its right tributaries—Jruchula, Katskhura, and Buja form 100–300 m depth canyons; they dissect the surface in deep wavy sectors surrounded by cliffy edges (Fig. 11.7). The karst forms are well-developed with various forms of sinkholes, corridors, dry gorges, and many caves into the slopes of the canyon. In the bottom of the lime remnants of beautiful shape and large boulders fallen from the cliffs are common. On the right bank of the Kvirila gorge, there is a "Katskhi Pillar"—the 40-m high denudative remnant erected at the Katskhi Village, on the top of which a church is built (Fig. 11.8). Karst vaucluses are common; the town of Chiatura is supplied by vaucluse streams. Most of the karst

Fig. 11.7 Chiatura (Kvirila River) canyon (*photo by* L. Tielidze)

Fig. 11.8 40-m high denudative remnant formed in limestone rocks (Katskhi Column) (*photo by* L. Tielidze)

caves were the dwellings of humans during the Stone Age, and in the historical period, they were used as fortifications.

The Kvirila-Dzirula interfluves are dissected by the tributaries, including the Dzirula tributaries—Dumala and Gemrula and the left tributaries of Kvirila—Sadzaliskhevi, Prone, Izvara, and others. On the left side, in the areas of distribution of Cretaceous limestones, karst forms are common. It is particularly notable for the areas of the villages of Shukruti, Nigozeti, Sveri, and other areas. Surface karst forms are visible in different kinds of sinkholes, connected by underground corridor at the Nigozeti Village.

In the Chiatura Plateau, there are cone-shaped hills that represent the Quaternary volcanic laccoliths of Perevisa and Goradziri. The Goradziri cones are visible in the relief near the Korbouli Village. The Makhaturi landslide is notable among the exodynamic forms, which have created a large limestone cirque and the hills composed of loose material. There are lakes or former lakes among the hills, notable Tepnadze Lake.

The small plateau of Bazaleti-Ghoresha is located in the relief built of limestones in the south of the Dzirula-Chkherimela interfluves. On the surface karst sinkholes are formed, and the edges of the plateau are cliffy. Out of the karst caves, Paleolithic humans lived in Deviskhvreli Cave.

In the north of the Dzirula massif, at the southern border of the Racha Range, the Sachkhere depression is located in the extended section of the Kvirila River gorge. The length of the Depression is 10–12 km and width 6–7 km. The relief building rocks—Oligocene and Miocene clays, sandstones, etc., have a certain role in the formation of its shape. The slopes of the depression are terraced. There are 2–4, 8–12, 20–22, and 40–50-m high terraces. The upper terraces are dissected by actions of landslides, while the lower one is broad and used in agriculture.

The denudative, strongly dissected plateau of the Imereti Highland is spread over the watersheds of the Dzusa, Kvirila, Dzirula and Chkherimela rivers, as well as in the Likhi Range. The relief is constructed from Paleozoic and Proterozoic granitoids and gneisses. Over a long time, the relief underwent the denudation while uplifting and sinking. At the end of the Palaeozoic, the peneplain was formed. In later periods, it was covered with new sediments and washed again. The modern medium and low mountainous relief were formed after the Pliocene (Gobejishvili 2011).

The smoothed surface is dissected by river valleys, the depth of which is 100–200 m in the west and 700–800 m in the east. In the Likhi Range, the peneplain is represented as a form of submeridional belt (20 km long, 5–6 km wide) and is cut by the tributaries of the Dzirula and Prone Rivers. The peneplain is visible as straight crests and is elevated to 1100–1300 m height. In the south side of the Dzirula-Chkherimela watershed, the peneplain is elevated to the highest point (1300–1500 m). Sometimes on the surface of the peneplain 20–40 m high denuded hills occur. The surface of the peneplain is covered with a powerful crust. The wreathing rind in the Likhi Range is represented by thin detritus and sand.

11.6 Likhi Range

The central and southern parts of the Likhi Range genetically and morphologically belong to the Dzirulula denudative Plateau. Its relief from the Chkherimela gorge to the Jvari Pass is constructed from Pre-Paleozoic old crystalline rocks. The crest is a peneplain and is dissected by erosive action of rivers. The northern part of the Likhi Range is built of Bajocian porphyrites. Hypsometrically the northern sector is higher. In the extreme north, the highest peak—Ribisa (2470 m) is located in the northernmost part of the range. Morphologically and genetically, the northern part belongs to the Greater Caucasus folded system.

The Likhi Range is an orographic border of western and eastern parts of Georgia. At the same time, the range is a watershed of the basins of the Black and Caspian seas. Artificial tunnels of Likhi Range are the main highways of the traffic movement.

References

Apkhazava I (1986) Georgian Soviet Encyclopedia, vol 10. Tbilisi, p 90 (in Georgian)
Devdariani G (1980) Qartuli Sabchota Enciklopedia (Georgian Soviet Encyclopedia), T. 5. Tbilisi, p 595 (in Georgian)
Gobejishvili R (2011) Relief of Georgia, monograph. Publishing House "Universali", Tbilisi (in Georgian)
Maruashvili L (1970) Saqartvelos fizikuri geografia (Physical geography of Georgia), monograph. Part 2. Tbilisi (in Georgian)
Maruashvili L (1981) Kavkasiis fizikuri geografia (Physical geography of the Caucasus), monograph. Publ. "Metsniereba", Tbilisi (in Georgian)
Mamisashvili K (1984) Qartuli Sabchota Enciklopedia (Georgian Soviet encyclopedia), T. 8. Tbilisi, p 657 (in Georgian)
Nemanishvili S (1982) River terraces of mountainous areas. vol 2. Tbilisi (in Georgian)
Tatashidze Z (1997) Dynamics of the reliefs of karst caves of Georgia and evolution of their slopes. Tbilisi (in Georgian)

Levan Tielidze is a Ph.D. of Ivane Javakhishvili Tbilisi State University in the field of geomorphology–glaciology–cartography and a senior research scientist at Vakhushti Bagrationi Institute of Geography. He is also a researcher-scholarship holder of the Department of Earth Sciences of the Georgian National Academy of Sciences. The field of his research is mountain cartography and glacial-geomorphological study of the mountainous areas in the Quaternary (Late Pleistocene and Holocene). He is a member of several international organizations. In 2014/2015, he conducted a research stay at the Climate Change Institute of the University of Maine, USA, and in 2015/2016–2017, at the University of Northern British Columbia, Canada. He is the author of about 40 scientific papers, maps, and eight monographs.

Ramin Gobejishvili (1941–2014) was a Doctor of Geographical Sciences, a geomorphologist–glaciologist. Until his last days, he was Professor of the Faculty of Exact and Natural Sciences, Ivane Javakhishvili Tbilisi State University and the chair of Geomorphology and Cartography. In 1996–2008, he was the president of the Geographical Society of Georgia, and in 2005–2007—the Director of the Institute of Geography. He was the author and co-author of about 150 scientific works and ten monographs.

Levan Maruashvili (1912–1992) was a prominent Georgian geographer. Doctor of Geographical Sciences (1954), Professor (1965). Honored Scientist of the Georgian Soviet Socialist Republic (1966). Honorary member of the Russian Geographical Society (1985). He produced over the 500 scientific and popular-scientific works in Georgian, Russian, and English languages.

Nino Chikhradze is a Ph.D. student and research scientist at the Department of Geomorphology and Geoecology of the Vakhushti Bagrationi Institute of Geography, Ivane Javakhishvili Tbilisi State University. Her research area is the use of photogrammetric methods in the study of coastal zone dynamics. In 2013–2014, she completed her Ph.D. internship at the University of Mino, Braga, Portugal. She is a Vice-Delegate of the European Speleological Federation and has published about 40 scientific works.

Eastern Georgia (Iveria) Intermountain Plain

12

Levan Tielidze, Ramin Gobejishvili, David Tsereteli,
Levan Maruashvili and Roman Kumladze

Abstract

In this chapter, the geomorphological characterization of the subzone of the Eastern Georgia (Iveria) intermountain Plain is presented. The orographic units are described in detailed, such as the Shida Kartli and Kvemo Kartli plains, as well as the orographic units including in the Tiripona Plain, Shida Kartli foothill mountainous-hilly relief, Kvernakebi Range (Hill), Mtkvari River gorge, Saguramo-Ialno Range, the Tbilisi depression, Marneuli and Gardabani plains, foothill belt of the Trialeti Range, Kumisi depression, and Iaghluja Hill. Also, the Iori Upland, Gombori Range, and Alazani Plain are described in detail. The text also provides the geological structure, tectonic conditions, and current exogenic processes of the study region. Each subchapter is accompanied by relevant images and maps.

Keywords

Eastern Georgia (Iveria) Plain • Shida Kartli Plain • Kvemo Kartli Plain • Tbilisi depression • Iori Upland • Gombori Range
Alazani Plain

12.1 Study Area

The relief of the Eastern Georgia (Iveria) Plain is divided by the Mtkvari River and its tributaries. In some literature, it is often known as the Mtkvari depression (Maruashvili 1970, 1981; Gobejishvili 2011). The relief is constructed from Tertiary sediments, only the Gombori Range and some smaller sectors are built of the suite of the Cretaceous period. The relief of the Iveria Plain has been formed with the interaction of differential tectonic movements and

L. Tielidze (✉) · R. Gobejishvili · D. Tsereteli
L. Maruashvili
Department of Geomorphology and Geoecology, Vakhushti Bagrationi Institute of Geography, Ivane Javakhishvili Tbilisi State University, Tbilisi, Georgia
e-mail: levan.tielidze@tsu.ge

R. Kumladze
Laboratory of Cartography and GIS, Vakhushti Bagrationi Institute of Geography, Ivane Javakhishvili Tbilisi State University, Tbilisi, Georgia

L. Tielidze
Department of Earth Sciences, Georgian National Academy of Sciences, Tbilisi, Georgia

R. Gobejishvili
Department of Geography, Faculty of Exact and Natural Sciences, Ivane Javakhishvili Tbilisi State University, Tbilisi, Georgia

R. Kumladze
Geological Mapping Division, Department of Geology, National Environmental Agency, Ministry of Environment Protection and Agriculture of Georgia, Tbilisi, Georgia

L. Tielidze (ed.), *Geomorphology of Georgia*, Geography of the Physical Environment,
https://doi.org/10.1007/978-3-319-77764-1_12

exodynamic processes. According to the orographic and morphogenetic features, the Iveria Plain is divided into several regions as follows: (1) the Shida Kartli Plain, (2) the Kvemo Kartli Plain, (3) the Iori Upland, (4) the Gombori Range, and (5) the Alazani Plain; they are divided into subregions by themselves (Fig. 12.1).

Kvemo Kartli and Alazani regions are plains, the relief of the Shida Kartli Plain is more complex and includes five subregions: the Tiripona-Mukhrani Plain, foothill mountainous-hilly zone of the Greater Caucasus, the Kvernakebi Range, the Mtkvari River gorge and the Saguramo-Ialno Range. The Iori Upland is a plain and wavy relief with the alternation of flatlands and anticline–mononclinal hills with height 1100–1200 m. The Gombori Range due to its tectonic nature is uplifted to 2000 m.

12.2 Shida Kartli Plain

The Shida Kartli Plain is a central region of the intermountain region. It is bordered by the southern slopes of the Greater Caucasus in the north, and by the Trialeti Range in the south. Its western border goes along the Likhi Range. In the east, it is bordered by the low hills of Saguramo.

The relief of the Shida Kartli Plain is mainly built from Tertiary and Quaternary sediment suites; only in the west are the ancient crystalline rocks (granites and crystalline shales) outcropped. The distinctive morphological and morphometric features of the Shida Kartli terrain, as well as its plains, terraced gorges and mountainous-hilly, allow distinguishing the following subregions: (1) the Tiriphona-Mukhrani Plain, (2) the hilly foothills of the Greater Caucasus, (3) the Kvernakebi Range (Hill), (4) the Mtkvari gorge, and (5) the Saguramo-Ialno Range.

12.2.1 Mukhrani-Tiripona Plain

The Mukhrani-Tiripona Plain was created by the accumulation of alluvial material brought by rivers during the relative sinking period of the intermountain depression. The inclination of their surface is defined by the direction of the following rivers: Didi Prone and Patara Prone, Liakhvi, Lekhura, Ksani, and Aragvi. The edges of the slight wavy terrain of the plain are complicated by the prolovial–deluvial deposits imported from the foothill hills and hillocks.

The Mukhrani—Tiripona Plain is divided into two parts: the Tiripona Plain in the west and the Mukhrani-Saguramo Plain in the east.

The Tiripona echelon-shaped Plain is dissected by tributaries of the Didi Liakhvi and Patara Liakhvi that meet at the city of Gori. The Tiripona Plain inclined toward Gori is built of Quaternary alluvial deposits the thickness of which is 50–60 m (Fig. 12.2). The Liakhvi River develops two 2–3 and 10–12 m terraces in the plain (Nemanishvili, 1982). The second extensive terrace is well represented in the relief. The western part of the Tiripona Plain (in the Prone gorge) is called the Doghlauri gorge, while the eastern part the Saamilakhvro gorge. The entire Tiripona Plain is spread at 60–65 km latitudinally. Its width is $\sim$30 km between Gori and Tskhinvali. The height of the Tiripona Plain is 550 m asl at Gori, while it is at 850 m at Tskhinvali; the inclination of the topographic surface is 10 m/km.

The Mukhrani Plain is separated from the Tiripona Plain by the Okami Highland (Fig. 12.3). The Mukhrani Plain is slightly inclined toward the east. The rivers of Ksani, Narekvavi, and Aragvi flow through its surface. The Mukhrani Plain also extends to the left side of the Aragvi gorge by the name of Saguramo Plain. The relief of the Mukhrani Plain is constructed from Pleistocene age alluvion, loam, and clay. The thickness of alluvial sediments is $\sim$100 m (Gobejishvili 2011).

The length of the Mukhrani Plain is $\sim$35 km and the width is 8–10 km. The lowest place at the Natakhtari Village (490 m), and the highest at the Ksovrisi Village (610 m). Within the plain, the Aragvi River develops into the Narekvavi River flowing along the Mukhrani Plain in the north and the Saguramo Plain is cross-sected by

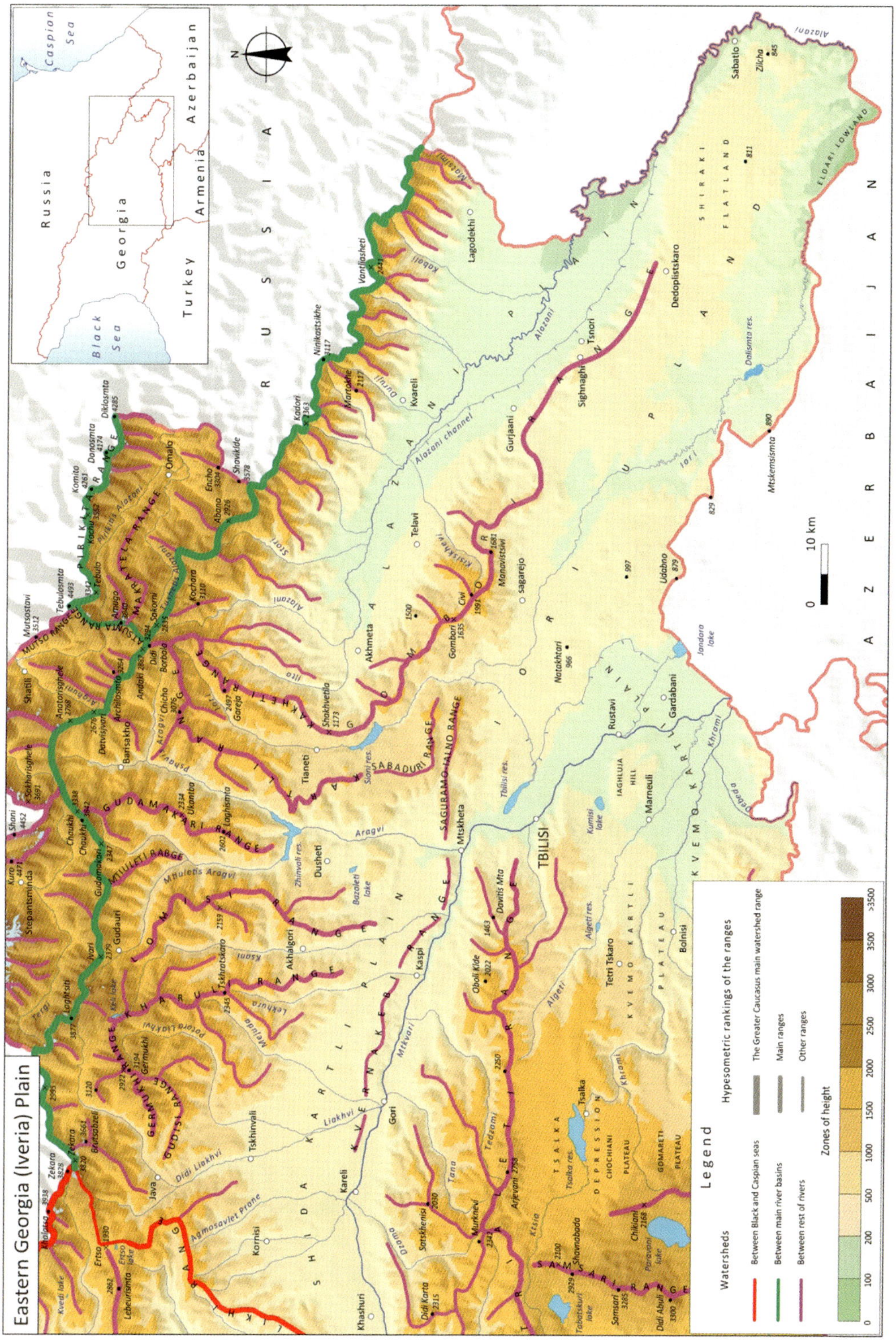

Fig. 12.1 The orography of the Eastern Georgia Plain

Fig. 12.2 Tiripona Plain (*photo by* L. Tielidze)

Fig. 12.3 Okami Highland—a divider of the Tiripona and Mukhrani plains (*photo by* R. Kumladze)

the Tedzami River that begins on the northern slope of the Ialno Range. The Mukhrani-Tiripona Plain is irrigated by the rivers, therefore, the agricultural sector is developed at a high level.

12.2.2 Foothill Mountainous-Hilly Relief

The foothill mountainous-hilly relief follows the Tiripona-Mukhrani gorge from the north. Its surface is mostly denuded and in the relief, it is sometimes represented as a pediment (step). The mountainous-hilly relief is built of marine and continental molasses of Tertiary age, such as conglomerates, clays, sandstones, and some marls and limestones. The hilly belt extends from the Suramula River gorge to the Iori River. The surfaces of the relief of the individual sectors are different from each other, but they have common morphological and morphometric features. The relative height of the mountainous hill is more than 250–350 m on average above the adjacent plain. The width of the hill is 8–10 km. The origin of the Relief origin is related to tectonic-erosive processes. In the east of the foothill belt, the Bazaleti and Ertso tectonical-accumulative depressions are formed. The Bazaleti flatland is located on the right side of the Aragvi River. It is built of $\sim$350 m thick Lower Pliostocene alluvial sediments. In its depression, the Bazaleti Lake is located at a height of 876 m asl.

In the easternmost part of the mountainous-hilly belt, on the right bank of the Iori River, the Ertso Flatland, is dissected by the tributaries of the Adzedzi River. They create debris cones of small thickness in the edges of the flatland. The length of the Ertso Flatland is about 10 km, width 5–6 km and height is 950–1050 m asl.

The foothill mountainous-hilly belt is cross-sected by the river valleys (Western Prone, Eastern Prone, Didi Liakhvi, Patara Liakhvi,

Lekhura, Ksani, and Aragvi), where terraces are common. It is noteworthy that the remains of the volcanoes in the river heads are not found in the terraces of 15–20 m height of the Ksani and Aragvi rivers, which indicates the young age of these volcanoes (Upper Pleistocene-Holocene) (Gobejishvili 2011).

The landslides are widespread in the foothill mountainous-hilly belt, including the Surami, Iremiantkari, and other landslides. In the Liakhvi River gorge, ice occurs in summer in the boulders near the Sabatsminda Monastery.

12.2.3 Kvernakebi Range (Hill)

The Kvernakebi Range follows the Tiripona-Mukhrani gorge to the south. It is constructed from Miocene age marine and continental molasses, such as conglomerates, clays, sandstones, and some marls and limestones. The hill has a monoclinal structure and is a cuesta in shape. The northern slope of the hill is sloped toward the Tiripona-Mukhrani depression, while the southern slope descends steeply to the bottom of the Mtkvari River gorge (Fig. 12.4).

The rivers of Shida Kartli (Liakhvi, Lekhura, Ksani, and Aragvi) antecedently cross the Kvernakebi Range and form a narrow, intricate through gorges of 200–400 m depth. With the erosive action of the rivers, the Kvernakebi Range is divided into four sections: the Ruisi Hill, the Kvernakebi Range itself, the Tslevi-Tkhoti Hill and the Skhaltba Hill.

The Ruisi Hill is located in the westernmost, between the rivers of Prone and Liakhvi. It is the smallest and lowest. Its length is 12–14 km, width 5–6 km, and height is 844 m asl. The crest is flattened (Fig. 12.5), the northern slope is sloped toward the plain, the southern slope is relatively steep and dissected by gorges.

Kvernakebi Hill, which is the highest and longest part of the range, is located between the Liakhvi and Lekhura rivers. Its length is 24–25 km and width 6–7 km; here is the highest peak of the range Mount Zegarda (1112 m). The northern slope of the hill is inclined, where small landslides are developed. In the central part of the northern slope is the saline Nadarbazevi Lake.

The southern slope of Kvernakebi Hill directed toward the Mtkvari River is steep and dissected by the ravine-gorges. In the cliffs of the southern slopes a group of artificial caves was

Fig. 12.4 Southern slope of the monoclinal Kvernakebi Range (*photo by* L. Tielidze)

Fig. 12.5 Ruisi Hill on the background of central Greater Caucasus (*photo by* L. Tielidze)

Fig. 12.6 Uplistsikhe—one of the ancient settlements of the Caucasus located on the southern slope of the Kvernakebi Hill (*photo by* L. Tielidze)

made, including one of the oldest settlements in the Caucasus and the city center (fortress-city) Uplistsikhe (Fig. 12.6), the main part of which is dated to the VI-IV centuries BC (Lezhava 1971).

Between the Ruisi and Kvernakebi hills, in the territory of the city of Gori, the erosion witness (remainder) is preserved, on which the Gori Fortress is now built.

The Tslevi-Tkhoti Hill is located between the Lekhura and Ksani rivers. The southern slope is steep and heavily ravined, while the northern one is sloped toward the Mukhrani Plain.

The Skhaltba Hill is the name of the eastern sector of the Kvernakebi Hill. It stretches ~11 km between the Ksani and Aragvi rivers. A morphological face characteristic to the cuesta-shaped hill is a high cliff of the southern slope, wherein the neogene conglomerates the artificial caves are made in the surroundings of the Shio Mghvime Monastery. The structural and morphological continuation of the Kvernakebi Range in the east is a higher Saguramo-Ialno Range.

12.2.4 Mtkvari River Gorge

The Mtkvari River gorge extends along the northern slope of the Trialeti Range from the Tashiskari Village to Mtskheta. The gorge is bordered by the branches of the Achara-Imereti Range in the west and then it extends along the southern slope of the Kvernakebi Range in the north (after Agara Village). The morphology of the gorge is determined by the nature of the structures; the river sometimes crosses the structures, sometimes it flows around it. The Mtkvari River is flowing along the synclinal and anticlinal structures at a long distance.

The Mtkvari River flows divaricated into a wide alluvial plain and forms interfluves from Tashiskari to the Prone River confluence. Both slopes of the gorge are terraced. The terraces are uplifted at high altitude on the front slopes of the Trialeti Range. On the left side, there are lower accumulative terraces. It should be noted that on the second terrace the Suramula River (to Prone) flows in parallel to the Mtkvari River.

From the Prone confluence to Mtskheta the Mtkvari River is flowing between the foothills of the Trialeti Range and the Kvernakebi Range in a large alluvial plain (Fig. 12.7). The gorge is asymmetric. In this part of the Mtkvari gorge, the terraces are visible, especially on the right side. The terraces show the uplifting rate of this area in the Quaternary, the five terraces at 2–4, 12–15, 20–25, 80, and 100 m height. From Dzegvi Village the narrow gorge of the Mtkvari River is cut between the Satskepela and Skhaltba ranges; sometimes this section is called Dzegvi's Klde-kari (Dzegvi rock-gate).

From Mtskheta to Avchala the Mtkvari gorge is antecedent; it crosses the end of the Trialeti Range.

12.2.5 Saguramo-Ialno Range

The Saguramo-Ialno Range stretches between the Aragvi and Iori gorges over a 32–33 km distance and borders the Tbilisi depression in the north. The highest point (1874 m) of the range is

Fig. 12.7 Mtkvari River gorge between the Trialeti Range and Kvernakebi Hill (*photo by* L. Tielidze)

Fig. 12.8 Saguramo Range behind the Tbilisi city (*photo by* G. Lominadze)

located in the east. In the north, the range is separated from the Sabaduri Hill by fault lines, and in the south from the Iori Upland. The Range is built of marine and continental molasses (conglomerates, clays and sands). The southern slope of the range is steeper than the northern. Morphosculpturally the range is divided into two parts (by Lelubani Pass and Tbilisi-Tianeti highway)—Saguramo and Ialno ranges.

The Saguramo Range (Mount Kukheti) is a latitudinal range of medium height and is the western extension of the Ialno Range (Fig. 12.8). It is built of Neogene sedimentary rocks (conglomerates, sandstones, and clays) (Changashvili 1984). There are mixed deciduous forests on the slopes of the range (relic Colchic subforest—Colchic holly (*Ilex colchica*), ivy (*Hedera helix*) and box tree (*Buxus colchica Pojark*), and the secondary subalpine meadows in its crest. The Zedazeni Monastery is built in the Saguramo Range.

The Ialno Range is an eastern continuation of the Saguramo Range and is named "Akriani Mountains" after old literary sources (according to Vakhushti Bagrationi "Ialoni"). The length of the range is ~ 18 km and the highest point is Mount Ialno (1874 m asl). The Range is built of Neogeneous sedimentary rocks (conglomerates, sandstones, and clays). The Ialno Range is a watershed, on the one hand, of the rivers of Tedzami (the left tributary of the Aragvi River) and Gorani (the right tributary of the Iori River), and on the other hand, of the rivers of Norioskhevi, Martkopiskhevi, and Satskheniskhevi. The slopes are covered with oaks, oak-hornbeams, and beeches.

12.3 Kvemo Kartli Plain

According to morphological and morphometric signs, the Kvemo Kartli Plain is represented by the following morphological sectors and subregions: the Tbilisi depression, Gardabani Plain, Marneuli Plain, the foothills of the Trialeti and Loki ranges, Iaghluja Hill, and Kumisi depression.

In the Tbilisi depression (Fig. 12.9), the Mtkvari gorge (from Avchala to Soghnalughi) cross-sects the endings of the Trialeti Range in the form of synclines and anticlines. These structures determine the morphological appearance of the right side of the gorge. The synchlinal flatlands of Dighomi, Saburtalo and Krtsanisi are located between the Mukhatgverdi, Lisi, Mtatsminda and Teleti anticlinal ranges. The lakes of the same name are located in the flatlands of the Lisi and Mukhatgverdi dissected relief, including Turtle Lake of exotectonic origin on the slope of Mtatsminda. The synclinal flatland is filled with alluvial material and overlaid with proluvial-deluvial sediments. The left side of the Mtkvari gorge is open and soft, and the river terraces create plains.

Within the city, the main right tributaries of the Mtkvari River are the Vere and Dighmistskali rivers. The Vere River confluences Mtkvari at Heroes' Square, while the Dighmistskali River at the bridge of Didube district. Within the city, the Vere River has its tributaries as streams that flow into the ravines, e.g., Varazi Ravine (flowing into the Vere River in the Vake district).

Fig. 12.9 Tbilisi depression (*photo by* L. Tielidze)

The region of Tbilisi and its surroundings are characterized by the development of paleogenic sediment-flysch and volcanogenic sediments of considerable thickness. In the most of the territories these sediments are covered by the strong Quaternary formations, though in the right part of the city, rocks are outcropped in the slopes of the ranges, in the dividing gorges and in the river valleys. The Quaternary sediments spread in the city area can be grouped by their genesis, composition, character, and properties as follows:

(1) Deluvial–prolovative clays, loams with sand middle layers and lenses, loess and gypsum sediments;
(2) Lacustrine-wetland sediments (Fig. 12.10);
(3) Alluvial terraces of the rivers of Mtkvari and Vere with gravel, gravel-sand, and sandy fillers (Jajanidze 2014).

Tectonically the region of Tbilisi is located in the eastern part of the central Tectonic zone of the Achara-Trialeti fold system. In general, this zone is characterized by a wide distribution of Middle and Upper Cretaceous volcanogenic-carbonate rocks and thick Eocene volcanogenic-flysch formations. The rocks are collected in the latitudinal large anticline folds of echelone or box shape. The Tbilisi region includes the Dighomi syncline, Lisi anticline, Saburtalo syncline, Tabori anticline, Krtsanisi syncline, and Teleti anticline that are sinking in the east.

Seven terraces are distinguished in the Tbilisi depression, on the surfaces of which the city districts are located.

The first beyond-floodplain terrace is 2–3 m high, well-expressed below Avchala and at the Orkhevi River confluence; the second terrace of 10–15 m high, is visible in the vicinities of Didube-Dighomi; the third, the Saburtalo-Avlabari terrace, with height of 40–50 m, is sharply expressed in both sides of the gorge; the fourth, the Arsenali terrace is 60–70 m high; the fifth is the Kukia terrace (110–120 m); the sixth is the Lotkini terrace (150–160 m) with a wide smooth surface with Lotkini settlement in it (Fig. 12.11) and the seventh terrace, which is the highest (200–250 m) and morphologically well-

Fig. 12.10 Lacustrine deposits in the territory of Tbilisi (*photo by* L. Gadrani)

expressed in the area of Makhata and Mukhatgverdi.

The relief in the depression of Tbilisi is entirely modified due to anthropogenic factors. From the natural processes in the depression of Tbilisi the landslide phenomena (in Mamadaviti, Mukhatgverdi, and Nadzaladevi) and mudflows (of small size) are notable. Particularly noteworthy is the disaster—flood and mudflow, occurred at night on 13th of July 2015, stipulated

Fig. 12.11 Settlement of Lotkini built in the sixth terrace (*photo by* L. Tielidze)

Fig. 12.12 Consequences of the disaster of 13 June, 2015 in Tbilisi (Tbilisi Zoo) (*photo by* L. Tielidze)

by the rock-avalanche type landslide in the Vere gorge and caused major damage to the capital (Fig. 12.12).

The relief of Tbilisi is strongly transformed by human activity. The surface of the depression is completely covered by the anthropogenic forms, such as housing, transport communications, etc.

The relief of the **Marneuli Plain** is represented by three terrace steps in the river gorges of Mtkvari, Mashavera, and Algeti. The plain is dissected by the same rivers. The bottom of the gorge is wide, the bed is branched and there are vast interfluves. The terraces are located at the relative heights of 2–4, 10–15 and 25–30 m. The age of the terraces is Pleistocene (Maruashvili 1970). It is constructed from marine sands and alluvial gravelites. There are pseudo karst forms (wells and caves) developed in the river terraces composed of clays.

The Gardabani Plain is located on the left side of the Mtkvari River. The surface of the plain is represented by several accumulative terraces of Mtkvari. The number and height of the terraces are the same as on the right bank of the Mtkvari River (2–4, 10–12, and 25–30 m). The Lochini River flows in the plain, the surface of which is covered with irrigation channel systems.

In the study area, **the foothill belt of the Trialeti Range** is constructed from Upper Eocene clays and sandstones, Middle Eocene volcanogenic formations, Lower Eocene limestone sandstones, marls, Maikop gypsum clays, and sandstones. This surface is dissected by small rivers (Enagetiskhevi, Asuretiskhevi, and Tsalaskuriskhevi) and is overlapped by the deluvial–proluvial layers.

The northern foothill of Loki covers the area from the Kveshi Village to the Shulaveri Village. It is built of the Upper Chalk Cretaceous volcanogenic suites. It is dissected by the tributaries (Shulaveristskali, Poladauri, and Talavristskali) of the Mashavera River and is characterized by wide-bottomed gorges with some steep hills. Hypsometrically it is located within the range of 350–800 m asl.

The Kumisi depression is located between the Trialeti Range and Iaghluja Hill, containing Kumisi Lake. Tectonics are of great importance in the formation of the depression, which is built

Fig. 12.13 Mud volcano in the Iori Upland (*photo by* R. Gobejishvili)

of the powerful suite of the lacustrine deposits, and prolovial–deluvial materials on its sides.

The Iaghluja Hill is a name for the low hilly relief, which extends in the Marneuli Plain and is constructed from Neogene molasses (conglomerates and sandstones). Morphotectonically, it creates an inverse hill (synclinal horst). The massif reaches a maximum height of 765 m and the Iaghluja foothill is located at 400–500 m asl. Some scholars believe that the main deepened part of the hill is the former riverbed (Maruashvili 1970).

12.4 Iori Upland

The Iori Upland is located in the southeastern part of Georgia. It is surrounded by the Mtkvari and Alazani depressions with the Ialno and Gombori ranges. The relief extends within the height of 90–1000 m asl, where the highest point is Mount Nikorastsikhe (1001 m) at Dedoplistskaro Town. The high peaks are Mount Demuri (990 m), Natakhtari (966 m), Didi Udabno (905 m), and Gareji's Udabno (879 m) in the west and Mount Zilcha (845 m) in the east. The morphological appearance of the terrain is created by narrow anticlinal hills of common Caucasian direction and vast plain-depressions. From sculptural forms the erosive gorges, riverine terraces, smoothed surfaces, gorges, badlands, pseudo-karsts, mud volcanoes (Fig. 12.13), artificial caves, and denudation witnesses are notable. The dry climate of the upland determined the specific shapes of landforms, and therefore, the upland is distinguished with the most interesting and peculiar shapes of landforms from other territories of Georgia.

The relief of the Iori Upland is mainly constructed from Miocene–Pliocene marine and continental molasses: conglomerates, clays, sandstones, middle layers of volcanic ash, and limestones and marls in the Middle Miocene marine molasses in some areas of high hills. In the area of Dedoplistskaro, the Upper Jurassic limestones and marls are found in the form of islands. The flatland foot is built of Quaternary alluvial–proluvial sediments.

Structural-denudative forms spread almost across the whole territory, represented by anticlinal hills, create the appearance of the relief. Structural-accumulative forms are formed in the synclinal depressions, as well as the structural-erosive forms related to mononclinal hills. Significant denudative and accumulative forms are as follows (small plain-flatlands): Didi Shiraki, Patara Shiraki, Taribana, Ole, Naomari, Udabno, and Kajuri; the hills are as follows: Amartuli, Demurdaghi, Gareja, Pirukughmarta, Kotsakhuri Range, Burdamta, Shvindiskeli, Katarkeli, Zilcha, Kachi, and Kalandara.

The structural-denudative hills worthy of note include the Kashi Range (Hill), located in the easternmost part, which is a watershed between the Alazani gorge and the Shiraki flatland (Steppe). It is asymmetric and its northeastern slope is dissected by dry ravines and steeply descends toward the Alazani gorge. The relative height of the slope is 400–500 m. The southwestern slope of the range gradually merges into the Shiraki Plain, where the relative height is 180–190 m asl.

Hypsometrically, the Kashi Range (Hill) is not high. The height of the peak located in the central part is 742 m asl. Due to the dry climatic conditions, among the modern exogenic processes the weathering, deflation and surface denudation and mudflow phenomena in dry ravines are intensive. In the southwest, the Shiraki flatland (steppe) is bordered by the Nazarlebi Hill and its continuation Kaladabra. At Mount Nazarlebi (876 m) begin the Didi Shiraki and Patara Shiraki watershed hill—Shuamta, the highest point of which is 810 m.

The relief of the Kotsakhuri and Berdosmta Hills are structural-denudative. The southern slope of the Kotsakhuri River is dissected by erosive processes and is uplifted at 400 m over the Iori riverbed. Its northern slope gradually transits into the Natbeura depression.

In the western part of the Iori Upland, the anticlinal hill of Katarkeli is located, uplifted from the Naomari and Ole synclinal depressions by 150–200 m. The upland is strongly dissected by dry ravines and creates a hilly relief. Structural-denudative forms of relief include low anticlinal hills of Natakhtari, Aklemiskeli, and the Mount Dobi. They are uplifted by 300–400 m from the Gardabani Plain and the Udabno synclinal depression and are heavily denuded. Vast dry gorges and depressions of dried-up lakes are characteristic to this area.

Structural-accumulative forms of the relief are common in the tectonic depressions and in the synclinal depressions of different sizes, which are often filled with powerful Quaternary sediments. The accumulative plain-flatlands of Didi Shiraki and Patara Shiraki stretch from the northwest to the southeast and are inclined in the same direction. Waterless Shiraki Plain is connected to the depressions of the Kasristskali and Jeirani plains, the relief of which is wavy with traces of erosion action. The thickness of the Quaternary deposits in the Shiraki depressions is ~100 m.

The Eldari Plain-Lowland of accumulative origin is located in the southeasternmost part of the Iori Upland. In the south, it is bordered by the Iori River, while the Komeri and Duzdali Hills in the north. In the peripheries, the relief of the Eldari Lowland is represented by the debris cones giving a wavy shape. It is built of Quaternary alluvial and deluvial sediments with a thickness of 100–250 m. The Eldari Lowland is a part of the Mtkvari-Araksi accumulative Plain-Lowland. With the maximum width of 6–7 km, it extends at ~25 km along the left bank of the Iori River within Georgia. It is almost a horizontal flat surface, the absolute height of which varies within 100–200 m and is the lowest place above sea level in eastern Georgia.

Additionally, the annual sum of precipitations in the Eldari Lowland is 250–300 mm and the lowland is considered the most dry place in Georgia, which is the only place where a true semihumid landscape is presented. On the gray and light brown soils the wormwood (*Artemisia absinthium*) semidesert vegetation has developed, against a background of salsola (*Salsola arbuscula*) and broomsedge (*Andropogon*). The Eldari Lowland is distinguished by a sharply negative balance of moisture (the rate of moisture is 0.2–0.3. Gobejishvili 2011). Due to the dry climate of the plain and the outcropped slopes (of the edge of the Iori Upland) leaned on it from the northeast, the region is completely devoid of a hydrographic network. There are temporary flows, though they are not able to cross the Eldaro Lowland and they leak into the soil. A significant part of the surface of the Elderi Lowland is completely outcropped and chapped from drought. The southeasternmost part of the lowland is known as "Bughamoedani". The Eldari Lowland is a winter pasture for sheep.

The vast tectonic depression of Tsitsmatiani-Ole is located between Sagarejo and Kachreti, which is divided into two parts by the Sagarejo highland. The southern part of the plain is dissected by the river of Lakbeti and its tributaries. The thickness of alluvial and deluvial sediments is several tens of meters. The relief of the northern part of the Tsitsmatiani Plain contains debris cones, which are formed by temporary and permanent flows from the Gombori Range. The thickness of accumulative deposits is 200–400 m.

On the side of the Iori River, in the vicinities of the Tsitsmatiani Plain, riverine terraces are formed. There are four terraces at the Lambalo Village with heights of 4–6, 20–25, 50–60, and 100–150 m. The terraces are composed of alluvial deposits and lacustrine deposits with peat layers.

In the middle sector of the Iori Upland, the structural-accumulative forms include the depressions of Vake, Didi Taribana, Patara Taribana, Natipari, and Naomari. The relief of the depressions is characterized by flat bottom and debris cones that are formed by dry gorges in the peripheries.

The Samgori Plain is located in the northwesternmost sector of the Iori Upland. Its surface is strongly peneplained, slightly wavy and dissected by dry ravines and gorges. There are several artificial hills—kurgans visible in the relief. In the eastern part of the plain, the terraces of the Iori River are widespread (in the area of Sartichala).

Structural-erosion forms are common in most parts of the Iori Upland and they are associated with monoclinal hills. These hills represent the limbs of anticlinal rings or strongly washed synclinal folds. The monoclinal hills are well represented in the southern part of the Iori Upland, in the Mtkvari-Iori interfluves. The southwestern slopes of the Eldari and other hills are straight, while the northern slopes are sloping toward the layers. The hills are built of Lower and Middle Sarmatian sandstones and strongly cemented conglomerates.

The northern slope of the Udabno Hill stretches monoclinally. It is built of Sarmatian and Meiotic color suite. The color suite is represented by red and green clays, sandstones, and conglomerates. It is noteworthy that the residual bones of the hipparion fauna and the human-like monkey's teeth (Gareja desert-Piteki) have been found in this suite (Gobejishvili, 2011). There is an erosion depression between the Udabno and Dodokari hills, which is constructed from color suite (Fig. 12.14). The depression is characterized by hilly relief. The bottom of the depression is 280–300 m lower than the adjacent hills.

In the south and west of the Shiraki Plain, the following monoclinal structural forms are spread: the Parasi, Komra, Kaltani, Umru, Duzdala, Shavimta, and other hills.

Similar forms are observed in Pantishara, Kumuriskhevi, Dekistskali, and Goristskali gorges. Soft clays and conglomerates contribute to the construction of monoclinal hills. The most interesting and strange morphostructural forms (in the shape of fortress, column, etc.) originate in the relief, including badlands (Fig. 12.15). Clay karst is very well distributed at the foot of the monoclinal hills in the Iori Upland. Their origin is associated with strong deluvial and proluvial deposits.

Rocky forms of fortress shape built of limestones are sharply distinguished in the relief of the surroundings of Dedoplistskaro. Such island hills are Ori Dzma, Nikorastsikhe and others, the heights of which reach 900–1000 m asl.

Fig. 12.14 The erosion depression between the Udabno and Dodokari hills built of color suite (*photo by* L. Tielidze)

Fig. 12.15 The badlands in the Iori Upland (*photo by* T. Bragvadze)

Smoothed surfaces (denudative forms) of the relief are spread all over the Iori Upland and are related to the crests of the ranges and hills.

The interesting landforms are the meanders and terraces spread in the gorges of the Alazani and Iori rivers. In the Iori River gorge, the meanders are particularly visible below the Lambalo Village; the plants in the floodplain of the river are spread chainwise. Meanders of this area differ from the meanders of the Alazani River as the meanders of the Iori River do not develop dry former riverbeds. In the Iori River gorge (upstream), the two-stair terrace of the Holocene age is located at 2–5 and 10–15 m: the second terrace is visible in the gorges of both rivers and their tributaries and its relative height is 25–35 m. The third and the fourth terraces are high terraces, they were formed in the Lower and Middle Pleistocene (Nemanishvili, 1982).

12.5 Gombori Range

Orographically the Gombori Range is a continuation of the subrange (Kakheti Range) of the Greater Caucasus, but according to location and geological construction, it belongs to the intermountain plain. The range runs from the north-west to the south-east. Its length from Shakhvetila Pass (1200 m) to the town of Dedoplistskaro (800 m) is ~90 km, the width of the range is 20–25 km and the highest peak of the range is Mount Tsivi (1991 m).

Morphometrical-morphologically the north-western part (Shakhvetila-Chalaubani) is average mountainous relief, with old remnants of fragments of the smoothed surfaces, valleys filled with box-shaped gorges, badlands, and landslides. Hypsometrically the southeastern sector of the Range is low mountainous and hilly. The northeastern slope of the Gombori Range is dissected by the tributaries of the Alazani River: Orvili, Khodasheniskhevi (Didrike), Turdo, Kisiskhevi, Shromiskhevi, Vantiskhevi, Shashianiskhevi, Chermiskhevi, Papriskhevi, and others. The southwestern slope is dissected by the tributaries of the Iori River: Gomboriskhevi, Lapiankhevi, Sagarejoskhevi, Chailuriskhevi, Lakbe, and others. Morphologically the gorges are V-shaped and box-shaped. The gorges and their watersheds are located in perpendicular to the Gombori Range (considering structural construction).

The Gombori Range is constructed from Mesozoic and Kainozoic suites and is divided into two distinct complexes from the tectonic and lithological point of view. The older Upper Cretaceous sediments occupy the central part of the structure and are represented by the sandstone-aleurolitic (in the lower part) and cluster-limestone (in the upper part) flysch: aleurolitic, sandstone, gravelite, and cluster-limestone turbidites, pelagic marls, flint argillites, and some alternation of heliotrophic limestones and marls, boulder-breccias and conglomerates.

Younger deposits are located in the heart of a huge anticline and are constructed from Neogene (Sarmatian), continental and marine molasses: sandstones, clays, conglomerates, and occasionally marls. On the edges of these deposits, the relief is built of Upper Pliocene molasses: conglomerates, sandstones, clays, and loams. The watershed crest of the Gombori Range is located asymmetrically toward the anticlinal structure, which is generally shifted to the southwest and folded.

The main appearance of the region resembles erosive landforms, represented by river gorges and their watersheds. The depth of the gorges reaches 500–1000 m and mostly they have a box shape; the wide alluvial gravels are created by the destruction of the material of the Tsivi suite. In the areas of landslide development, the gorges are narrow bottomed.

The cliffy slopes of the Tsivi suite, especially in the southwestern part of the range, are strongly washed with rain water and form a badland relief. The classical example of distribution of such relief is Lapiankhevi, Sagarejoskhevi, Chailuriskhevi (left tributaries of the Iori River), also the heads of the right tributaries of the Alazani River, as well as the Turdo and Kisiskhevi mudflow basins (Fig. 12.16). Two morphological types are distinguished in the lithologically homogeneous and non-homogeneous facies of the Tsivi suite among the badlands of the Gombori Range. The badland surfaces formed in the non-homogeneous facies have stairs-shaped profiles, while in a homogeneous setting, they are thinly dissected.

12.6 Alazani Plain

The Alazani-Agrichai Plain (Georgia-Azerbaijan) is a unified structure, with a total length of ~200 km. Alazani Plain is located between the Kakheti Caucasus and Gombori Range. In Georgia, the Alazani Plain stretches ~110 km in length and 28–30 km in width (Fig. 12.17).

Fig. 12.16 The mudflow basins on the slopes of the Gombori Range (*photo by* Z. Turmanidze)

Fig. 12.17 Alazani Plain (*photo by* L. Tielidze)

From a distance the Alazani Plain leaves the impression of a flat plain; in fact, the absolute height of the surface of the plain is highly variable and varies from 175 to 550 m (the Pankisi gorge). Due to the slight inclination of the Alazani Riverbed, the river flows slowly and forms sharply visible meanders with well-expressed loops (Fig. 12.18). The surface of the plain rises

Fig. 12.18 Meanders of the Alazani River in the Alazani-Agrichai Plain. Google Eart imagery 18/5/2009

from the riverbed toward the bottom of the ranges and is covered by powerful debris cones on both sides. The area of some cones are tens of square kilometers with heights of 200–300 m.

From a geological point of view, the Alazani Plain is a peculiar tectonic unit of the territory of Georgia. This is an intensive sinking zone, a live continental geosyncline with intensive accumulation of sediments. Its formation began from the Pliocene and is continued to the present day. This is confirmed by the fossil soil in the boreholes and the existence of a thick layer containing cultural horizons, powerful debris cones of the rivers flowed down from the Greater Caucasus and the Gombori ranges with deformation signs (Maruashvili 1970). The foothill relief leaned against the plain is constructed from the sediments of Upper Jurassic and Cretaceous periods and represents the remainder of the sunken zone, which is cut down by fault lines.

In these foothills of the Greater Caucasus, the absolute heights reach 500–800 m. The belt is cross-sected by the gorges of the following rivers flowed down from the Greater Caucasus: Stori, Lopota, Didkhevi, Chelti, Duruji, Avaniskhevi, Shorokhevi and Lagodekhiskhevi. Besides the meridional gorges, short alongside gorges are common (Chagurgulastskali and Bursa). The accumulative relief of the Alazani Plain is complicated by secondary morphological forms, such as the erosive forms with small depth and rare erosive forms, that are cut in debris and do not create a joint network.

The right side of the Alazani Plain is constructed from the Tsivi suite sediments (cobbles, gravelites) brought by mudflows from the Gombori Range. Large debris cones are formed by the Telaviskhevi, Turdo, Kisiskhevi, Vantiskhevi, Shromiskhevi, Papriskhevi, and others.

In the present bed, the Alazani River flows serpentine in the east of the Sighnaghi meridian and forms meanders. Within Kakheti the Alazani's bed basically coincides with the axis of depression (graben). In the surroundings of the Iori Upland the Alazani River flows along the foot of the upland, which is conditioned due to the absence of mudflow basins from the right side. The rivers flowed down from the Greater Caucasus have inclined the Alazani thalweg to the right.

References

Changashvili M (1984) Georgian Soviet Encyclopedia, vol 8. Tbilisi, p 620 (in Georgian)
Gobejishvili R (2011) Relief of Georgia, monograph. Publishing House "Universali", Tbilisi (in Georgian)

Jajanidze T (2014) Engineering-geological analysis and evaluation of closed depressions of Vake, Delisi, Saburtalo and Dighomi. The summary of the dissertation thesis presented for the Doctor's academic degree. Tbilisi (in Georgian)

Lezhava G (1971) The cave-city Uplistsikhe architecture, friend of the monument № 27/28, Tbilisi (in Georgian)

Maruashvili L (1970) Sakartvelos pizikuri geograpia (Physical geography of Georgia), monograph, Part 2. Tbilisi (in Georgian)

Maruashvili L (1981) Kavkasiis pizikuri geograpia (Physical geography of the Caucasus), monograph (in Georgian)

Nemanishvili S (1982) River terraces of mountainous areas, vol 2. Tbilisi (in Georgian)

Levan Tielidze is a Ph.D. of Ivane Javakhishvili Tbilisi State University in the field of geomorphology–glaciology–cartography and a senior research scientist at Vakhushti Bagrationi Institute of Geography. He is also a researcher-scholarship holder of the Department of Earth Sciences of the Georgian National Academy of Sciences. The field of his research is mountain cartography and glacial-geomorphological study of the mountainous areas in the Quaternary (Late Pleistocene and Holocene). He is a member of several international organizations. In 2014/2015, he conducted a research stay at the Climate Change Institute of the University of Maine, USA, and in 2015/2016–2017, at the University of Northern British Columbia, Canada. He is the author of about 40 scientific papers, maps and eight monographs.

Ramin Gobejishvili (1941–2014) was a Doctor of Geographical Sciences, a geomorphologist–glaciologist. Until his last days, he was Professor of the Faculty of Exact and Natural Sciences, Ivane Javakhishvili Tbilisi State University and the chair of Geomorphology and Cartography. In 1996–2008, he was the president of the Geographical Society of Georgia, and in 2005–2007—the Director of the Institute of Geography. He was the author and co-author of about 150 scientific works and ten monographs.

David Tsereteli (1905–1976) was a Doctor of Geographical Sciences, Professor (1965). Until his last days he was deputy director of Vakhushti Bagrationi Institute of Geography. His research area was Quaternary geology, glaciology and paleogeography. He was the author and co-author more than 100 scientific works and several monographs.

Levan Maruashvili (1912–1992) was a prominent Georgian geographer. Doctor of Geographical Sciences (1954), Professor (1965). Honored Scientist of the Georgian Soviet Socialist Republic (1966). Honorary member of the Russian Geographical Society (1985). He produced over the 500 scientific and popular-scientific works in Georgian, Russian, and English languages.

Roman Kumladze is a Ph.D. student and research scientist at the Laboratory of Cartography and Geographical Information System (GIS) of the Vakhushti Bagrationi Institute of Geography, Ivane Javakhishvili Tbilisi State University. His research areas are geodynamic processes and geomorphological mapping.

Achara-Trialeti-Loki Mountainous Terrain

13

Ramin Gobejishvili and Levan Tielidze

Abstract

This chapter presents the geomorphological characterization of the subzone of the highland of Southern Georgia. There are detailed descriptions of orographical units such as Achara-Trialeti-Loki mountainous terrain, Achara-Imereti (Meskheti) and Trialeti ranges, Borjomi gorge, Achara depression (Acharistskali River gorge), Shavsheti and Arsiani ranges, Akhaltsikhe depression, Khrami-Loki medium mountainous region and its subregions—the Bedena mountain massif, Khrami River gorge, Kviriketi-Lukuni subregion, and the Loki Range. The text also provides geological structure of the study area, its tectonic conditions and current exogenic processes. Each subchapter is accompanied by appropriate images and maps.

Keywords

Highland of Southern Georgia • Achara-Trialeti-Loki mountainous relief Achara depression • Borjomi gorge Shavsheti Range • Arsiani Range • Akhaltsikhe depression • Khrami-Loki medium mountainous region

13.1 Study Area

The mountain system of Achara-Trialeti-Loki is stretched latitudinally from Batumi to Tbilisi, for about 260 km, while it has a southern direction from the heads of the Algeti River to the Loki Range. The western sector of the subzone is wider (65 km) than the eastern (30 km). The highest peaks of Mepistskaro (2850 m) and Shavi Klde (2850 m) are located in the Achara-Imereti and Trialeti ranges. The Mtkvari River divides the Achara-Trialeti Range into two parts—the Achara-Imereti and Trialeti ranges.

The north and northwest slopes of the Achara-Imereti Range are characterized by abundant atmospheric precipitation that stipulates the intensity of erosion processes. The crest of the Achara-Imereti and Arsiani ranges are the watersheds of the Caspian and Black sea basins, while the Shavsheti, Trialeti and Loki ranges are the watersheds of the Chorokhi and Mtkvari rivers.

The Achara-Trialeti and Loki medium mountainous subzone are divided into separate

R. Gobejishvili · L. Tielidze (✉)
Department of Geomorphology and Geoecology, Vakhushti Bagrationi Institute of Geography, Ivane Javakhishvili Tbilisi State University, Tbilisi, Georgia
e-mail: levan.tielidze@tsu.ge

L. Tielidze
Department of Earth Sciences, Georgian National Academy of Sciences, Tbilisi, Georgia

R. Gobejishvili
Department of Geography, Faculty of Exact and Natural Sciences, Ivane Javakhishvili Tbilisi State University, Tbilisi, Georgia

L. Tielidze (ed.), *Geomorphology of Georgia*, Geography of the Physical Environment,
https://doi.org/10.1007/978-3-319-77764-1_13

regions according to morphological, morphometric, and genetic signs; they are as follows: the Achara-Imereti Range, Borjomi gorge, Trialeti Range, Khrami-Loki medium mountainous region, Shavsheti Range, Achara depression (Acharistskali River gorge), Arsiani Range and the Akhaltsikhe depression (Fig. 13.1).

13.2 Achara-Imereti Range

The Achara-Imereti mountain range is the broadest in the southern highland and is clearly visible orographically; its length is about 182 km from the Mt. Erge (892 m) to the Mt. Kokola (1341 m), by direct line, the length is about 151 km and the width is 40–45 km in the Persati meridian. The range is of submeridian (SW-NE) direction (west sector), from Mt. Erge to the Mt. Sakornia (2755 m), then to the Kvishkheti Village, it is of general latitudinal direction. The highest peak is Mepistskaro (2850 m). The submeridian (western) sector of the range is a watershed between the right tributaries of the Acharistskali River and the Chakvistskali, Kintrishi and Natanebi river basins. The latitudinal watershed separates the basins of the left tributaries of the Mtkvari and Supsa-Rioni rivers. In this watershed, the following peaks are located: Sakornia (2755 m), Mt. Zoti (2676 m), Khakhlamo (2635 m), Didmaghali (2588 m), Lomismta (2187 m), and Dedabera (1838 m). In this sector of the range, there are passes at 2100–2300 m height. The Zekari Pass (2182 m) is particularly noteworthy, containing the Abastumani-Sairme-Baghdati highway.

Across the country, the Achara-Imereti Range is named differently. In the west, it is called the Achara-Guria Range, in the central part—Meskheti and Achara-Akhaltsikhe Range and in the east—the Akhaltsikhe-Imereti Range (Nizharadze and Jibuti 1957). In some works of Al. Javakhishvili (Javakhishvili 1926) and in the Geographical Atlas of the Georgian SSR published in 1964, it is known as the Meskheti Range.

The relief of the Achara-Imereti folded structure is constructed from Eocene volcanogenic-sedimentary rocks: basalts, andesites, massive and thick layered breccias, tuffs, tuffo-sandstones, and marls. In the east of Zekari Pass, in the northern part of the crest, the Upper Cretaceous thin layered red and pink limestones and marls are outcropped. Eocene diorites are represented in small segments. Folded structures are complicated by faults and overthrusts (Gobejishvili 2011).

The Achara-Imereti Range is divided into three main parts according to the genetic signs of morphological, morphometric and landforms; they are the watershed crest, the northern and southern slopes. The watershed crest has a wavy morphology of 100–200 m height; only in some sectors, individual mountain massifs are uplifted above the surface. Pleistocene glaciers made great impact in the formation of relief. The glacial forms of corries, moraines, and small troughs are created in the peaks higher than 2400 m asl, such as Khino, Taginauri, Sakornia, Zoti Mountain, Mepistskaro, and Didmaghali (Tielidze 2017). It is noteworthy that the northern slopes of the peaks are more rocky and steep than the southern slopes.

The western sector of the Achara-Imereti Range is a watershed of the Chakvistskali and Acharistskali river basins, often called the Chakvi Range. In the north, it is parallel to the Kobuleti Range (the watershed of the Kintrishi and Chakvistskali river basins). The Range is morphologically visible and stretches to the Black Sea in the vicinity of Tsikhisdziri.

Another branch, known as Sapileti-Goraketi, stretches from the Bodishi Pass in the northeast to the southwest and creates the watershed between the right tributaries of the Acharistskali River. This range retains a 2000 m height up to Mt. Sapileti, then gradually reduces and ends at the confluence of the rivers of Acharistskali and Chvana.

The northwestern slope of submeridian (NE-SW) direction from Mt. Sakornia is dissected by the Natanebi, Kintrishi, Chakvistskali and Korolistskali rivers, and their tributaries. The depth of the erosion section is 600–1200 m and the slope width is 6–12 km. There are several flattened steps in different heights of the river watersheds of the rivers; their heights range between 1200 and 2200 m asl.

From the Mt. Sakornia, the Achara-Imereti Range is in an eastern direction. Its northern slope

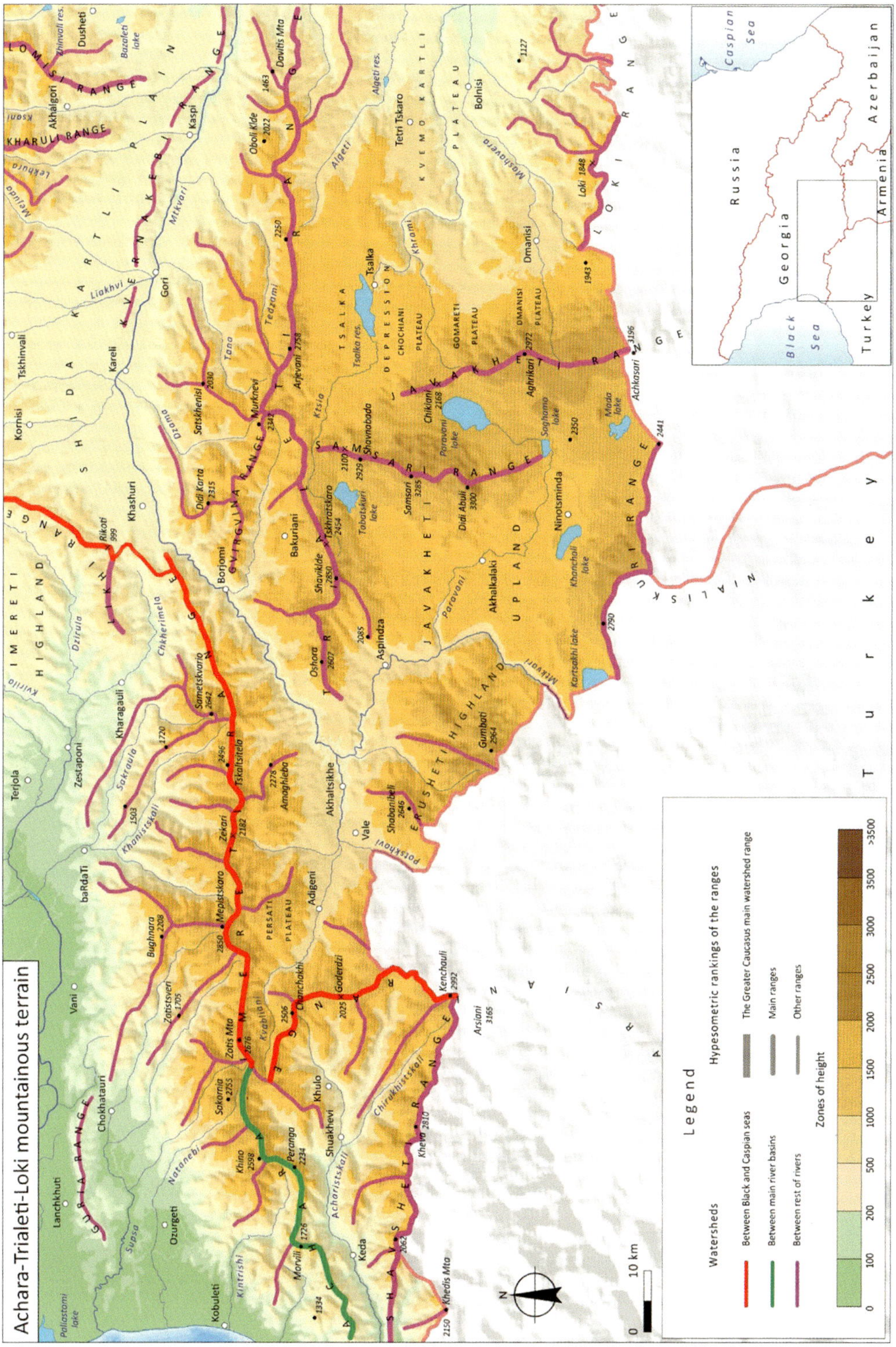

Fig. 13.1 The orography of the Achara-Trialeti-Loki mountainous terrain

is dissected by the tributaries of the Supsa and Khanistskali rivers, whose watershed is a Laboriti subrange. On the western slope of the subrange in the Supsa River basin, the flattened surfaces are located at the heights of 1100, 1650, 2200, and 2600 m asl (Tskhovrebashvili 1978). The forms created due to the selective erosion are common in the Moghobila subrange. In the Laboroti subrange, especially, in the area of the Mt. Mepistskaro, the old glacial (Pleistocene) forms remain, such as cirques and weakly expressed troughs.

The Supsa River forms a V-shaped deep gorge in the northwestern direction; the depth of the gorge is 800–1000 m and rapids are common in its bed. Several terraces are observed in the expanded sectors with relative heights of 6–10, 60, 90–100, and 180 m (Nemanishvili 1982). The Khanistskali River and its tributaries—Kershaveta, Tsablar-iskhevi, Sakraula, and Laitura cross the structures of the range mainly in perpendicular and develop deep gorges with strongly inclined slopes. Sometimes the shape of the latitudinal gorges is asymmetric; the depth of the cut is 1000–1200 m and rapids are common in the bed. Because of the morphological and morphometric characteristics, some of the sectors of the gorge are almost inaccessible; this is especially true for the middle sector. In the watersheds of the river basins, which are at the same time the branch-ranges of the Achara-Imereti Range, the ancient flattened surfaces are preserved (terraces are preserved only in the lower parts of the gorge). The eastern sector of the Achara-Imereti Range is dissected by the short, rather deep and narrow gorges of the left tributaries of the river of Chkherimela.

The southern slope of the western sector of the Achara-Imereti Range is far behind the northern slope. It is a north wing of the Achara depression. It is characterized by the action of the snow avalanches, landslides, and erosion processes. There are also the well-visible flattened surfaces on its northern slope

On the southern slope of the Achara-Imereti Range, the fragments of the steps of flattened surfaces are representable by quantity and relief appearance. These resemble shapes on the northern slope of the range, including the number of steps created by the main fragments and also their high spread rate.

There are four clearly visible steps of the flattened surfaces on the southern slope of the Achara-Imereti Range (Tskhovrebashvili 1978), the average absolute height of which differ from similar characteristics of the corresponding steps on the other side of the range by only three or four meters (Table 13.1).

We are highlighting this similarity because we consider them one of the most important features in identifying the genesis of the flattened surfaces. On the southern slope, the accumulation surfaces associated with volcanic processes are common, giving a hint in dating some of the steps, to this surface, which is represented in the Persati Plateau.

It is also noteworthy that the fragments of the highest flattened surfaces are observed in two or three places on the southern slope of the Achara-Imereti Range, while they are much more widely represented on the northern slope of the same range, as noted above. In addition, most of them have a straight surface or are more inclined toward the north and northwest.

These factors provide the basis to suppose the following:

Table 13.1 Flattened surface of the Achara-Imereti range

Flattened surfaces	Average absolute height of the group of the fragments in the northern slope of the Achara-Imereti Range (m, asl)	Average absolute height of the group of the fragments in the southern slope of the Achara-Imereti Range (m, asl)	Difference in heights (m, asl)
I	2510	–	–
II	2185	2155	30
III	1972	1947	25
IV	1660	1616	44
V	1253	1208	45

(1) Main arteries of the hydro-net located on the northern slope of the Achara-Imereti Range should have the same common direction (northern and west–northwest) of the flow as in the geological past;
(2) Streams on the northern slope of the Achara-Imereti Range have cut the original orographic axis of the range in the most part, due to which the main watershed is relocated to the south by 5–10 km;
(3) Surfaces on both slopes are genetically similar and the leading role, in this case, belongs to water erosion.

Among the fragments of importance on the southern slope of the Achara-Imereti Range the following are notable:

The representable fragment of the first surface is located near the Mt. Chubukhnari; it is wavy, hilly and inclined toward the northeast.

Among the fragments of the second surface worth of attention are those located in the (1) upper stream of the Aliberi River, (2) to the left side of the Kvabliani River along the Mt. Shavnabada meridian, and (3) between the Tsvinta and Naomari mountains; their heights are more than 2000 m and they have plain surfaces.

Among the most visible fragments of the third surface are the forms in the Tsinubani River basin near the Atskuri Village, which are latitudinally elongated and are slightly hilly plains.

The giant fragment of the Persati Plateau deserves special mention; although its average height is 2200 m asl, it is included in the third stage. The point is that the substrate of the Persati Plateau is located under the 150–200 m thick lava layers (Tskhovrebashvili 1978) and, therefore, the absolute height of the Upper Miocene denudation substrate covered by them should be within 1900–2000 m. This is confirmed by the altitudinal index of the watersheds and the fragments of the flattened surfaces located in the east and west of the plateau.

Only one fragment of the fourth surface can be considered as representative. This fragment is located in the west of Patara Abastumani and is a slightly sloped plain. It is oriented toward the northeastern direction.

Describing the general characteristics of the main fragments of the fifth surface, it should be noted that they exceed the previously mentioned clearly visible steps of the second surface by morphological clarity and even by number of fragments. Among those steps, the forms located at Mt. Terati and the Tsriokhi Vllage, are of special attention, being morphologically visible and both are denuded volcanoes, incompatible with the structural peculiarities of the place. Good morphological manifestations are also characteristic of the fragments of the surface above, found at Dagva Mountain in the east of the latitude of Mt. Kintrishi, at Didi Mtirala River, at Mt. Markhani, and near the Akhaldaba Village.

The southern slopes of the latitudinal sector of the Achara-Imereti Range border the Akhaltsikhe depression, and the Mtkvari River gorge. The relief is dissected by the left tributaries of the Mtkvari River. The slope is narrow, the depth of erosive section is 200–400 m, and the gorge lacks terraces. Mudflows are common in some gorges: Kvabiskhevi, Kurtskhana, and Tsopianiskhevi at Borjomi). The fragments of erosive-denudative and structural genesis of flattened surfaces are observed in the subranges of the watershed of rivers.

On the southern slope of the Achara-Imereti Range (in the central sector) is the Persati Plateau which is morphologically and genetically interesting (Fig. 13.2). It is constructed from the andesite-dacite lavas of the Lower Pliocene age. Its surface is slightly fragmented and is mainly sloping toward the west. The length of the plateau is about 11 km and the width is 7 km. The maximum heights and peaks of the plateau are located in its eastern part and edges—the Usakhelo (2450 m), Persati (2372 m), and Saburtalo (2258 m) peaks. The Plateau is cliffy in all sides. The surface of the plateau is erosive-denudative origin and has not experienced the Late Pleistocene glaciation. Some scientists believed that the volcanic boulders in the lower located villages were glacial (Maruashvili 1970), which is not confirmed by the our research. These boulders are displayed by means of

Fig. 13.2 Persati Plateau (*photo by* G. Lominadze)

landslide processes and rock avalanches (Gobejishvili 2011).

13.3 Borjomi Gorge

The sector from Atskuri to Tashiskari on the Mtkvari River gorge is known as Borjomi gorge; its length is 40–42 km. The common direction of the gorge is northeastern. This sector of the Mtkvari River is a complex crossing gorge, with a visible V-profile and its slopes are complicated with erosive gorges and terraces (Fig. 13.3). Bakuriani lava flow ends at Borjomi. The height of the bottom of the gorge is 890 m at Atskuri and 720 m a.s.l. at Tashiskari.

The relief of the Borjomi gorge is constructed from Middle Eocene suites (mainly the alternation of the layered colorful tuffs of the basaltic composition of tuffs, argillites, limestones, and marls), as well as basalts, andesites, trachytes' massive and thick layered volcanogenic breccias, tuffs, and tuffo-conglomerates. In the area of Borjomi-Rveli, the relief is built of Paleocene limestones, marls, and sandstones (flysch suite). On the right side of the gorge, the Upper Pliocene–Pleistocene andesites, and andesite-dacites are common (Gobejishvili 2011).

The slopes of the gorge are dissected by the gorges descending from the Achara-Imereti, and Trialeti ranges. It is worth mentioning the Kvabiskhevi River, which flows from the Mt. Lomismta (2187 m) on the southern slope of the Achara-Imereti Range and conflows the Mtkvari River at Dviri. There are narrow cavity-like sectors cut in the gorge. The narrow sectors are related to the andesite lavas of the Trialeti Range, and the extended ones with the Middle Eocene tuffo-breccias and tuffo-sandstones. Other left tributaries have similar morphology. The right tributaries—Dviri and Nedzvistskali are deeply sat on the slopes of the Trialeti Range developing the V-shape gorge. The rockslide occurring in late Pleistocene on the edge of the Dabadzveli Plateau is noteworthy, blocking the Mtkvari River gorge and forming a temporary lake. Signs of this phenomenon are still visible in the relief (existence of rocky material on the left side of the gorge and spreading 10 m thick lacustrine deposits to the Dviri Village.

In the Borjomi gorge, the river terraces are visible in fragments (I-II terraces of 5–7 and

Fig. 13.3 Borjomi gorge (*photo by* G. Lominadze)

10–15 m height) and occupied by settlements and arable lands.

13.4 Trialeti Range

The Trialeti Range is the eastern half of the Achara-Trialeti mountain system, which extends to the sublatitudinal direction from the Akhaltsikhe depression to Tbilisi at ~180 km. The width of the range is 20–32 km, of which the main part is a well-developed northern slope. The widest is in the Mt. Arjevani meridian. The highest peak of the mountain range is Shavi Klde (2850 m). Among other peaks, the following are notable: Oshora (2608 m), Sanislo (2748 m), Sakveliosmta (2806 m), and Arjevani (2758 m). Despite such hypsometry, the depth of the erosive fragmentation of the Trialeti Range in comparison to the Achara-Imereti Range is relatively low, which is caused by a high level of local erosion basis. The range has several important subranges; the Gvirgvina subrange stretches from the peak of Murknevi (2342 m) to the west up to Borjomi, which borders the Gujareti River gorge in the north. In the east, the Trialeti Range is divided into several subranges (Satskeplia, Mskhaldidi, Mtatsminda, and Teleti), the ends of which reach Tbilisi. The Kodiani (the northern slope), Oshora, and Veli (in the southwest) subranges are separated from the Trialeti Range in its western sector.

Geologically, the Trialeti Range is built of Paleocene and Eocene volcanogenic-sedimentary suites, alkaline and lime-alkaline basalts and andesites, thick layered volcanic breccias, tuffs, tuffo-sandstones, and gravelites. The relief around Mt. Satskhenisi and in the eastern part of Mt. Arjevani is constructed from Cretaceous sediments, such as limestones, marls and sandy limestone, tuffits, tuffo-conglomerates, and sandstones (carbonate flysch) (Gobejishvili 2011). The volcanic lavas (andesites and andesite-dacites) are common in the Tori-Gujareti region and the Eocene volcanic diorites in the Dzama gorge. The direction of the geological structures, which build the relief, is reflected in the morphology of the range.

The northern slope of the Trialeti Range in a morphological and genetical point of view is conventionally divided into western and eastern sectors. The border lies on the Gvirgvina subrange. Together with erosive-tectonic processes, volcanism and landslide processes played a great role in the formation of the northern slopes of the Trialeti Range in the west of Mt. Murknevi

(2342 m). The relative height of relief from the Mtkvari gorge side is 1500–2000 m, and slope width is 18–20 km. This sector of the Trialeti Range is the highest. The erosive gorges of Nedzvistskali, Borjomula, Dviri, and Tiseli are of meridian direction and the Gujareti gorge has a latitudinal direction. The Gujareti gorge is deeply cut by 800–1000 m; the slopes of the Gvirgvina and Trialeti ranges are dissected by its short tributaries.

The Tori-Tadzrisi depressions are formed in the syncline of the same name, built of the Upper Eocene, Maikop suite-like clays, sandstones, and thin middle layers of marls, with active associated landslide areas. Landslides are widely represented in the area of Kodiana Hill. Landslide cirque and tongues are well-developed in the relief. The Dgvari landslide is active in the Tadzrisi depression, causing serious damage to population. The Tadzrisi and Tori depressions at a height 1800–1900 m asl, on the wavy, containing lakes such as Kakhi and Tsero lakes.

Volcanic formations play a prominent role in the appearance of the relief by their morphological and morphometric indices, in particular, they create a rocky, cliffy, and plain surface plateau-like relief.

The volcanic formations of Borjomi-Bakuriani and Gujareti form lava flows. The Gujareti lava flow begins from the areas of the Didi Toneti Village. The volcano eruption center is Saghravi, which at present is a hill of 70–80 m relative height. The lava flow runs along the bottom of the Paleo-Gujaretistskali gorge. This lava flow is divided into mesas due to erosive action of the Gujaretistskali tributaries.

The lava flow of Bakuriani descended into the Mtkvari gorge and extended below at 1 km. In the lower part, the lava flow has the form of a boot. The edges of the lava flow are cut by the rivers of Borjomula and Gujaretistskali at 60–70 m depth and form asymmetric gorges. The lava erupted from the Tsikhisjvari and Mukhera hills. The hills of Saghravi, Tsikhisjvari, and Mukhera are located in one fault line.

The erosive actions of the Dzama, Tana, Tedzami, Kavtura, and their tributaries are important in the formation of the eastern sector of the northern slope of the Trialeti Range and in the dissection of the topographic surface. The gorges are cut at a depth of 500–600 m from the surfaces of the subranges (watersheds) of Ghorisnamkali, Satskhenisi, Tsereti, and Satskepela. The meridian gorge sectors are V-shaped and narrow, while the sectors of latitudinal direction are relatively wide.

The Trialeti Range has branches in the east and ends in the Tbilisi depression. In the relief, the geological structures of the Trialeti Range—anticlines and synclines—are expressed with subranges and gorges or depressions among them. Morphologically, the anticlines are the Satskepela, Mskhaldidi, Mtatsminda (Fig. 13.4), and Teleti Ranges, while the synclines are the Dighomi, Saburtalo, and Krtsanisi gorge-depression. The Dighmistskali and Vere river gorges change from the middle mountainous erosive gorges into low mountainous hilly gorges. A flattened surface is visible in the terrain at a height of 1100–1200 m asl, containing Turtle Lake in the exotectonic depression in the northern slope of the Mtatsminda Hill and Lisi Lake in the Lisi flatland.

The southern slope of the Trialeti Range stretches from the Minadze Village to the Soghanlughi Village. The main part of the southern slope is characterized by small width (2–8 km) and low relative altitude (100–400 m) due to the impact of volcanic terrain of Southern Georgia. In the area of the Tskhratskaro Pass, the volcanic lava is directly leaned in the crest. The western sector of the south slope is relatively wider. The relief is dissected by the erosive action of the Oshora, Ota, Chobareti, and Baraletistskali rivers. The central and eastern sectors of the southern slope are dissected by the heads and left tributaries of the Ktsia and Algeti rivers at small depth.

13.5 Achara Depression (Acharistskali River Gorge)

The Achara depression is bordered by the slopes of the ranges of Achara-Imereti (in the north), Shavsheti (in the south), and Arsiani (in the east).

Fig. 13.4 Ending of the anticline Mtatsminda subrange at the territory of Tbilisi (*photo by* L. Tielidze)

Below, it transits to the Kolkheti (Achara) hilly relief. Orographically, it is locked by a shape of hollow, which causes differentiation in natural conditions (with less atmospheric precipitation in Achara). Due to the sharp inclination of the slopes, there are actively developing exogenous processes. Anthropogenic factors are important in transforming relief.

The deep gorges of the tributaries of the Acharistskali River and sharply expressed watersheds represent the orohydrographic carcass of the Achara depression. The watersheds are the subranges on the southern slopes of the Achara-Imereti Range; they are of southwestern direction, while the subranges of the Shavsheti Range are of northwest direction. The Acharistskali River gorge mainly follows the syncline axis.

The depression is constructed from Paleogenic structures. The largest part of the Acharistskali River gorge is presented by the Middle Eocene volcanogenic flysch, tuffs, and tuffogenic rocks. The crest and the southern slope of the Achara-Imereti Range are constructed from the suites of porphyritic tuffs and tuffo-breccias of Upper Eocene age. Bellow the Tskhmori Village, the Acharistskali River gorge is formed in the same suites. Paleogene suites are folded and their complex creates the extensive syncline of Achara.

The depression is characterized by medium mountainous erosive relief; absolute heights vary from 100 to 1000 m (bottom of the gorge) to 2000–2700 m (the edges of the depression). There are several erosive steps in the depression in the form of pediments. The main villages of Khulo and the mountainous Achara depression are built on the pediment of 800–1000 m height (Fig. 13.5).

Intensive exodynamical processes are characteristic for the Acharistskali River basin (within the depression) causing great damage to the population. Particularly, active are landslides, rock avalanches caused by erosion, and snow avalanches.

13.6 Shavsheti Range

Orographically, the Shavsheti Range belongs to the medium height ranges of latitudinal direction. It begins on the right side of the Chorokhi River

Fig. 13.5 Achara depression, town of Khulo (*photo by* R. Kumladze)

(at the mouth of the rivers of Machakhela and Acharistskali) and joins the Arsiani Range at the peak of Kenchauli. Its length is ~77 km, width 20–25 km, the average height is 2100 m and its highest peak is Mt. Kheva (2812 m). This range is a watershed of the Acharistskali River's left and Chorokhi River's right confluences.

The Shavsheti Range is built mainly of Eocene volcanogenic rocks—pyroclastic layers, andesite-basalts, tuffo-breccias, and slightly terrigenous sandstones. The intrusive diorites are outcropped in the Merisistskali River gorge. The range is morphologically asymmetric. In the northern part of the range, there are well-defined flattened surfaces in the form of steps at heights of 1500–1600, 1900–2000, at 2200–2600 m asl.

Morphologically and geologically, the Shavsheti Range is a continuation of the Ponto Range, as indicated by the hypsometry (3440 m) and morphology of the Karchkhali massif in the west of the range.

There are transverse subranges in the north of the Shavsheti Range descending to the Acharistskali gorge forming the watersheds between the left tributaries of the Acharistskali River. It is worth mentioning the Ghoma-Tsinavela Range that creates the watershed between the rivers of Akavreta (Merisistskali), and Tbeti. The Ghoma, Tsinavela, and Sakoniati mountains are higher than 2000 m. At Mt. Gerati, the Shavsheti Range has one more subrange—the Medzibna-Milisi Range, which is relatively low and is the western end of the Shavsheti Range. On the northern slope of the central sector, the Shavsheti Range has one more subrange of Namtsvarevi, which is a watershed of the other left tributaries of the rivers of Tbeti and Chirukhistskali.

There are three passes in the Shavsheti Range. Among them, the Kvirala Pass is the most accessible, located at an altitude of 2222 m asl. The subranges, descended opposite to each other from the Achara-Imereti and Shavsheti ranges, greatly narrow the Acharistskali gorge and in some places create almost closed depressions.

Formations of the ancient glaciations such as corries, moraines, and short troughs are preserved in the northern slope of the main watershed crest in the higher part of the Shavsheti Range. The southern slopes of the range located in the territory of Turkey are dissected by the heads of the right tributaries of the Chorokhi River—the Imerkhevi and Machakhelistskali rivers.

13.7 Arsiani Range

The Arsiani Range is a watershed of the Mtkvari-Chorokhi river basins, and therefore, the watershed of the Black and Caspian sea basins. At present, the main part of the Arsiani Range is located in the territory of Turkey. The northern part of the range—from the Mt. Kenchauli (Kanlidaghi) to the Achara-Imereti Range located in the territory of Georgia; the length of this area is about 55 km. The highest peaks of the Arsiani Range are over 2400 m asl. From the north to the south, the notable peaks are Chanchakhi (2506 m), Zambora (2420 m), Kldistavi (2452 m), Ghrmani (2541 m), Prenauli (2635 m), and Kenchauli (2986 m).

The northernmost sector of the range is constructed from Middle Eocene volcanogenic marine sediments—tuffs, tuffo-breccias, and tuffo-sandstones. The southern part of the Arsiani Range, in the heads of the river of Potskhovi to Mt. Arsiani is constructed from Pliocene age andesite-dacite lavas, known as the Goderdzi suite. On the northern slopes of the Arsiani Range, 10–15 km away from the Goderdzi Pass, fossil trees and leaves are found in the volcanic ash of the Goderdzi suite. It is known as "Flora of Goderdzi" (Fig. 13.6). There are numerous water sources at the junction of the lava and tuffogenic deposits, which are the beauties of the Arsiani Range and at the same time, they intensify the dissection of the slopes of the range.

Two types of terrains are distinguished in the area of the Arsiani Range, built of the Goderdzi suite. The first covers the crestal part over 2200 m asl; the relief is weakly dissected, wavy, and developed on andesite-dacite substrate. Individual uplifted volcanic cones are visible here. Morphologically, all the high dome-like peaks are of volcanic origins. On the slopes of some of the high peaks, the Late Pleistocene glacial forms are visible, such as corries and sometimes, moraines.

The second type of relief is the mountain-gorge. The gorges are quite deeply cut, their form is V-shaped. Erosion processes in the slopes of the Arsiani Range cause intense dissection. The eastern slope is dissected by the tributaries of the rivers of Potskhovi and Kvabliani, while the western slope is dissected by the tributaries of the Acharistskali River and the tributaries of its left tributaries—Skhalta and Chirukhistskali. Unlike the rivers of the eastern slopes, the rivers of the western slope are characterized by a short profile. The Arsiani Range is an orographic, climatic, and landscape barrier between the Achara and Samtskhe-Javakheti regions.

Fig. 13.6 Fossil trees in the volcanic ash of the Goderdzi suite (The Arsiani Range) (*photo by* R. Gobejishvili)

From the western subranges of the Arsiani Range, the Furtio (the watershed of the Chirukhistskali and Skhalta rivers), Samkria (the watershed of the Skhalta and Acharistskali rivers), and Karati (the watershed of the Chvanistskali and Ghorjomi-Acharistskali rivers) ranges are important and from the eastern subranges—the Grmani Range (the watershed of the

Dzindze-Kvabliani and Grmanistskali rivers). The Arsiani Range does not have important eastern subranges in the territory of Georgia.

13.8 Akhaltsikhe Depression

The Akhaltsikhe depression is located in the Mtkvari River gorge and morphostructurally belongs to the Achara-Trialeti folded system. In the north, the depression is bordered by the southern slopes of the Achara-Imereti Range, in the east by the eastern end of the Trialeti Range, in the south by the Erusheti Highland, and in the west by the eastern slope of the Arsiani Range (Fig. 13.7). The depression is open toward the Borjomi gorge. Hypsometrically the depression is limited by the 1500-m asl contour line. The depression extends latitudinally from the Mlashe Village to the Atskuri Village for a distance of ~42 km (in direct line). In the direction of meridian, it stretches 12–14 km (in the center of the depression). The lowest point (890 m) in the depression is at the Atskuri Village. Morphologically, the depression consists of two parts—the bottom of the depression and the hilly belt.

Geologically, the Akhaltsikhe depression is a syncline, the limbs of which are the slopes of the Achara-Imereti Range and Erusheti Highland. The syncline axis is aligned to the bottom of the Akhaltsikhe depression. Morphologically the syncline extends to the west as the Achara depression, and in the northeast in the form of the Tadzrisi-Tori depressions. Thus, the depression is of a tectonic origin, in the modification of which the erosive-accumulative action of the rivers is important.

In the central part, the relief of the depression is constructed from Oligocene shallow marine sediments, such as clays, sandstones, conglomerates, aleurolites, marls, and brown coal layers. In the peripheries, it is constructed from Upper Eocene carbonate and gypsum clays with thin layers of marls and sandstones (Maikop series), as well as middle Eocene alkaline basalts with massive and thick layered volcanic breccias, tuffs, and lava layers. Miocene–Pliocene continental sediments are found in separate islands, such as tuffs, volcanic breccias and conglomerates, tuffo-diatomites, and diatomites (Goderdzi suite).

The geological structure of the depression is important in the formation of morphosculptural

Fig. 13.7 Akhaltsikhe depression (*photo by* L. Tielidze)

landforms. Namely, the narrow, cutting gorges with steep slopes are related to the distribution of lava layers. In the reliefs constructed from Oligocene sediments, the bottom of the gorge is expanded and alluvial floodplains are formed; the slopes of the gorge are terraced. The intensive denudation of the slopes of the depression and the development of soft relief are related to the Maikop series. Active and stable landslides are common on the slopes. It is worth mentioning the active landslide of Adigeni. The large stable landslides are at the bottom of the Persati Plateau, on the surface of which the only individual sectors are active.

Erosive-accumulative action of the rivers are important in the formation of the appearance of the relief of the Akhaltsikhe depression. The Mtkvari River crosses the depression in the eastern sector; often changes direction regarding structures; the riverbed is enlarged in the syncline structures is represented by floodplains and interfluves, while the narrow, short sectors with steep slopes are formed when cutting the anticlines. The Kvabliani River, with well-developed vast gorge, floodplains, and interfluves, flows along the bottom of the depression (Fig. 13.8). The Potskhovi River crosses the depression first from the south and then, at the confluence with the Kvabliani, passes along the bottom of the depression. The Otskhe River crosses the depression from the north. Erosive-accumulative terraces are well developed in the river gorges. Twelve terraces are located in the central part of the depression. The height of the first terrace is 5–7 m, and the 12th terrace is 550–600 m (Gobejishvili 2011).

Fig. 13.8 Kvabliani River gorge (*photo by* G. Lominadze)

13.9 Khrami-Loki Average Mountainous Region

The Khrami-Loki average mountainous relief occupies the vast area of Kvemo Kartli. Orographically, the Mashavera River gorge divides it into two parts or subregions: the middle Khrami mountainous subregion and the Loki Range. This section is a link between the Meskheti-Trialeti and the southeastern slope of the south Lesser Caucasus.

The middle Khrami mountainous subregion is divided into several massifs: the Gomeri Massif in the north of Tetritskaro Town; the Bedena—Tavshishveli massif in the west from Tetritskaro Town; and the Kviriketi and Lukuni (Shindliari) massifs at Dmanisi.

The Khrami River separates the Tavshishveli and Gomeri massifs from the Kviriketi and Lukuni massifs. These latter two are separated from each other by the Shavtskala thalweg.

The Mashavera River is the border between the Loki Range and the middle Khrami mountainous sector. The Khrami-Loki region is an independent geosyncline, a part of the Artvini-Bolnisi Block. The terrain is built of Jurassic, Cretaceous, and Eocene age inclined folded suites, underneath of which, in two places the Paleozoic granitoids are outcropped. The most widely distributed are Upper Cretaceous formations volcanogenic and limestone facies. The Eocene deposits build the Loki and the Lukuni massif. The Khrami crystalline massif is cut by the Khrami River below Tsalka Town, while the Loki massif is cut by the Poladauri River. The tectonic structure of the region is complex; often structures are directly reflected in the reliefs as horst massifs. The tectonic uplifting in the west formed mountainous relief, while the eastern sinking coincides with alluvial plains.

The overall morphological appearance of the Khrami-Loki region was formed by the interaction of volcanic, tectonic, erosive, lithological and denudative factors. Against the background of neotectonic movement, the erosive processes formed mountainous relief of medium height. The volcano lavas created the meso and microforms—the plateaus of flattened inclined surface, canyon-like and asymmetrical gorges; because of them the gorges were locked and lakes were formed. Several subregions are distinguished here.

The Bedena mountain massif. The Gomeri and Tavshishveli ranges come out of the mountain massif of Bedena. They are separated from each other by the lava flow of Bedena and Kvemo Kartli. The crystalline massif of the left side of Khrami enters this region. The massif is dissected by the left tributary of the Khrami River (Kldeisistskali, Chivchavi, and Torne) and right tributary of the Algeti River. Bedena massif is stretched latitudinally. The highest peaks Iaila (1951 m) and Bedena (1875 m) are located in the watershed of the rivers of Algeti and Khrami. The surface of the massif is sloped to the south. The massif is constructed from Cretaceous volcanogenic suite with outcrops of Paleogenic dacites in some places. In the southwest of the Bedena Range, the Tavshishvela Range is separated (1788 m), which is a watershed of the Khrami and Torne river basins. The depression of the same name is located in the Torne River basin.

The Khrami River gorge. The Khrami canyon is cut into the granitoids of the crystalline massif, Cretaceous limestones, and volcanogenic suites from the Trialeti Village to Samshvilde. In the relief of Trialeti, there are visible small plateaus of the andesite-basalt lava flows of Tkemliani and dolerite lava flows of Kakliani. The Khrami gorge is of zigzag form, cut at 500–700 m depth. In this sector, the gorge thalweg reduces from 1300 to 600 m. Only the first terrace is easily visible, containing farmlands. In the right side of Khrami, in the rocks of the Gomareti and Kakliani plateaus, there are false terraces, the genesis which is associated with avalanche-landslide processes. There are many rocky microforms in the area of the gorge that are formed in the Cretaceous suite.

The Kviriketi-Lukuni subregion consists of two massifs. The Kviriketi massif is located between the Gomareti Plateau and Khrami gorge, with the highest point 1744 m, mainly built with Paleozoic granitoids. The Lukuni (Shindliari) massif is located between the Khrami and

Fig. 13.9 Mashavera gorge-canyon in the area of Dmanisi (*photo by* L. Tielidze)

Mashavera gorges. From its highest peak (Shindliari 1896 m) two hills are distinguished. In the north, the watershed of the Khrami and Mashavera river basins is the Darbazi (Tandzia) Hill and in the east the Kakliani Hill. The Lukuni massif is constructed from Eocene volcanogenic suites and its branches with Cretaceous volcanogenic formations. The watershed hills are characterized with flattened surfaces (Tandzia, Ormosheni, and Shorsholeti). The Gomareti and Dmanisi volcanic lavas have a role in the formation of the river gorges of this massif.

The relief is medium mountainous, dissected by the Mashavera and Khrami tributaries. The Mashavera River gorge is a canyon shape in the area of Patara Dmanisi that is a result of cutting lavas (Fig. 13.9). The gorge is asymmetric in its upper and lower parts; the left part is cut into lava and forms the cliff, while the right part is formed in the Cretaceous volcanogenic suites.

The Loki Range. Only the northern slope of the Loki Range enters into Georgia extending to the bottom of the Mashavera gorge. The range is dissected by the right tributaries of this river. Heights of the peaks of the range exceed 2000 m asl (Lalvari 2545 m and Loki 2142 m). The range is built of the volcanogenic suites of the Cretaceous age; under these suites, in the Pinezauri and Poladauri river basins, the old crystalline rocks of the Loki massif are outcropped. The relief is of medium mountainous and erosive, sometimes with denudative flat lands (villages of Opreti, Damaia, etc.) and flat bottoms of the gorges (Debeda, Bolnisistskali, and Talavristskali). The morphological trace of shifting the hydrographic network in the different sectors of the Loki Range is visible in dry gorges and passes.

References

Gobejishvili R (2011) Relief of Georgia, monograph. Publishing House "Universali", Tbilisi (in Georgian)

Javakhishvili A (1926) Geography of Georgia, vol 1—geomorphology (with Atlas), Tiflis (in Georgian)

Maruashvili L (1970) Saqartvelos fizikuri geografia (Physical geography of Georgia), monograph, Part 2. Tbilisi (in Georgian)

Nemanishvili S (1982) The river terraces of mountainous areas, vol 2. Tbilisi (in Georgian)

Nizharadze N and Jibuti N (1957) Autonomous Republic of Achara of the Soviet Republic of Georgia, Batumi, 264 pp (in Georgian)

Tielidze L (2017) Late Pleistocene and Holocene Glaciation. Chapter in Glaciers of Georgia, Springer, Geography of the Physical Environment, https://doi.org/10.1007/978-3-319-50571-8-6

Tskhovrebashvili S (1978) Geomorphology of Achara-Trialeti Mountain Range, monograph. Tbilisi (in Georgian)

Ramin Gobejishvili (1941–2014) was a Doctor of Geographical Sciences, a geomorphologist–glaciologist. Until his last days, he was Professor of the Faculty of Exact and Natural Sciences, Ivane Javakhishvili Tbilisi State University and the chair of Geomorphology and Cartography. In 1996–2008, he was the president of the Geographical Society of Georgia, and in 2005–2007—the Director of the Institute of Geography. He was the author and co-author of about 150 scientific works and ten monographs.

Levan Tielidze is a Ph.D. of Ivane Javakhishvili Tbilisi State University in the field of geomorphology–glaciology–cartography and a senior research scientist at Vakhushti Bagrationi Institute of Geography. He is also a researcher-scholarship holder of the Department of Earth Sciences of the Georgian National Academy of Sciences. The field of his research is mountain cartography and glacial-geomorphological study of the mountainous areas in the Quaternary (Late Pleistocene and Holocene). He is a member of several international organizations. In 2014/2015, he conducted a research stay at the Climate Change Institute of the University of Maine, USA, and in 2015/2016–2017 at the University of Northern British Columbia, Canada. He is the author of about 40 scientific papers, maps, and eight monographs.

14 Volcanic Highland of Southern Georgia

Levan Tielidze, Ramin Gobejishvili, Levan Maruashvili, Alexander Javakhishvili and Nikoloz Astakhov

Abstract

In this chapter, the geomorphological characterization of volcanic highland of southern Georgia is presented. The orographic units are considered in detail, such as the Erusheti Highland, Mtkvari River upper gorge, Javakheti Upland with its subregions (the Akhalkalaki Plateau, the northern hilly low mountainous subregion, and the southwestern hilly subregion), Nialiskuri, Samsari and Javakheti ranges, Tsalka depression and the volcanic plateaus of Bedena, Kvemo Kartli, Chochiani, Gomareti, and Dmanisi. The text also provides geological structure, tectonic conditions, and current exogenic processes in the study region. Each subchapter is accompanied by appropriate images and maps.

Keywords

Volcanic highland of southern Georgia Erusheti Highland • Javakheti Upland Nialiskuri Range • Samsari Range • Javakheti Range • Tsalka depression • Volcanic plateau

14.1 Study Area

The volcanic highland is located in southern Georgia, containing volcanic cones, lava plateaus, canyons, and depressions as the main forms of the relief. The highest peaks of the volcanic highland are the Didi Abuli (3301 m) and Samsari (3285 m) peaks.

Together with volcanic processes, the tectonic, erosive, gravitational, and nival processes have also participated in the formation of the relief. According to the orographic and morphogenetic signs, the subzone of the volcanic highland is divided into several regions, such as the Erusheti Highland, the upper gorge of the Mtkvari River, Javakheti Upland, Nialiskuri Range, Samsari Range, Javakheti Range, Tsalka depression, and volcanic plateaus (Fig. 14.1). These regions include subregions.

L. Tielidze (✉) · R. Gobejishvili · L. Maruashvili · A. Javakhishvili · N. Astakhov
Department of Geomorphology and Geoecology, Vakhushti Bagrationi Institute of Geography, Ivane Javakhishvili Tbilisi State University, Tbilisi, Georgia
e-mail: levan.tielidze@tsu.ge

L. Tielidze · A. Javakhishvili
Department of Earth Sciences, Georgian National Academy of Sciences, Tbilisi, Georgia

R. Gobejishvili · A. Javakhishvili
Department of Geography, Faculty of Exact and Natural Sciences, Ivane Javakhishvili Tbilisi State University, Tbilisi, Georgia

L. Tielidze (ed.), *Geomorphology of Georgia*, Geography of the Physical Environment,
https://doi.org/10.1007/978-3-319-77764-1_14

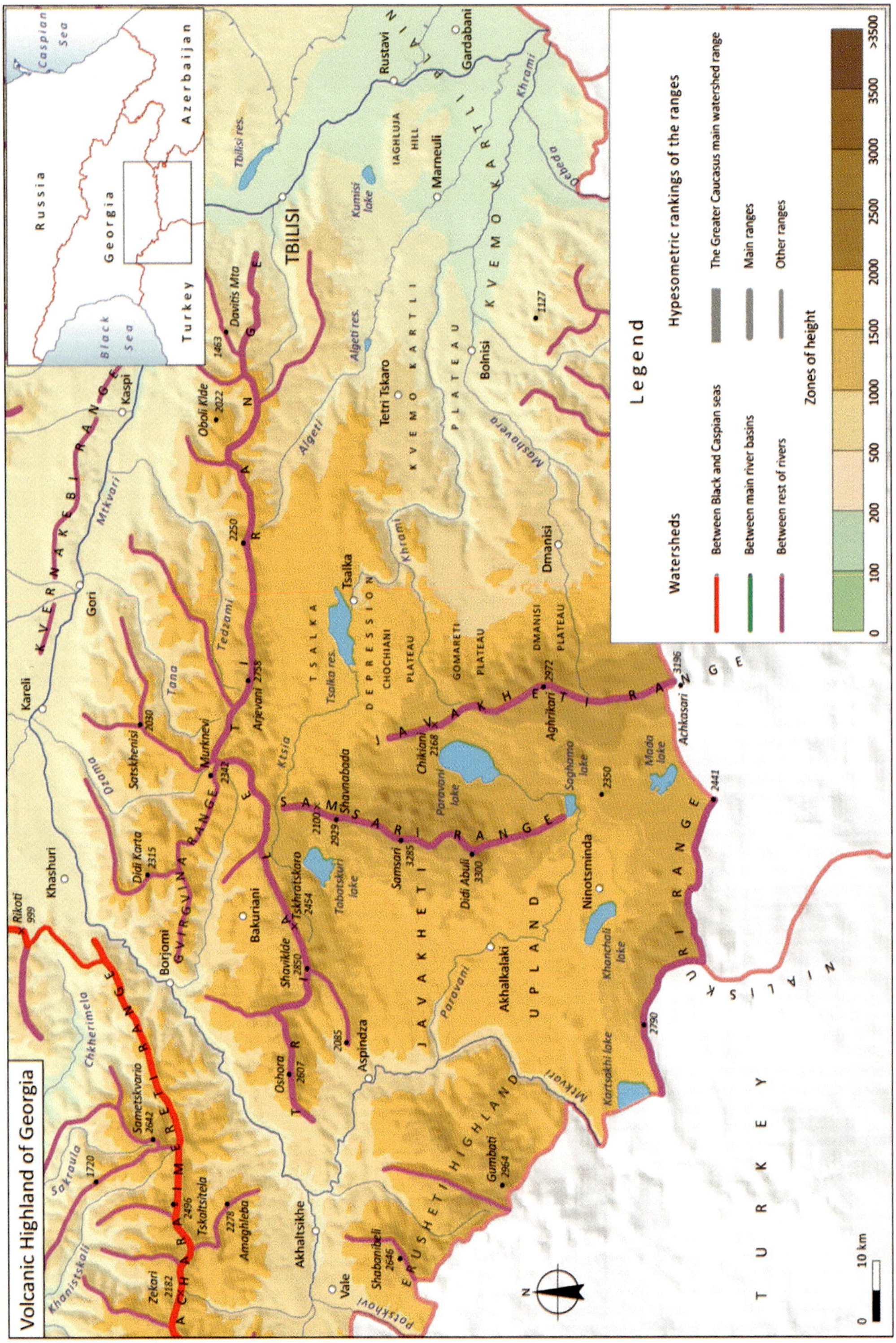

Fig. 14.1 The orography of the Volcanic highland of Georgia

14.2 Erusheti Highland

The Erusheti Highland is located in the southwest of Georgia and only part lies in the territory of the country (its length is approximately 60 km, and the width 25–30 km). The Erusheti Highland has a transitional form between the mountain range and upland in terms of general morphology. The highest point of the Erusheti Highland is the dome-like Mount Gumbati (2964 m). The heights of the highest peaks range within 2300–2800 m (Tlili, Shuakaravi, Shabanibeli, and Sholosheti). The bottom of the northern slope of the highland is located at 1000–1200 m. The Khaniori Pass (2435 m) includes the Akhaltsikhe-Artaani road.

The Erusheti Highland is mainly built of Goderdzi suite, which is compressed in the form of inclined folds and consists of lavas, tuffo-breccias, and tuffs. In places, effusive material contains fresh water layers and lenses, deposited in Tertiary age lakes, accumulating diatomite algae (Gobejishvili 2011), for example, the Kisatibi diatomites in the Uraveri River gorge. In the gorges of the Potskhovi and Mtkvari Rivers, Eocene sediments are outcropped under the Goderdzi suite. Alluvial, glacial, deluvial, and other continental sediments are represented fragmentary.

The main genetic land types of the Erusheti Highland are tectonic-volcanogenic, erosive and old glacial. The tectonic-volcanogenic terrain is found in the fragmented wider watersheds. Morphologically, it is expressed with dome-like massifs and stepped smooth surfaces. The erosive relief in the foothills is represented by average mountain-valley relief. The depth of the erosive intersection is 300–800 m. In the massifs higher than 2400–2800 m asl, old glacial forms can be found locally, such as cirques and strongly deformed troughs (at the head waters of the Uravelistskali River) (Tielidze 2017). The depressions with lakes and swamps are distinguished in the relief and the avalanche–landslide forms are common at the edges of the denudative steps.

In the relief of the Erusheti Highland the Shuakarva Range is in the west, the Vani Range in the central part and the Sholosheti Range in the east. Among these watersheds, the tributaries of the Uraveli and Khevistskali rivers flow and dissect the terrain.

The Shuakarva Range stretches about 18 km to the north. The watersheds often have a plateau shape in the erosive gorges. Its high peaks are the Shuakaravi (2562 m) and Shabanibeli (2646 m) peaks. The western slope is dissected by the Potskhovi River tributaries, and the eastern slope by the Uraveli River tributaries.

The Uraveli River basin is divided into several morphological parts: the Uraveli upper stream, the Uraveli depression, and the Inja River gorge.

The Uraveli hollow is located in the lower stream of the gorge, where the rivers of Uraveli and Inja join. The width of the hollow is 5 km, length 8 km; its bottom is located at the height of 1080–1100 m asl and the depth of the hollow is 400–800 m. The bottom of the hollow is terraced. The terraces are located on the right bank of the Uraveli gorge (1–2 m, 5–6 m, 10–12 m, and 15–20 m). The old terraces are fragmented above the fourth terrace and are found in the form of hills. The hollow is located in the sublatitudinal synclinal depression, built of Eocene and Miocene volcanogenic suites. The Kisatibi diatomite mine is formed in the lacustrine sediments of the Goderdzi suite in the western part of the hollow.

In the hollow, the step surfaces, old and new landslides are associated with the cliffs of the volcanogenic plateaus. The Uraveli River heads are erosive and its gorges are deeply cut. The erosive gorge of the Inji River is similar to that of the Uraveli River.

In the east of the Erusheti Highland, between the Inja and Sholosheti hills, lies the Sholo tectonic-denudative hollow, in which the Khevistskali River flows with its tributaries. The hollow is oval, developed in the Goderdzi suite and located at 1850–2000 m asl. The abundant underground water in this area leads to the swamping of some sections. The bottom of the hollow is filled with alluvial–prolovial type loose deposits, with active peat mining. There are terraces of 1–2 m and 5–8 m height in the Khevsistskali gorge, in the area of the hollow.

14.3 Upper Gorge of the Mtkvari River

Above the Minadze Village, the Mtkvari River flows through a deep gorge canyon, the right side of which is cut at the edge of the Akhalkalaki Plateau, and the left side—on the eastern slope of the Erusheti Highland. The height of the thalweg at the confluence of the Uraveli River is 950 m asl, and 1360 m at the Turkish border (from which it enters Georgia). Above Aspindza Village the Mtkvari gorge is built of Neogene Goderdzi suite, which is covered by the Quaternary dolerite lavas of the Akhalkalaki Plateau on the right side of the gorge and partly, on its left side. The Goderdzi suite builds the high slopes of the Erusheti massif, while the Mtkvari gorge is formed in Eocene volcanogenic and sedimentary suites. The Quaternary deposits on the bottom of the gorge are presented by alluvial and deluvial materials.

The morphological shape of the gorge is created by the compatibility of the forms created by the major erosive processes with the volcanic, exotectonic, accumulative and anthropogenic forms. The deep gorges with steep and fragmented slopes, curved bottom, and step-like forms are created under the influence of the erosive processes. Such morphology of the gorge has led to the development of exotectonic forms, in particular, the large landslides are widespread in the extended sections, on the one hand, and in the Eocene terrigenous suite below Khertvisi, on the other hand. Anthropogenic forms are expressed by artificial terraces (this region is well-known in Georgia in this respect) and whole groups of caves. Morphologically the gorge is divided into three sections: upper section from above the Tmogvi Village up to the border of Turkey; the middle section from Tmogvi to the Saro Village; and the lower section from Saro to the Minadze Village.

The gorge above Tmogvi is a canyon. The slopes are steep above with cliffs and inclined below (Fig. 14.2). From the upper part of the inclined slopes built of Goderdzi suite, the dolerite boulders are falling down and accummulating in its foot. In the upper stream, the width of the gorge is 2.5–3.0 km. The bottom of

Fig. 14.2 A canyon in the upper gorge of the Mtkvari River (*photo by* L. Tielidze)

the gorge is occupied by its bed and the fragments of the lower terraces. There are many complexes of artificial cavities in this section of the gorge such as Vardzia (Fig. 14.3) and Vani cavities, etc.

The gorge is a canyon from Tmogvi to Saro, only at a distance of about 5 km, in the Nakalakevi hollow, is it relatively expanded and the width of the gorge is about 5 km. The leading role in its expansion belongs to erosion and avalanche–landslide processes. Below the Nakalakevi hollow the gorge is still a canyon. The depth of the sides of the gorge is 400–500 m to the Akhalkalaki Plateau and 700–800 m to the Erusheti Highland. There are agricultural terraces arranged on both slopes of the gorge.

The lower section of the Mtkvari gorge from Saro to Minadze is a V-form, due to the Middle Eocene tuffs, tuffo-breccia, sedimentary rocks and Quaternary basaltic, and andesite-basaltic lavas. Alluvial terraces (including at a relative height of 400–600 m) are visible on the open slopes.

The landslide phenomena play an important role in the morphology of the gorge. On the left bank, from the Sholosha Village to the Rustavi Village the landslides are related to the Upper Eocene sediments and occupy the 2 km long belt. The serrature of the eastern part of the Erusheti Highland is stipulated by landslides. Creation of the beautiful Atskvita Lake is also related to them on the right bank of the Kuntsiskhevi River.

Landslide forms are well developed above Khertvisi, especially in the Nakalakhevi depression, which was formed due to the landslide of the Akhalkalaki Plateau. In the bottom of the Mtkvari gorge, the weathered boulders fall down from the cliffs built of dolerites into the landslide cirques produced within the tuffs and tuffo-breccias of the Goderdzi suite, and descend to the bottom of the gorge. The result of landslide is the Tsunda Lake, which some researchers believe is a crater of volcanic explosion (Gobejishvili 2011). Larger landslides are Bertakvana and Dakhvandra.

The Paravnistskali gorge is interesting and morphologically similar to the Mtkvari gorge (the Akhalkalakistskali and Mtkvari rivers of Javakheti), which is a canyon from Khertvisi to Akhalkalaki.

14.4 Javakheti Upland

The Javakheti Upland is the most extensive and complex geomorphological region of the volcanic highland of southern Georgia. The region is bordered by the northwestern part of the

Fig. 14.3 The XII–XIII century Georgian monument of art, the rock cave monastery ensemble of Vardzia (*photo by* L. Tielidze)

Trialeti Range in the north, the Javakheti and Samsari ranges in the east, the Nialiskuri Range in the south, and the upper stream of the Mtkvari River in the west. Therefore, the Javakheti Upland occupies the central part of the highland of southern Georgia. The main part of the upland is located within 1500–2000 m asl.

The modern relief of the upland is formed by the impact of volcanic, tectonic, erosive, accumulative, and anthropogenic processes. The main morphological appearance of the upland was created by volcanic activities. The tectonic deformation of the old lava surface led to the existence of the mountain-gorge and mountain-hollow terrain in its edges. Erosion dissected part of the upland by the canyon-like gorges. Lacustrine and riverine deposits enlarged the area of the plains, which are formed by the Quaternary lavas (Maruashvili 1970). Anthropogenic forms are found on the canyon slopes in the form of caves.

The Javakheti Upland can be divided into several main subregions such as the Akhalkalaki Plateau, northern hilly low mountainous subregion, and southwestern hilly subregion.

The Akhalkalaki Plateau, extends from the Mtkvari upper gorge to the Javakheti Range in the east. The Akhalkalaki Plateau has a strong plateau shape only in the northwestern part, which is bordered by the Samsari Range in the east, while the southeastern part has quite different relief; the shape of the plateau is distorted by turbulence lacustrine depressions and hilly highlands. These two parts of the Akhalkalaki Plateau will be considered separately.

The plateau, in a narrow sense, is a mountainous plain with a height of 1500–1800 m, stretching 35–40 km in length in a meridian direction, and 25–30 km in width. It is constructed mainly by the Quaternary dolerite lavas and lacustrine–riverine sediments that were accumulated in the earliest depressions of the ancient relief. The Goderdzi suite is outcropped here, on the slopes of the canyons on the one hand, and in the hilly highlands on the other hand, which are both located both in the eastern edge of the plateau. The age of dolerites and continental sediments is Upper Pliocene or Lower Pleistocene. The plain is dissected with erosive canyons of Paravnistskali River and its tributaries—Chobaretistskali, Baraletistskali and Murjakhetistskali.

The depth of the erosive section reaches a maximum (400–500 m) in the lower stream of the Paravnistskali River and decreases in the upper stream; at Akhalkalaki its depth is 200 m (Fig. 14.4). Here, in the canyon walls artificial and natural caves are found. The homogeneous relief of the plateau is disturbed by the peak of Amiranisgora, a camp-like massif (absolute height of 1883 m, relative to 150 m) at Akhalkalaki, built of dacite extrusives. In the northeastern part of the plateau a group of "Sakaraulo" hills are erected between the villages of Kachio and Ikhtila, the relative height of the highest is 150 m. Similar highlands are also found in the south of the canyon in the region of the villages of Kumurdo and Gogasheni, where their relative height does not exceed 50–70 m.

An essential morphological element of the plateau is the Baraleti Plain, located in the north of the Paravani canyon, which is drained by its right tributary, the Baraletistskali. The above-mentioned plain is completely flat and partially swamped, a result of lacustrine–riverine accumulation. The Baraleti Lake existed in the Lower and Middle Pliestocene and was filled with lacustrine and riverine deposits of tens of meters (Gobejishvili 2011). The length of the plain is approximately 6 km. The large villages of Baraleti, Grtila, Merenia, Bughasheni, Gomani, Zakvi and Kistela are located in the edges of the plain.

The southeastern part of the Akhalkalaki Plateau is occupied by a surface of variable altitude; in the east, it is bordered by the Javakheti Range, in the south by the Nialiskuri Range, and in the west and partially in the north by the Samsari Range. This surface is built of geological formations similar to the Akhalkalaki Plateau, though it is somewhat different in character of the relief; here, the Quaternary lavas do not form vast cover and do not make the leveling of the relief (Goderdzi suite) of Neogene

Fig. 14.4 Paravani River canyon at Akhalkalaki (*photo by* L. Tielidze)

substrate. Young basaltic and andesite-basalitic flows merged the forms of the depression elements of the relief and created small plain areas.

These lavas mainly erupted in the area of Emlikli Peak (the southern part of the Javakheti Range) and flowed to the west on the dissected surface of the Goderdzi suite (Gobejishvili 2011). A relatively large plain location is formed by Quaternary lavas, which are located between the villages of Jigrasheni and Orlovka, Gorelimi and Efremovka, Poka and Gandza. Most of tpocahe depressions created by lavas are occupied by the lakes of Paravani, Saghamo, Mada, and Khanchali (Fig. 14.5), the levels of which are located at 1920–2110 m asl. The Paravani and Khanchali lakes are of mixed, tectonic-volcanic genesis. The surfaces were folded before and in the Quaternary, as a result of the lava flow, and modified into the lacustrine hollows. In this part of the plateau, the following rivers flow: Paravnistskali and its left tributary Bughdashenestskari, cutting the plateau to the lower depth.

The northern hilly, low mountainous edge of the Javakheti Upland extends to the northeast from the Khizabavra Village to the Nariani hollow.

The Zemo Ktsia synclinal gorge between this hill and the Tskhratskaro-Sakvelo Mta section of the Trialeti Range is a monocline of the Goderdzi suite. The hollow of Tabatskuri Lake is also of a tectonic origin. It is a synclinorium (or semi-synclinorium) located between Tabatskuri Mountain Range and the northern part of the Samsari Range.

The southwestern hilly section of the Javakheti Upland is located at the border of Turkey. The Goderdzi suite, which builds this region, creates the shapes such as in the northern part of the upland. They are expressed with anticlinal hills, synclinal gorges, and hollows. There are clearly visible three mutual parallel hills: Azmani and Shua Murakvali in the north, and Nialiskuri in the southwest. All of them are stretched in the latitudinal direction and are separated by synclinal depressions.

Fig. 14.5 Saghamo Lake (*photo by* L. Turmanidze)

14.5 Nialiskuri Range

The Nailakuri Range extends in the east for about 35 km from Mt. Gekdaghi to Mt. Okiuzdaghi (2442 m). On its crest passes the southern state border of Georgia to the west with Turkey and to the east with Armenia. The width of the range in the middle part reaches about 17 km, while a large part belongs to the better developed northern slope. Its peaks are Gekdaghi, Uchtapalari (3011 m), Ortuldaghi, Okiuzdaghi, and the row of unnamed peaks with an absolute height of 2450–3000 m and a relative height of 350–1000 m. Their height is reduced from the middle part to the flanks. These peaks have the form of a dome; their origin is tectonic and not volcanic—they represent brachy-anticlinal complications, not the central volcanoes. The highest peak of the range within Georgia is Sambortsva at 3003 m, and the average height of the range is 2670 m.

The slopes of the Nialiskuri Range are dissected by the dense but shallow erosion streams of Kodala, Murjikhevistkali, Bughdashenistskali, and the lakes of Khanchali and Mada. The river gorges of the southern slopes look like a canyon. In many places, especially in the east, the rivers flow between the low and flat swampy banks, especially in the above river basins. The glaciation in the Nialiskuri Range was of about the same range as in the Javakheti Range.

Morphologically, the Nailiskuri Range can be divided into two different parts—the western and the main part that cover almost 2/3 of the range. It is characterized by less inclination and a flat watershed crest. The soft relief of the northern slope is disturbed only by the El-Daghi Peak (2497 m), which is uplifted in the south of the Akhalkalaki Plateau.

The western part, which is located between the Shaitan-Kali Gorge and the Gekdaghi Peak, is completely different from the above-described section of the range. The slopes are more inclined here, and the crest is clearly visible, with Ziareti and Gekdaghi—the peaks of dome-like massifs (in the north of the Zurzuni, within the borders of Turkey). Due to the different inclination of the slopes in both parts, their width is different: the width of the northern slope is only 2 km in the Khozapini Lake basin, while the same slope of the ridge in the east reaches 10–14 km in the Kodala, Kirkh-Bulakhi, and Khanchali River basins.

Nialiskuri Range is built mainly with lava deposits. It is noteworthy that the range is poorly studied from a petrographic point of view, which is related to its location. The Nialiskuri Range is composed of trachyte-andesite and andesite-dacite lavas. The western section of the Nialiskuri Range from the Gekdaghi peak to the Zurzuni Village is an anticline, directly reflected in the relief.

14.6 Samsari (Abul-Samsari) Range

The Samsari Range is the highest orographic unit in the highland of southern Georgia. It is a watershed of the Paravani and Ktsia river basins. The range stretches from the north to the south at 40–42 km, with width 20–22 km (Fig. 14.6). The volcanic massifs erected from the smooth surfaced pedestal of the Akhalkalaki Plateau and Paravani Lake depression and the individual cones are located in the meridian direction and create two rows. The volcanic cones located in the west are distinguished by high altitude (2900–3300 m) and are erected from the Akhalkalaki volcanic plateau at the relative altitude of 1000–1200 m. The highest peak is Didi Abuli (3301 m), the other high peaks are Samsari (3285 m), Godorebi (3188 m), Karakuza (3194 m), and western Shaori (2921 m). The volcanoes in the eastern meridian row are relatively low, they are Grigoli (2770 m), eastern Shaori (2757 m), Tsiteli Mta (3099 m), and Shavi Mta (3047 m). Their relative height from the Paravani Lake (2074 m) is 700–1000 m. Morphologically the volcanoes of the Samsari Range are divided into five groups: Tavkvetili, Shavnabada, Samsari, Didi Abuli, and Patara Abuli. They are separated from each other by 1.0–1.5 km wide passes.

The western slopes of the range intrude in the northwestern part of the Javakheti Upland, and the eastern slope of the Tsalka and Paravani hollows. In the west, it is worth mentioning the lava stream Pampula, the width of which is 5–6 km, length 20 km, and capacity 100–150 m. The lava stream flowed in the Paravani hollow formed the largest lake in Georgia, Paravani Lake, elevation 2074 m.

The geologic structure of the Samsari Range almost entirely involves the volcanogenic and volcanogenic-sedimentary formation of Neogene–Quaternary age, with several blocks constructed with Middle Eocene age tuffs and tuffo-breccias. The volcanic processes in the Samsari Range are associated with the faults of meridian direction. The dynamics of the eruptions of central type developed here determines the composition of the volcanic material and the form of volcanic mechanism. The three stages of magma activity are distinguished here. In the Late Miocene–Early Pliocene, a large volcanic material erupted, known as Goderdzi suite. The lava streams were andeste-dacite. In the next stage, in Late Pleistocene, the local eruptions occurred from the central mechanisms followed mainly by andesites, and in Holocene only a few eruptions occurred (Gobejishvili 2011).

The morphological appearance of the Samsari Range is created by the forms of tectonic, volcanic, erosive, glacial, and nival genesis. Anthropogenic forms can be found such as artificial caves (the Samsari River gorge and the Shaori volcano). Older volcanic shapes in the range (built of Goderdzi suite) are modified by denudation and Pleistocene glaciation. Young volcanic cones and domes have retained a

Fig. 14.6 Samsari (Abul-Samsari) Range (*photo by* Z. Turmanidze)

volcanic appearance. Nival and erosive relief are expressed by different forms.

In addition to tectonics and volcanism, in the formation of the modern terrain of the Samsari Range, the leading role belongs to the nival-glacial processes that were active in the entire Late Pleistocene and Holocene. At the modern stage, there is no nival-glacial zone on the range; subnival processes emerge fragmentary in the individual high peaks.

Small valley, cirque-valley and cirque type glaciers developed on the slopes of the high peaks of the Samsari Range during the Late Pleistocene glaciation. The valley type glaciers of the Samsari and Abuli massifs were distinguished by relatively large size, with their cirque area of 6–10 km^2, and the length 5–8 km. The glaciers' tongues were covered with boulder material. The traces of the Würm glaciers are clearly visible in the forms of cirques, moraines, and rock glacier deposits (Tielidze 2017). The "Sea of rocks", rock flows and other nival-glacial sculptural forms are well represented in the relief.

The Samsari Range pedestal (foundation) is developed on both sides; it is represented in the west by the Akhalkalaki Plateau and the east side is covered with volcanic cones. There is a two-cone massif of Tavkvetili in the north of the volcanic meridian row of the range. This massif is erected between the Ktsia gorge and the Tsalka-Tabatskuri highway. The cones have a truncated shape. Here, the eastern Tavkvetili is higher, looking down at the Nariani swampy plain in the Ktsia gorge. The relative height between them is 530 m.

There are two cone massifs of meridian direction in the south of Tavkvetili. The cones are located along the fault line. Despite the high altitude of western Shavnabada (2929 m), there is no trace of glaciation, which should be explained by its young age (Holocene).

The whole group of old and new volcanoes is located in the central part of the Samsari Range. The volcanoes are oval shaped in the relief with old glacier cirques in the center. The highest peak is the Samsari Peak (3285 m). In the Miocene–Pliocene relief, two rows of volcanoes are sharply distinguished. The peaks of Samsari, Karakuza, etc., constructed with Miocene–Pliocene age dacites, are located in the west, while the young Late Pleistocene–Holocene age volcanic cones of the Shavi Mta, Tsiteli Mta, Mesame Dzmobili, Shaori, and others are located in the east. The relief of the old volcanic cones is heavily modified due to the active influence of glaciers in the Pleistocene. Samsari cirque, a typical pyramidal peak was formed during this period. The trace of the Würm glaciation is mostly well-known in this section of the range. The young cones of the Tsiteli Mta and Shavi Mta have not experienced the impact of the glaciers, while barancos are developed on their slopes.

The Didi Abuli is the highest in the Samsari Range; this massif includes Grigoli, Godorebi, Ivantepe, and western Shaori. It is separated from the Samsari massif by the Sabatkne Keli at 2300–2600 m, in the western part of which lies the Levani Lake (Fig. 14.7). The Ivantepe and western Shaori volcanic cones have two steps. The trace of glaciation is visible in the Didi Abuli and Mt. Godorebi in the northeast. A strongly degraded middle moraine indicates that the Würm glaciers, descended from their slopes, and joined each other.

The southernmost massif of the Samsari Range is a volcanic cone of the Patara Abuli massif (2800 m). The massifs of Didi Abuli and Patara Abuli are separated from each other by the Poka-Abuli corridor. There is a highway through this corridor between Poka and Akhalkalaki. The lava erupted from Mt. Patara Abuli, spreads 6 km in the east to the Paravani River. This contributed to the creation of Paravani Lake.

The relief of the Akhalkalaki Plateau is constructed with the lavas erupted from the Samsari Range (Miocene–Pliocene), including the Mtkvari gorge. There are many lakes of volcanic origin in the Samsari Range (Apkhazava 1975).

14.7 Javakheti Range

The Javakheti Range (in some literature it is named the Kechuti Range or the Wet Mountains) stretches to the east and in parallel of the Samsari

Fig. 14.7 Levani Lake, located on the western slope of the Samsari Range (*photo by* Z. Turmanidze)

Range; its southern part with the highest peaks (Gegladaghi, Akhchala) is located in the territory of Armenia. Within Georgia, the range extends about 50 km meridionally from the western part of the Tsalka depression (Burnasheni Village) to the southern state border. Its width is about 25 km and it has an oval shape projection. The range is a watershed of the Paravnistskali and Ktsia river basins at a long distance of its extension from Mt. Dali to Mt. Legli. On its eastern slopes, the Chochiani, Shavtskala (Karabulakhi) and Mashavera rivers originate, as the right tributaries of the Khrami River. No significant rivers flow down from the western slope. The eastern slopes of the range intrude in the lava plateaus of Chochiani, Gomareti, and Dmanisi, as well as in the Lukuni and Kviriketi mountainous massifs, and in the west in the hollows of the Paravani, Saghamo, and Mada lakes. The highest peak within Georgia is Legli at 3156 m. Heights of other peaks range between 2300–3000 m (Mt. Dali, Chikiani, Shambiani, etc.). The lowest mountain pass is the Chikiani (Tiklatashi) Pass (2164 m).

In the geological structure of the Javakheti Range, as well as in that of Samsari Range, there are two main effusive formations: Neogene—Goderdzi suites and Quaternary—Zurtaketi suites. Tectonically, the Javakheti Range is located in the fault line. The rocks that construct the range were erupted in this line and experienced bending, on the axis of which the central-volcanic structures appeared.

The Javakheti Range is different from the Samsari Range with less fragmentation of the pedestal and with almost a complete lack of the continuous chain of effusive cones and domes located there, which result in few rocky and stony places. The passes of the range have a large plateau-like character and the peaks are rounded.

The relief of the Javakheti Range consists of forms of tectonic-volcanic, volcanic, erosive, lesser old glacial, periglacial, and other genesis. The tectonic-volcanic type of terrain forms the geomorphologic skeleton of the entire region and is expressed with considerably sloped lava covers that coincide with the slopes of the watershed crests. Volcanic forms are represented by separate cones and domes that do not form the entire chain. Neogene and Quaternary formations can be distinguished between these massifs. The old volcanic structures of the Javakheti Range, as well as the Neogene volcanoes of the Samsari Range, are strongly transformed by the impact of exogenous processes. It is noteworthy that the peaks of the Javakheti Range got the rounded, gentle forms due to weak Pleistocene glaciation and strong denudation. Erosion forms are characterized by a relatively small width and depth. The narrow gorges are better developed on the eastern slopes with depth 150–200 m, in the gorges of the Khrami River tributaries. The erosive network of the Javakheti Range is generally distinguished by less density than is typical of mountain-erosive regions.

The old glacial forms are less developed in the Javakheti Range than in the Samsari Range; they are not found on the western slopes at all, while the glacial forms are spread only in the middle part of the eastern slope, namely, in the upper parts of the Shavtskala River (Karabulakhi). The old glaciers are represented by strongly transformed corries and troughs. Such distribution of glacial relief is the result of interaction of climatic and orographic conditions; it is also caused by the hypsometry of the middle part of the range and the snow transferring from the western slope to the eastern as a result of winter anticyclones. Periglacial formations are less common. Anthropogenic forms are presented by caves on the slopes of the narrow gorges.

In the south of the Mt. Dali, the plateau-like Chikiani Pass (2164 m) is represented in the crest of the Javakheti Range. To the south of the pass, in the watershed of the Paravani Lake and Chochiani River, the dome-like Chikiani mountain massif (2415 m) is uplifted, and cut by oxide veins, from which the prehistoric Georgian population was producing a glass of volcanic material.

14.8 Tsalka Depression

The Tsalka depression is bordered by southern slope of the Trialeti Range in the north, by the Samsari Range in the west, by the Javakheti Range in the south, while the east side is mostly open and ends at the slopes of the Bedena Range. Its length is 40–42 km and the width is 18–19 km on the Arjevani meridian. The depression extends from 1500 m (the height of the bottom) to 1800–1900 m asl. Today, the main part of the bottom of the depression is covered by the Tsalka water reservoir—the largest water reservoir in Georgia. The Ktsia River flows into the depression from the west, while the Khrami River flows out from it in the east and develops the Dashbashi canyon.

The relief is constructed with Pliocene-Pleistocene volcanic lavas—dolerites, basalts, and andesites. Bottoms of the hollows are covered with young alluvial–proluvial sediments. Flatlands are synclines and hills are anticlines.

The leading role in the formation of the relief of Tsalka depression belongs to tectonic-volcanic, erosive and accumulative processes. The interaction of endogenous and exogenous processes led to the modern face of the relief of the Tsalka depression. Generally, the Tsalka depression is dissected by smaller hills, such as Tsalka, Beshtasheni, Bareti, and Korsu.

The Tsalka Hill is located in the northern part of the depression. A hill of anti-synclinal genesis begins at Mt. Tavkvetili and extends to the village of Arjevan-Sarvani at 35–36 km distance with the relative height of 150–200 m. The volcano lava blocked the rivers flowing down the Trialeti Range and formed the flatlands of Rekhi, Tarsoni, Kaburi, Tedzisi, and Livadi, the bottom of which is covered by proluvial–alluvial sediments. Subsequently, the rivers at Gumbati, Ashkola, and Kariaki cut the anticlinal hill and created the canyons. In these areas, the anticlinal genesis of the hills is clearly visible. The rivers have divided the Tsalka Hill into several sections (Gobejishvili 2011).

In the south of Tsalka Hill the synclinal flatlands of Beshtasheni and Kariaki are separated from each other by the Beshtasheni Hill. The lakes are located at the bottom of the flatlands. In the south of Beshtasheni is the bottom of the Tsalka depression, with the water reservoir in its central part. The depression extends to the east as a narrow belt, while in the west it is wider and is clearly visible in the relief. At the bottom of the depression, the Ktsia River flows and divides it into two equal parts. The bottom of the depression is constructed with the alternation of Pleistocene lacustrine and alluvial sediments. The southern slope of the depression is constructed with the lavas that flowed down Mt. Bula (2661 m) and Mt. Dali (1996 m) and its slopes are dissected by small ravines and gorges.

14.9 Region of the Volcanic Plateaus

The volcanic plateaus are spread in the east of the Javakheti Range and their origin is mainly associated with the lava flows from the Javakheti Range. The following subregions are distinguished here: the Bedena-Kvemo Kartli, Chochiani, Gomareti, and Dmanisi plateaus.

The Bedena-Kvemo Kartli Lava Plateau is located in the center of the relief. It is the remnants of the giant lava flow from the central side of the Javakheti Range. The Bedena Plateau located in the northwest is separated from the Kvemo Kartli lava plateau by the lava slope. The entire lava plateau is about 48 km long and ends at Marneuli. Hypsometrically, it begins at 1700–1800 m and ends at 400–500 m. The west of the plateau is the Algeti and Khrami interfluve, while the east is the Algeti-Mashvera interfluve. The latter is divided into two parts by the Khrami River canyon. The Bedena-Kvemo Kartli Plateau is formed in the Upper and Lower Pliocene by basalt layers with dolerites erupted several times. From the west and southwest, the plateau is surrounded by the massifs built of Bedena and Lukuni Cretaceous volcanogenic and carbonate rocks.

The Bedena Plateau is located in the south of the Bedena Range and is divided by the heads of the left tributaries (Kldeisistskali, Torne and Chivchavi) of the Khrami River. A shape of a trough is formed in the Bedena Plateau that some scientists consider the part of the Paleo-Khrami gorge (Maruashvili 1970); it was then filled up by dolerite lavas and was uplifted by tectonic movement (Gobejishvili 2011). From the bottom of the trough or from the Bedena Plateau the lava wedge is intruded in the south, which reaches the Kldeisi Village in the form of the broad steep-edged plane. In this area, there is a natural well at a height of 1450 m in the dolerites, where Vakhushti Bagrationi indicates the presence of ice (Khorkhebi) in summer (Vakhushti 1941).

The Kvemo Kartli Lava Plateau extends from west to east at about 38 km with the width of about 10 km. The plateau is divided into two parts morphologically by the Khrami River, the larger Tetritskaro Plateau in the north and the Disveli Plateau in the south.

In the surroundings of the plateau, the Khrami canyon stretches for a distance of 20–22 km from the Torne River to the Nakhiduri Village. The width of the upper part of the canyon is 0.4–1.7 km and the depth is 100–300 m (Fig. 14.8); it reaches the maximum depth at Nakalakari. The castle-town of Samshvilde is located on the cape-like cliff formed among the Khrami and Chivchavi canyons.

The length of the Tetritskaro section of the lava plateau is 37–38 km from west to east and the width is 7 km. The height of its surface decreases from 1500 m (Nadarbazevi Village) to 450 m (Marneuli). The western part of the plateau is fragmented by the left tributaries of the Khrami River, forming the Chivchavi River. In the west of the Tetritskaro Plateau the Nadarbazevi hollow is located with the lake in the stage of senility. The eastern section of the plateau is buried in the alluvial sediments of the Marneuli Plain.

The Disveli Plateau is far behind the Tetritskaro Plateau in size and height. Its surface is almost horizontal and is characterized by smooth terrain.

The Chochiani Plateau is dissected by the right tributaries of the Khrami River and its tributary the Chochiani River. The relative height of the plateau is 1400–1600 m, its surface cut at 300–500 m depth by the Khrami and Chochiani gorges. There are closed caves, lakes, and marshes in the southern part of the Chochiani Plateau. The Chochiani Plateau is divided into three parts: the southern, northern, and middle parts.

The Gomareti Plateau is located in the basin of the Shavi Tsklebi (Karabulakhi) River, the right tributary of the Ktsia River. The Shavi Tsklebi River divides the plateau into two parts; its eastern edge has a form of a cliff in the Khrami gorge. Plateau height in the north-east is 1250–1350 m and in the southwest 1400–1500 m. It consists of several layers of dolerite lavas overlapping each other and the associated lacustrine sediments, which are mainly the weakly cemented fine-grained and straight layered sandstones of tuffogenic type. Landslides

Fig. 14.8 Khrami gorge-canyon (*photo by* R. Gobejishvili)

are common for the edges of the plateau, which is caused by erosion of the lacustrine sediments deposited under the lava layers. The huge dolerite creeps down and forms the false terraces in the Khrami gorge.

The southwestern part of the plateau is covered by debris cones of the streams and mudflows flowing from the Javakheti Range that give its surface a significant inclination. These debris cones with the thickness of several tens of meters are represented by boulder conglomerates.

The Dmanisi Plateau is located in the south of the region. The shape of the plateau is a triangle; it narrows along the Mashavera River indicating that it follows the Paleo-Mashavera gorge. The Dmanisi Plateau starts at the foot of the Javakheti Range (1600–1700 m) and ends in the vicinities of the Didi Dmanisi Village (900–950 m) (Fig. 14.9). The Dmanisi Plateau consists of dolerite lavas, which erupted in the Lower Quaternary and Middle Quaternary in the southern part of the Javakheti Range and flowed down the Mashavera River gorge; there are also lacustrine-tuffogene outlets of sandstones.

The Dmanisi Plateau surface is step-like, where three steps are distinguished: the first step is from Didi Dmanisi and Garinjuki to the settlement of Dmanisi within 900–1250 m; the second step is in the vicinities of the Ormozani (Orozmani) Lake (at 1350–1450 m), and the third one is at 1600–1700 m. On the lower step of Dmanisi, in the cape of the rivers of Mashavera and Pinezauri, the ruins of the old city Dmanisi can be found as well as the remains of the dwelling with the bones of the first human in Eurasia, dating back to 1.8 million years (Vekua et al. 2002). In this place, the Mashavera River crosses the lower section of the Dmanisi Plateau and develops quite deep canyon. The Mashavera River gorge is asymmetric above Patara Dmanisi; it is developed on the right edge of the main carbonate and Dmanisi lava stream.

Fig. 14.9 Dmanisi volcanic plateau (*photo by* L. Tielidze)

References

Apkhazava IS (1975) Oziora Gruzii [The lakes of Georgia], Monograph. Tbilisi (in Russian)

Gobejishvili R (2011) Sakartvelos reliepi [Relief of Georgia], Monograph. Universali, Tbilisi (in Georgian)

Maruashvili L (1970) Saqartvelos fizikuri geografia [Physical geography of Georgia], Monograph, Part 2. Tbilisi (in Georgian)

Tielidze L (2017) Late Pleistocene and Holocene glaciation. In: Chapter in glaciers of Georgia. Springer, Geography of the Physical Environment. https://doi.org/10.1007/978-3-319-50571-8-6

Vakhushti B (1941) Aghtsera sameposa sakartvelosa [Description of the Kingdom of Georgia], Tbilisi (in Georgian)

Vekua A, Lortkipanidze D, Rightmire GP, Agusti J, Ferring R, Maisuradze G, Mouskhelishvili A, Nioradze M, Ponce de Leon M, Tappen M, Tvalchrelidze M, Zollikofer C (2002) A new skull of early Homo from Dmanisi, Georgia. Science 297(5578):85–89. https://doi.org/10.1126/science.1072953

Levan Tielidze is a Ph.D. of Ivane Javakhishvili Tbilisi State University in the field of geomorphology–glaciology–cartography and a senior research scientist at Vakhushti Bagrationi Institute of Geography. He is also a researcher-scholarship holder of the Department of Earth Sciences of the Georgian National Academy of Sciences. The field of his research is mountain cartography and glacial-geomorphological study of the mountainous areas in the Quaternary (Late Pleistocene and). He is a member of several international organizations. In 2014/2015, he conducted a research stay at the Climate Change Institute of the University of Maine, USA, and in 2015/2016–2017 at the University of Northern British Columbia, Canada. He is the author of about 40 scientific papers, maps and eight monographs.

Ramin Gobejishvili (1941–2014) was a Doctor of Geographical Sciences, a geomorphologist–glaciologist. Until his last days, he was Professor of the Faculty of Exact and Natural Sciences, Ivane Javakhishvili Tbilisi State University and the chair of Geomorphology and Cartography. In 1996–2008, he was the president of the Geographical Society of Georgia, and in 2005–2007—the Director of the Institute of Geography. He was the author and co-author of about 150 scientific works and ten monographs.

Levan Maruashvili (1912–1992) was a prominent Georgian geographer. Doctor of Geographical Sciences (1954), Professor (1965). Honored Scientist of the Georgian Soviet Socialist Republic (1966). Honorary member of the Russian Geographical Society (1985). He produced over the 500 scientific and popular-scientific works in Georgian, Russian, and English languages.

Alexander Javakhishvili (1875–1973) was one of the oldest geographer and anthropologist in the Soviet Union. He was the founder of Georgian geographical school. His qualifications and awards include Doctor of Geographical Sciences (1937), Academician of the Georgian National Academy of Sciences (1944), an honorary member of the Soviet Union Geographical Society, and winner of a State Prize of the Georgian Soviet Socialist Republic.

Nikoloz Astakhov (1911–1993) was a Doctor of Geographical Sciences, Professor (1965). He was researcher at the Department of Geomorphology of the Vakhushti Bagrationi Institute of Geography. His research area was structural geomorphology.

Part IV

Exogenic Processes Occurred in Georgia

15 General Characteristics of Exogenic Processes

Ramin Gobejishvili and Levan Tielidze

Abstract

This chapter discusses the distribution of various exogenic processes in the territory of Georgia. In the beginning, the following processes are discussed, such as accumulation, gravitational, karst, and erosive-denudative processes. The map of exogenic processes of Georgia is presented. Individual paragraphs are dedicated to the processes such as landslides, mudflows, and snow avalanches. The latest subchapter reviews flooding on the Vere River occurred in Tbilisi on June 13, 2015.

Keywords

Exogenic process · Landslide · Mudflow
Snow avalanche · Flooding

R. Gobejishvili · L. Tielidze (✉)
Department of Geomorphology and Geoecology, Vakhushti Bagrationi Institute of Geography, Ivane Javakhishvili Tbilisi State University, Tbilisi, Georgia
e-mail: levan.tielidze@tsu.ge

R. Gobejishvili
Department of Geography, Faculty of Exact and Natural Sciences, Ivane Javakhishvili Tbilisi State University, Tbilisi, Georgia

L. Tielidze
Department of Earth Sciences, Georgian National Academy of Sciences, Tbilisi, Georgia

15.1 Introduction

The relief of Georgia is characterized by diverse and complex morphology, morphometry, and morphography. The complexity of the relief is stipulated by its geotectonic nature, and the diversity by altitudinal zoning and peculiarities of exodynamic processes. Exogenic processes are characterized by altitudinal zonality, though its development is defined by the lithology of structural rocks and atmospheric precipitation. Leading exogenic processes include erosion, accumulation, weathering, nival-glacial, and karst processes, snow avalanches, landslides, and mudflows. It should be noted that in some areas, several exogenic processes act simultaneously, as one process determines another one (Gobejishvili and Tsereteli 2012a) (Fig. 15.1).

In the zone of 0–200 m asl, accumulation processes with lateral erosion (Kolkheti and Eldari lowlands) prevail. In the second zone, which extends within 200–1000 m and is about 35% of the total area, the relief is a low mountainous and hillock built of Tertiary and Quaternary sediments. Danudative-erosive and gravitational processes are predominantly developed. The medium mountainous zone (1000–2000 m) occupies about 35% of the entire territory of Georgia, where erosion, gravitational and denudative processes prevail in the region; karst processes are also found in the same zone. In the medium and high mountainous zone

L. Tielidze (ed.), *Geomorphology of Georgia*, Geography of the Physical Environment,
https://doi.org/10.1007/978-3-319-77764-1_15

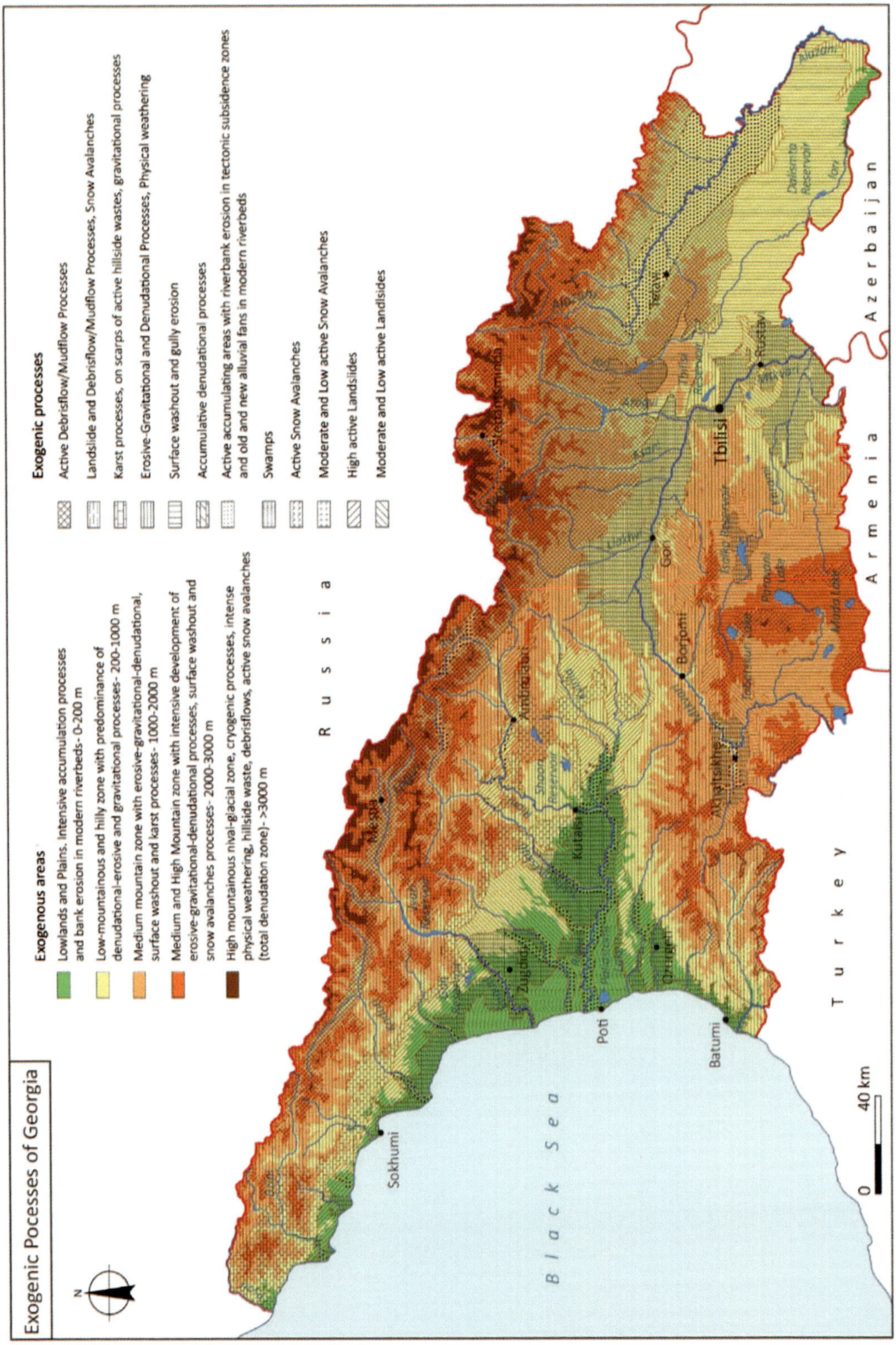

Fig. 15.1 A map of exogenic processes of Georgia

(2000–3000 m), the erosive-gravitational-denudational processes prevail with intensive development of snow avalanches. Above 3000 m the nival-glacial processes are developed with weathering, debris fall and snow avalanches. Relief in this zone is outcropped and exogenic processes occur at all points, therefore, this zone is called a total denudation zone.

Negative consequences caused by **landslide-gravitational** phenomena are distinguished among the natural hazards in the territory of Georgia. Because of them, more than 1.5 million hectares of land is damaged by various degrees. There are more than 2000 settlements in the danger zone among the documented more than 50 thousand landslide-gravitational phenomena. Through their danger risk zone pass, the Samgori-Batumi, Baku-Supsa, Baku-Tbilisi-Ceyhan oil pipelines, and Transcaucasian gas pipeline, while approximately 25% of the total length of highways is in the area of impact of these phenomena, where up to 500 landslide bodies are documented (Gobejishvili and Tsereteli 2012b).

Even under the background activity of landslide-gravitational phenomena, the damage inflicted on Georgia's economy is determined in tens of millions of dollars, and in extreme activation conditions often reach the billion limit.

All kind types of landslides known in the engineering geodynamics are formed in the territory of Georgia, starting from the simplest ones, the deformation of which do not go beyond the aeration zone and ending with those with depth of tens of meters (Fig. 15.2). Also, the areas (from hectare to hundreds of hectares) and volumes (within a few thousands of cubic meters and tens and hundreds of cubic meters) of individual landslides vary within a large range.

Georgia belongs to one of the complicated regions of mountainous areas with petrogenic nature of **mudflows**, the degree of damage to the area, the frequency of recurrence and the negative results caused by them (Gobejishvili and Tsereteli 2012c). The mudflow phenomena are found in almost all landscape-climatic and morpho-geological zones of mountainous territory, ranging from hillock, hilly and low mountainous areas to alpine-nival, where up to 3000 mudflow-forming water canals are observed (Fig. 15.3). The area of their negative impact covers hundreds of settlements including Tbilisi, Telavi, Kvareli, Sagarejo, Lagodekhi, Tsageri, Borjomi, Lentekhi, Oni, Mestia, Akhaltsikhe, Adigeni, and Mtskheta; highways and railways, oil and gas pipelines, irrigation channels, and farmlands.

The large development of mudflows is caused by the spread of sensitivity to these rock phenomena in large areas, as well as the erosion relief, climatic conditions, and human stress on geographical environment. The mudflows caused by heavy rains (~65%) prevail in distribution over those caused by landslides (~30%) and glaciers (~10–12%).

Fig. 15.2 Block slide in the Mulkura River valley near the Artskheli Village (*photo by* L. Tielidze)

Fig. 15.3 Mudflows basin in the Mulkura River valley (*photo by* L. Tielidze)

Snow avalanches annually inflict significant damage to the mountainous regions of Georgia and endanger people's lives. Important transport communications and settlement areas are located in the snow avalanche risk areas. However, snow avalanches rarely result in isolation of populated areas due to the blocking of roads (Sukhishvili et al. 2012). The snow avalanches in western Georgia are more frequent than in eastern Georgia, due to a greater inclination of slopes along with abundant amount of atmospheric precipitation (Fig. 15.4).

The massive descending of avalanches in the abundant snowy winter of 1970–1971 took the lives of 39 people, 42 in the winter of 1975–1977, and 80 in winter 1986–1987. The damage caused by the avalanches of these years has amounted to 60, 200, and 500 million dollars. Mountainous highways are avalanche dangerous, as well as the electricity transmission lines located in the mountains, agricultural, recreational, and other objects. Descending avalanches caused human victims and destruction of buildings in 77 settlements out of 348 avalanche-dangerous settlements, inflicted destruction and damage to buildings in 148, while 123 settlements are potentially avalanche-dangerous with the high likelihood of descending avalanches in these settlements. Many settlements are avalanche-dangerous in the regions of Mestia (61), Dusheti (49), Khulo (46), Lentekhi (35), Java (25), Shuakhevi (24), Akhalgori (17), Keda (17), Kazbegi (13), and Gulripshi (11). The river basins of Acharistskali (24%), Enguri (18%), Rioni (13%), Aragvi (13%), and Liakhvi (11%) stand out with a large number of avalanche-dangerous settlements.

While zoning the territory of Georgia by the rate of avalanche danger, the strongest (11.1%), medium (32.7%) and weak (12.2%) avalanche-dangerous regions are distinguished (Salukvadze and Kaldani 2012). There is no danger of avalanches: in western Georgia in the Kolkheti Lowland and the narrow strip along the Black Sea in Achara and Apkhazeti; in eastern Georgia in the Akhaltsikhe and Tsalka depressions, Shida Kartli, Kvemo Kartli and Alazani plains, and the most part of the Iori and Javakheti uplands.

Karst processes are related to the distribution of massive limestones of the Cretaceous period. They continuously follow the peripheral strip of the southern slopes of the Greater Caucasus, where low and high mountain karsts are distinguished. The karstification process is in progress in both underground hollows and on the surface of relief. On the flanks of limestone massifs, the debris fall is active and associated with debris cones. The Arabika and Bzipi massif, Askhi, Khvamli, Shaori depression, and others are distinguished by a great variety of karst forms.

Fig. 15.4 Snow avalanche in western Georgia (*photo by* L. Tielidze)

Erosion-gravitational and denudative processes are widespread and are associated with sedimentary and volcanogenic suites of Jurassic and Tertiary ages. Due to the active erosion processes, deep and wide gorges and sometimes wide depressions are developed (in the Bzipi, Kodori, Enguri, Tskhenistskali, Rioni, and Kvirila gorges) in the Lower and Upper Jurassic rocks; narrow deep gorges are associated with the spread of volcanogenic suites. The erosion and ravine forming is in progress in the Iori Upland, which is built of Tertiary suites. The denudative and accumulative processes are underway in the intermountain plain, namely, in the hillock and the plain areas (Kvemo Kartli and Shida Kartli and Kolkheti hillock strip, Akhaltsikhe and Tsalka depressions).

Active accumulation processes should be noted separately, the distribution of which are related with absolute and relative sinking areas, such as Kolkheti, particularly along the riverbeds, the bottom of the Akhaltsikhe depression, the Tskhenistskali gorge, the mountain Racha, the Racha-Lechkhumi depression, and the Tergi gorge (Fig. 15.4). Active accumulation processes are underway across the entire Alazani Plain. Active

Fig. 15.5 Active accumulation process in Racha region near the Chiora Village (*photo by* L. Tielidze)

accumulation along the Alazani riverbed is accompanied by lateral erosion, and the powerful debris cones are formed in the gorges of tributaries. In the areas of lateral erosion distribution (Alazani Plain, Kolkheti Lowland, Racha, Akhaltsikhe depression, and Kvemo Kartli), the rivers develop meanders or flow in bayous (Fig. 15.5).

15.2 Vere River Disaster (13/06/2015), Consequences and Analysis

The Vere River flows in the form of several small streams in its head, and then gradually increases, gathering the waters of the ravines flowing down the slopes of the Teleti Range and joining the Mtkvari River in Tbilisi with already enlarged runoff.

Regarding hypsometry, the Vere River basin lies in three geomorphological zones, such as the plain terrace relief, low mountainous foothills, and medium mountainous mountain-gorges. The terrain feature and forms of morphological types of all geomorphological zones are defined by tectonic conditions, different sensitivity of structural rocks to the erosion-denudation processes, virgation of rocks folding, their inclination in different directions and tectonic disorder with many regional and local faults, on the basic substratum of which the diverse terrain is formed by erosion-denudation and gravitational processes.

In June 13–14, 2015, heavy rainfall in the Vere River basin (by the data of Tbilisi meteorological station, 49 mm precipitation was observed in 3–4 h) (NEA 2015) caused a sharp rise in water level of Vere and its tributaries and

Fig. 15.6 Vere River disaster consequences, June 13–14, 2015 (*photo by* L. Tielidze)

large-scale generation/activation of mudflow processes. It was followed by the disruption of the roads of Tskneti-Betania, Tskneti-Akhaldaba and Kojori-Manglisi. It also caused damage to the houses, various infrastructural objects, buildings and the zoo located in the lower landmarks of the Vere River gorge in Tbilisi (Figs. 15.6 and 15.7).

During the disaster, in the territory of Tbilisi, in the Vere River bed, the mudflow stream was four times inundated that was caused by creation of temporary barriers in the water tunnels; the barriers were made of large erosion deposit material, especially of drifted mass of tree plants, and at certain sections of the riverbed and floodplain the river water level raised by 17 m (Kereselidze et al. 2015). As a result of the inundation, the water stream overflowed into the territory of the Heroes Square and flooded the underground passages. The water discharge on June 13, 2015 (according to the passing track) made 468 m^3/s by calculated various empirical formulas. The disaster took the lives of 19 people.

A crucial role in the disastrous events developed in the Vere River basin have been played by the anomalous rising of groundwater levels due to excessive atmospheric precipitation and saturation of slope-building base rocks and also the alluvial–proluvial and lacustrine clay-loamy soils. In April–June, a large amount of atmospheric precipitation was observed, namely, more than half of the annual norm (506 mm)—320 mm (NEA 2015), a significant part of which percolated into the soil and caused its high degree of watering, which contributed to fast generation/activation of landslide-mudflow

Fig. 15.7 Vake-Saburtalo damaged highway after the Vere River flooding, June 14, 2015 (*photo by* L. Gadrani)

processes. The excessive groundwaters were discharged in the form of plural frontal outflows in the natural and artificial (including newly formed) outcrops in the studied territory, during 2–3 days after the Vere disaster.

References

Gobejishvili R, Tsereteli E (2012a) Exogenic processes. National Atlas of Georgia. Cartograpia, pp 50–51 (in Georgian)

Gobejishvili R, Tsereteli E (2012b) Landslide danger. National Atlas of Georgia. Cartograpia, p 52 (in Georgian)

Gobejishvili R, Tsereteli E (2012c) Mudflow danger. National Atlas of Georgia. Cartograpia, p 53 (in Georgian)

Kereselidze D, Alaverdashvili M, Tsintsadze T, Trapaidze V, Bregvadze G (2015) What happened on June 13, 2015 In the Vere River watercatchment basin? Tbilisi, 44 p (in Georgian)

National Environmental Agency (NEA) (2015) The general analysis of the natural disaster hazards in the territory of Tbilisi and the preliminary assessment of the situation caused by the disaster occurred in Vere Basin on 13–14 June of 2015 (in Georgian)

Salukvadze M, Kaldani V (2012) Snow avalanches. Chapter in the National Atlas of Georgia. Cartograpia, p 97 (in Georgian)

Sukhishvili L, Megrelidze I, Westen CJ (2012) Snow avalanche risk assessment. Atlas of natural hazards & risks of Georgia. CENN/ITC, Tbilisi, Georgia. ISBN 978-9941-0-4310-9 (in Georgian)

Ramin Gobejishvili (1941–2014) was a Doctor of Geographical Sciences, a geomorphologist–glaciologist. Until his last days, he was Professor of the Faculty of Exact and Natural Sciences, Ivane Javakhishvili Tbilisi State University and the chair of Geomorphology and Cartography. In 1996–2008, he was the president of the Geographical Society of Georgia and in 2005–2007,the Director of the Institute of Geography. He was the author and co-author of about 150 scientific works and ten monographs.

Levan Tielidze is a Ph.D. of Ivane Javakhishvili Tbilisi State University in the field of geomorphology–glaciology–cartography and a senior research scientist at Vakhushti Bagrationi Institute of Geography. He is also a researcher-scholarship holder of the Department of Earth Sciences of the Georgian National Academy of Sciences. The field of his research is mountain cartography and glacial–geomorphological study of the mountainous areas in the Quaternary (Late Pleistocene and). He is a member of several international organizations. In 2014/2015, he conducted a research stay at the Climate Change Institute of the University of Maine, USA, and in 2015/2016–2017 at the University of Northern British Columbia, Canada. He is the author of about 40 scientific papers, maps, and eight monographs.

Index

L. Tielidze (ed.), *Geomorphology of Georgia*, Geography of the Physical Environment,
https://doi.org/10.1007/978-3-319-77764-1

B

H

I

L

N

Q

R

Printed by Printforce, the Netherlands